ALL ABOUT VACUUM TUBE GUITAR AMPLIFIERS

Kendrick Books
531 Country Road 3300
Kempner, Texas
76539-5755

ISBN: 978-0-9641060-3-1

Art Direction and production: R. K. Watkins / Corporate Design and Graphics
Printed in the United States of America

Disclaimer: Tube amplifiers contain high voltages which may be lethal, even if the
amplifier has been off for some time. We do not recommend that you open your amplifier or try to perform any repair operations unless you are properly trained in electronic servicing. Gerald Weber and Kendrick Books accept no responsibility for ac-
cidents resulting in personal injury or destruction of property. Again, there are
large voltages present in your amplifier that can kill, even with your amplifier unplugged from the wall.

DEDICATION

I dedicate this book to my four wonderful daughters; in order of their birth: Daphne D'Ann Weber, Ashley Aline Weber, Miriam Caitlin Weber and Helen Kendrick Weber. My heart is filled with love for you.

ACKNOWLEDGEMENT

There are so many people that contributed in one way or another to the completion of this book. In no particular order:

Jill Kendrick Weber—my soulmate, best friend, secretary, and wife. For her unselfish support and total commitment.

Billy F. Gibbons—my tone hero that defined "Texas Tone" way back in 1970 at that lil' ol' teen club in Groves, Texas—"The Townhouse."

Robert and Jeanne Watkins—for their support in making this work into a real book. Neither this book nor any of my other books could have been possible without you.

Art Thompson—for writing the foreword of this book and for publishing my writings in *Guitar Player* Magazine.

Ken Fischer—for his unselfish contribution and encouragement in helping me learn about tube amp circuitry in the first place. During the late 80's and early 90's, he was my mentor. May he rest in peace.

Terry Oubre—for his friendship, fabulous guitar playing and his unbelievable ear that was so essential in helping me to make distinctions of great tone.

Larry Acunto—for publishing many of the articles from which this book was compiled, in his *20th Century Guitar* Magazine.

Alan and Cleo Greenwood—for publishing many of the articles from which this book was compiled, in their *Vintage Guitar Magazine*.

All Kendrick Employees—for providing me the freedom to work on this book.

Everyone Else—that participated in this book.

Ms. Hansen—my 9th grade, 23-year old English teacher at Thomas Edison Jr. High School in Port Arthur, Texas. Had it not been for her and her wardrobe of miniskirts, I would never have paid such close attention in English class and may have never learned the difference between a noun, a verb, an adjective or an adverb.

Anyone Else—I may have forgotten to acknowledge.

PREFACE

Besides playing guitar professionally for over 44 years, I have spent the last 20 years helping guitarists to get great tone with their vacuum tube guitar amplifiers. Besides designing and building Kendrick amplifiers and guitars, writing monthly advice columns in three guitar magazines, and servicing thousands of tube guitar amplifiers; I have also hosted and led Vacuum Tube Guitar Amplifier seminars and Amp Camps in major cities to help aspiring techs to learn servicing, overhauling and designing tube guitar amplifiers. Amazingly, I am still learning new things about tube amps and seeing anomalies with tube amp designs that I haven't seen before. It seems there is always something new to learn. I am not an engineer, nor have I ever had any formal training on electronics. I am just a guy that has played guitar for 44 years, overhauled 15,000 tube amps, built over 4,000 tube amps and spent a lot of time playing, listening, and evaluating tube guitar amplifier designs.

My writings are always based on what people ask me about and what seems to come up consistently with guitar players. Every five years or so, I compile these writings, organized them and condense them into a 500+ page book. The book you now hold in your hands is my fourth such book. Like many grapes are pressed to make wine, the writings of more than 5 years have been edited and condensed into this book. I hope you enjoy reading it and learn a lot from reading it, but even more I hope that you use the information to get the great tone that inspires you to play your guitar on a higher level of consciousness. When you sound better, you play better. And when you play better, it is easy to go out of your mind. And that is where the true inspiration lives.

Besides this book, my works include:

Book 1 — A Desktop Reference of Hip Vintage Guitar Amps
        Foreword by Ken Fischer
Book 2 — Tube Amp Talk for the Guitarist and Tech
        Foreword by Billy F Gibbons
Book 3 — Tube Guitar Amplifier Essentials
        Foreword by Joe Bonamassa

DVD 1 — Tube Guitar Amplifier Servicing and Overhaul — 4 hours
DVD 2 — Understanding Vacuum Tube Guitar Amplifiers — 4 hours

FOREWORD

Texas is famous for a lot of things that warrant having the word "big" in front of them, and that includes guitar tones. From Freddie King to Billy Gibbons to Eric Johnson to the Vaughan Brothers, the sound of electric guitar has evolved to a high order in the Lone Star state. It seems only fitting then that an amp builder from Texas named Gerald Weber has played such a big role in the boutique amp business with his Kendrick line of tube-powered combos and heads. Featured in the first shootout that *Guitar Player* magazine did of this new breed of amps, the Kendrick 2410 4x10 combo was lauded for its quality, sound, and for being the only production amp of its type at the time to feature circuitry that was handwired in the same fashion as Leo Fender's classic tweed Bassman. Gerald Weber obviously paid his dues studying the inner workings of tube-powered guitar amps, and he would introduce many more amplifiers in the following years, as well as speakers, studio preamps, stompboxes, guitars, and a series of books and videos dealing exclusively with the subject of tube amps. Weber's fourth book, *All About Vacuum Tube Guitar Amplifiers*, draws on his extensive experience as an amp maker, repair tech, and educator, to cover a multitude of things you need to know about maintaining, repairing, and modifying guitar amps. It's the field guide to what makes American and British tube amps tick, so if you worship great tones produced by glowing glass bottles, make this book your bible.

—Art Thompson,
Guitar Player Magazine

INTRODUCTION

All About Vacuum Tube Guitar Amplifiers

DEMYSTIFYING THE INNER WORKINGS OF A TUBE GUITAR AMP

Thinking back to the mid-sixties to the very first time I ever saw the inside of a tube amp, which was a Gibson Rhythm King 2X12, I must say to describe the experience as total overwhelm would have been an understatement. There were resistors and capacitors everywhere and wires going top to bottom and bottom to top. I had to stop and think that it must've been some sort of Einstein that developed such a complex apparatus. How did they know how to do such workings and how do these workings affect the overall tone of the amp? A few months later, when I traded my Gibson in for a blackface Fender Pro, being the inquisitive mind that I am, I had to take that amp apart and look at it too. Was this some sort of Rubick's cube of tone that only the chosen few could understand? Who came up with all this confusing stuff?

For most of us, the complexities of the inner workings of a tube guitar amp remain a mystery, yet it is not very complicated at all if you understand a few principles.

I have to laugh when I think of the amp technicians out there that try to make it seem even more mysterious as if it is rocket science.

PRINCIPLE NUMBER 1: THEY ARE ALL BASICALLY THE SAME

The key word here is "basically." Think of a vintage automobile. The vintage cars are all basically the same. They all have a motor, a transmission, four wheels, a steering wheel, seats, a battery, etc. Sure there is difference in performance and outward appearance, but basically they are all the same. In that same way, vintage tube guitar amps are all basically the same.

I operate what is arguably the largest vintage amp restoration/overhaul facility in the world and I see virtually every type of amp that has ever been built. How many times do we use schematics to overhaul, repair or service and amp? Almost never. Sure, I have schematics if someone sends in an amp that has been severely modified and the owner requests the amp returned to stock circuitry; but the truth is that schematics are rarely needed for the technician because the amps are basically the same. Once you understand that, one can easily distinguish the simple sections inside.

WHAT ARE THE BASICS?

There are four basic parts to any tube amp. When I tell you these four parts you will probably start laughing about how simple it is. Drum roll... ahem...

And the four parts are: Preamp section, Output section, Power Supply and Speaker. I will pause a minute for you to get back in your chair. Now wasn't that hard?!

POWER SUPPLY

Let's start by looking at the power supply. Go back to the vintage car example. They all have a motor. My first car was a 1957 Volkswagen. It had four cylinders and was air cooled. I later bought a Buick Skylark that had 6 cylinders with a water cooled system and then traded it for an Oldsmobile Cutlass with 8 cylinders. Yes, they were all different and the Cutlass could have eaten the lunch of the other two; but the fact remained: they all had a motor.

Every tube guitar amp has a power supply. The wall voltage in the USA is 120 volts A.C. In most of Europe, it is 230 volts A.C. A guitar amplifier needs several different voltages to supply different components. For example, it will need heater voltage to light up the tubes. A tube doesn't work unless it is hot, so there is a heater inside of each tube to warm the insides. In most cases, the heater requires 6.3 volts A.C.

The amp will also need high voltage D.C. to run the plates of the tubes. Somewhere around 200 volts D.C. will be used on the preamp tubes and possibly 400 volts D.C. or more will be used on the power tubes. This high voltage D.C. is generally referred to as B+ voltage.

Besides the high voltage D.C. for the B+ voltage and the 6.3 volts

A.C. for the heater voltage, if an amp is designed with a fixed bias output stage, then it will also need a negative D.C. voltage supply.

We need all these different voltages, some of them A.C. and some D.C., but we only have 120 volts A.C. coming from the wall. A power transformer is used that has one primary and several secondaries. The secondaries are wound so they have the correct turns-ratio with respect to the primary to either step-up the voltage or step it down as needed. When D.C. voltage is needed, a rectifier is used to change A.C. (electrons moving in both directions) to D.C. (electrons moving only one direction). In the case of D.C. power, filter capacitors are used to smooth out the D.C. so that it is rid of the rippling that occurs when A.C. is rectified to D.C.

Now there are different types of power supplies. You could have one with a power transformer or without a power transformer. You could have one that uses one or more diodes to rectify the wall 120 volt A.C. electricity into higher voltage D.C. electricity. You could have a tube-type rectifier to change the wall voltage into D.C. Some amps even used two rectifier tubes for this same purpose. But the fact remains: all tube guitar amps have some sort of power supply. The purpose of the power supply is to take the standard 120 volt A.C. wall voltage and change it into voltages needed for the tubes to operate. The tubes will need high voltage D.C. and lower voltage heater voltage (which can be either A.C. or D.C.) and possibly a lower voltage heater supply for the rectifier tube and possible a negative D.C. voltage supply to bias the output tubes. Of course, if the amp used a cathode bias output stage, it would not need a negative D.C. bias voltage supply. Likewise, if the amp is solid-state rectified, it would not need a rectifier tube heater voltage.

There are different styles of D.C. rectification that could be used in a power supply. For example, there is something called a half-wave rectifier (used on the mid sixties Fender 6G15 stand alone reverb unit). This type of rectifier only uses half of the 60 Hertz wave coming from the wall electricity. There is also a full-wave rectifier which, as the name implies, uses the entire or full wave coming from the wall voltage. This is used on all Fender amps and most others. Sometimes the full-wave rectifier is configured in such a way that it uses a non-center-tapped power transformer, in which case it is called a full-wave bridge rectifier.

If you know that every amp has a power supply, then it is very easy to identify what components make up the power supply. If it has a power transformer and most will, the power transformer is the heart of the power supply. Then one can look for either diodes or a tube to determine if the rectification is tube or solid state. Obviously, a rectifier tube in the amp is a dead give away that it is tube rectified. Common rectifier tubes are the 5AR4, 5U4G, 5Y3, and 5V4. There are other types that are not as common, such as the 5R4, 5Z3, 6X4 and many others. But the fact remains that any amp will have some type of power supply and that supply will have some type of rectification.

Most of us have seen powertransformerless amps. These generally use some type of 12 volt preamp tube (12AX7 for example), a 35W4 rectifier tube (this rectifier tube requires a 35 volt filament supply) and a 50C5 output tube (which requires a 50 volt filament supply). If you add up the filament voltages needed; the 12 volts, 35 volts and 50 volts together; you get 97 volts. But the wall voltage is 120 volts. So, there is a ballast resistor connected to eat up the other 23 volts. This way, the tubes plus the ballast resistor are put in series so that the circuit will plug directly into the 120 Volt A.C. wall receptacle without requiring a power transformer.

Most of the Silvertone, "amp in the guitar case" amps use this type of circuit. The rectifier simply rectifies the wall voltage. The circuit was actually published in the back of the R.C.A. tube manual, so many companies used this as a low wattage (about 3 or 4 watts) design. If you look at an amp that uses the aforementioned tubes, it is a dead giveaway what it is. This type of amp can be powered with D.C. or A.C. voltage since there is no power transformer. The design is a shock hazard since the wall voltage A.C. grounds directly to the chassis (which is ultimately connected to the strings on your guitar!), and for this reason, amps are no longer designed this way.

THE PREAMP

When you play a "cowboy" E chord on a single-coil pickup, as-suming the pickup is adjusted correctly, you will have approximately ⅛ of a volt (125 millivolts) output coming from the pickup. This miniscule voltage is not near enough to drive a speaker. So every amp uses a preamp to take the miniscule "pickup voltage" and amplify it.

Your guitar has a magnetic pickup that consists of a coil of wire wrapped around one or more magnets. When the string vibrates near the magnetic pickup, a small A.C. voltage is induced in the coil of wire. A single-coil pickup may put out, approximately ⅛ of a volt, and a humbucker about ¼ volt, while a P-90 pickup could put out as much as ½ volt. This voltage lacks power and can not drive a speaker. In fact, if you did connect a speaker directly to a pickup, the speaker impedance is so low that the pickup would "think" it was shorted out. Zero ohms is a dead short, so for a 5,000 ohm pickup, an 8 ohm speaker looks almost the same as a dead short circuit!!

When you plug your guitar cord to an amp, you are simply connecting the output of the pickup to a one megohm resistor. The one megohm resistor is connected across the pickup to complete the circuit. One end of the resistor is connected to the chassis ground of the amp, and the other end is connected to the grid of the preamp tube.

Other than a heater circuit to warm the preamp tube, the preamp tube (usually a 12AX7) has three parts: the cathode, grid and plate. Say those words out loud: Cathode, Grid, and Plate. See that wasn't so hard was it? I don't want you to start thinking this is complicated because we are using three words that you don't normally use everyday. High voltage (around 200 volts D.C.) from the power supply is placed on the plate of the tube. The cathode of the tube is connected to ground through a small cathode resistor. Electrical current flows into the tube from the cathode and out of the tube through the plate. The grid of the tube is what will control how much actual current flows; depending on how positive or negative it is in relationship to ground. The tube is like a valve and the grid is what controls the valve. Now remember, the grid is connected to the one end of a one Meg ohm resistor that is connected to the guitar's pickup via a patch cord. When a string is plucked, an alternating voltage is produced that matches the movement of the string. Remember alternating current goes both directions—plus and minus. If you are playing an "A" 440 on your guitar, the string is moving back and forth 440 times per second. That means the alternating current coming from the pickup is also changing polarity from plus to minus at the rate of 440 times per second. That also means the grid is allowing more or less current to flow at 440 times per second. Since the tiny voltage of the pickup is controlling the 200 volts on the tube, there is some real amplification going on here. In fact, with a 12AX7, the amplification factor is 100.

That means for every change in grid voltage of one volt, there will be a 100 volt change in plate voltage. If you are using a single coil pickup guitar that is putting out ⅛ of a volt, then the signal voltage output of that 12AX7 will be 12.5 volts A.C.

But the plate of the preamp tube now has high voltage D.C. and signal A.C. voltage mixed together on it. How do we separate the A.C. signal voltage from the D.C. power supply voltage so we can pass the signal on through the amp to the next stage? We use a coupling capacitor (also called a blocking capacitor) which passes A.C. but blocks D.C. This allows us to take the signal voltage, which is A.C., and separate it from the D.C. voltage that was placed on the plate to make it work in the first place. The A.C. signal voltage can now be connected to a volume control or tone controls or another stage of gain.

So we have looked at one stage of gain in the preamp circuit. Most amps have multiple stages. If you want to use tone controls, there will be a loss of signal; so the signal is routed into another preamp tube to bring the signal voltage back up. Perhaps a volume control is used to adjust the exact volume of the preamp. Perhaps a reverb circuit or other tone control options are added to the preamp section.

Regardless of how many stages of gain, there will not be enough power in a preamp to drive a speaker. A preamp may put out enough voltage to drive a speaker, but it lacks sufficient current to drive a speaker. This is where another type of tube is used to add power.

THE OUTPUT STAGE

It takes wattage to drive a speaker and wattage equals voltage time current. The preamp has already amplified the voltage, now we need something to amplify current and that is where the output stage comes in. In electrical terms, power (wattage) equals voltage times current. With a preamp tube, the primary characteristic is an increase in gain (same as an increase in A.C. signal voltage). With a power tube, the primary characteristic is an increase in current. You already have the needed gain from the output of the preamp circuit. You need plenty of current added to that gain in order to develop enough power to drive a speaker. Power tubes are designed to add current and not increase the gain. This is why you never see a power tube such as a 6V6 used as a preamp tube.

The output signal from the preamp section is connected to the grids of the power tubes. Just as the grid of the preamp tube controls the flow of current in the preamp circuit to produce voltage gain, the grid of the power tubes (which is connected to the output of the preamp section) controls how much current flows through the output tubes to produce wattage.

Even though the power tubes add power to the gain that is already present from the preamp, the output impedance of a power tube is still much too high to drive a speaker. An output transformer is used for two reasons. First it matches the impedance of the speaker to that of the output tubes, and second; it blocks D.C. and keeps it from getting to the speaker. High voltage D.C. could blow a speaker instantly. A transformer, like a blocking capacitor, will allow A.C. signal voltage to pass while blocking the D.C. D.C. cannot flow through a transformer. Think of the output transformer as a type of transmission. You have a high impedance engine (think very high rpms with low power) and you are trying to move a heavy load up a hill (the speaker represents a heavy load because it requires much power).

Of course, there are many different styles of output stages. You could use only one tube, in which case, the output stage is said to be single-ended. You could use two tubes and the most common method of configuring two tubes would be push-pull. You could use four tubes, again push-pull. If the tube or tubes never cut off during a cycle of the input signal, the tubes are said to be operating in Class A. If the tubes cut off less than 180 degrees of each input cycle, then they are said to be operating in Class AB. If there is only one tube, it is operating in Class A because with one tube, you can't cut it off and expect for there to be any output.

THE SPEAKER

At this point, we have signal that has the voltage amplified and the current amplified. Together this gives us power. That power has been transformed to the correct impedance and now it is ready to be sent to the speaker.

The speaker simply takes the electrical energy and converts it into mechanical energy. Different speakers have different characteristics because the actual components that make up the speaker (magnet material, cone type, paper type, style of basket, the voice coil, spider, top plate, pole piece and back plate) will color the sound. Also, speakers can come in various impedances; 8 ohm being the most common, then 16 ohm then 4 ohm.

A speaker is basically a metal basket with a moveable paper cone. Attached to the apex of the cone is a coil of wire. This coil is called a voice coil. The voice coil is wound on a bobbin that is very precise. A magnet and corresponding pole piece is situated on the speaker frame such that the voice coil can interact with the magnetic field created by the magnet. As the amplifier sends powerful A.C. signal to the speaker's voice coil, the voice coil becomes an electromagnet; however, the polarity of this electromagnet is changing at the same rate as the guitar string vibrating. The paper cone vibrates as the magnetic field created by the voice coil is attracted and repelled by the permanent magnet. Since the polarity of the signal sent to the speaker is alternating at the exact same rate as the guitar string, a sound is made that is the same as the guitar, except it is much louder.

CONCLUSION

There is method to this madness. When an amp is designed, the circuit is placed inside a chassis; the power supply is almost always as far as possible away from the preamp as possible. Why is that? Because the power supply has 120 volt A.C. in it and if you get it close to the preamp, the 60 Hz hum can couple (by stray capacitance or stray inductance) to the preamp and be amplified. This would make for a very noisy amp. So if you open up an amp and see the power cord, it will be going to the power supply. Ninety-nine times out of a hundred, the preamp will be on the opposite side of the chassis. And the output stage will be in between the two. Also, the input jacks will nearly always be next to the preamp. Why is that? Because the input signal needs to go to the preamp section and it would only make sense to design them there.

Amplifiers can be designed with more features, multiple channels, multiple speakers, more gain, more power, stereo output stages, and other variations; but when you break it down, they all work using the four basics described.

All About Vacuum Tube Guitar Amplifiers

ALL ABOUT VACUUM TUBES AND HOW THEY WORK

It all started on February 13th, 1880. Thomas Alva Edison had failed over 10,000 times to create an incandescent lamp that would last more than a few minutes. He had tried all sorts of experiments but it seemed that no matter what he tried, the filament in the light bulb would burn up and black soot would form on the inside of the bulb. He needed a way to keep the filament from burning up and something to attract the soot so it wouldn't cover the inside of the glass. He tried placing a metal plate inside a glass bulb whose air had been removed to form a vacuum. This plate was electrically separate from the filament. He noticed that when the metal plate was given a more negative charge than the filament, electrons were not attracted from the filament to the plate. And conversely, when the metal plate was given a more positive charge than the filament, electrons were attracted from the filament to the plate and electrical current would flow through the vacuum. He found that the current flowing to the plate would increase rapidly with increasing voltage, and filed the first US patent for an electronic device on November 15th, 1883 for a voltage regulating device using his "Edison Effect." Oddly, Edison did not think this discovery was very significant!

It would be over twenty years later before the British physicist Sir John Ambrose Fleming, a university professor working as an engineering consultant for many technology firms of his day, including Edison Telephone and the British "Wireless Telegraphy" Company, discovered that the "Edison Effect" could be used as a type of electrical "check valve." Later known as the diode, meaning two active components, the device allowed electrical current to flow in only one direction from filament to plate, enabling rectification of alternating current. Fleming patented the diode over 100 years ago on November 16, 1904.

Because all tube guitar amplifiers use D.C. current which only flows in one direction, diodes can be found in the power supply of every tube guitar amp as they are used to change (rectify) A.C. wall voltage that flows both directions into D.C. voltage that flows in only one direction. Actually, most guitar amplifiers use a vacuum tube rectifier with two plates which is called a full-wave rectifier. Think of it as two diodes in one glass envelope to economize physical size.

Fleming Diode.

Two plates are used so that both halves of the 60 Hertz wall electricity can be converted into D.C. electricity. Common examples of these full-wave rectifier tubes are the 5AR4, 5U4, 5Y3, and 5V4. These all use a 5 volt filament (with the American tube nomenclature, the first digit of the name is how many volts are required to heat the filament). On solid-state rectified tube guitar amplifiers such as a blackface Fender Twin Reverb or a Marshall 100 watt, instead of a tube rectifier, one or more solid-state diodes such as the 1N4007 or the 1N5399 are used. The solid-state diode, like a rectifier tube, is a type of electrical "check valve" that only allows electrical current to flow in one direction.

But even after the patent of the diode in 1904, amplification was not possible. It would be a few years later when a pre-teen schoolboy working for Lee De Forest would discover that a grid of wire placed between the filament and plate of the diode, could control how much current would flow from filament to plate. This "control grid" acted something like the handle of a valve and was actually made from a piece of wire that was bent back and forth to make a grid. If the grid was charged negatively, less current would flow. As it was charged positively, more current would flow. As the voltage applied to the grid was varied from negative to positive, the number of electrons flowing from the filament to the plate would vary accordingly. Thus the grid was said to "control" the plate current, hence the name "control grid." This discovery brought forth the possibility of amplification. Lee De Forest patented the device in 1907 and called it an Audion. The device is now known as a triode, meaning three active components. Later it became common to use the filament to heat a separate electrode called the cathode, and to use this cathode as the source of electron flow in the tube rather than the filament

Audion Diode.

itself. Today, most triodes have this separate metal electrode (cathode), situated over the filament and connected to the circuit. This works better because the filament can be constructed from metals that are best for making filaments and the cathode can be made from different metals that are best for emitting electrons. Also, when using an A.C. voltage to power the heater, hum is minimized. So the three active components of the triode are cathode, grid and plate. The filament is not counted when there is a separate cathode because it is just used as a heater to heat the cathode and is not connected to the signal path. Heaters are necessary because electrons will not flow from the cathode unless it is hot.

In the preamp circuit of a guitar amplifier, the control grid of the triode is actually connected to the guitar pickup via the input jack of the amplifier. As the guitar string vibrates, an A.C. signal voltage changes from positive to negative—thus causing more or less electrons to flow from the cathode to the plate in the triode. The most popular type of triode used in a guitar amplifier is the 12AX7, which is actually two triodes in one glass envelope to minimize physical size and conserve space.

The non-linear operating characteristic of the triode caused early

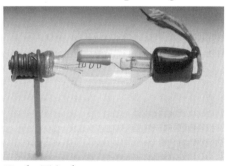

Early Triode.

tube audio amplifiers to distort the signal in a bad way. To remedy this problem, engineers plotted curves of the applied grid voltage and resulting plate currents, and discovered that there was a range of relatively linear operation right in the middle of the operating parameters of the tube. In order to use this range, a negative voltage had to be applied to the grid to have the tube idle in the middle of its range when no signal was applied. This concept is called grid bias. Once set, the plate current, with no actual signal applied, is called the "idle current." Preamp tubes such as the 12AX7 require very little grid bias and almost always develop this by placing a small resistor (usually 1500 ohms) between the cathode and ground. When current is drawn through this resistor, a small voltage develops on the cathode of the tube, thus making the cathode slightly positive with respect to the grid. The tube "sees" this the same as having the grid negative with respect to the cathode. Using the resistor on the cathode to achieve bias is called self-bias or cathode bias and is used on nearly all preamp tubes. Output tubes are sometimes biased this way and sometimes there is simply a negative voltage power supply that injects a constant negative voltage on the grid. The act of adjusting this negative voltage to get the best idle point is called biasing.

Back in the early days, batteries were used to provide the various voltages needed in an electronic tube device. There were two or three different voltages needed. In America, the filament/heater voltage supply batteries were called the "A" batteries or the "A" supply. In Britain, this battery was called the low tension supply. In America, the "B" batteries provided the plate voltage and because this was always a high positive voltage, the name evolved into "B+" supply in America and HT or high tension in Britain. And with amplifiers that used a negative voltage supply to provide bias voltage, the term "C-" supply or "C" battery was used. In Britain, these were called the grid bias battery.

Later, when A.C. wall power was used instead of batteries, sometimes a "C-" battery was still used because the grid bias battery, which draws almost no current, could sometimes last for a couple of years!

To increase the efficiency of the tube and to prevent unwanted parasitic oscillation in radio circuits, it was discovered that the addition of a second grid, located between the control grid and the plate and called a screen grid, could solve these problems. A positive voltage, lower than the plate voltage, was applied to this screen grid. The electrons would be attracted towards this screen, but after they got that far, they would see a higher voltage and more mass on the plate and go to the plate instead of the screen. This four component tube is called a tetrode and the four active elements are: cathode, control grid, screen grid, and plate. Sometimes tetrodes are used as power tubes in guitar amplifiers. For example, Marshall JTM45 amplifiers used the KT66. The KT66, KT77 and the KT88 are tetrode versions of the 6L6, EL34 and 6550 power tubes respectively. The "KT" in the name stands for "kinkless tetrode." The tetrode, though more efficient than a triode, has some new problems of its own because electrons can strike the plate hard enough to knock out other electrons. This is called secondary emission. In a triode, these electrons are simply re-captured by the plate; however in a tetrode, they can be captured by the screen grid, thus reducing the plate current and the amplification of the circuit! Another problem of this effect is that under severe overload, the current collected by the screen grid can cause it to overheat and melt, destroying the tube. That is why it is very important when using a tetrode to use a large screen resistor to limit screen current. If you don't, then too many electrons will go to the screen grid instead of the plate and the screen grid will burn up. Even a vintage Fender Champ with low plate voltage will burn up the screen of a KT66 if the Champ lacks a screen resistor.

So in 1928, Tellegen and Hoist of the Philips Company, in Holland, invented the suppressor grid to suppress the secondary emissions. The suppressor grid is placed between the screen grid and the plate, but is kept at the same electrical potential as the cathode which is negative compared to the plate. When electrons, which are also negative, bounce off the plate, they see the negative charge on the suppressor grid and they are repelled back to the plate. Opposites attract, and likes repell. This new tube is called a pentode because of its five active elements; namely: cathode, control grid, screen grid, suppressor grid, and plate.

Almost all guitar amplifiers use pentodes for the output tubes. Some common examples are the EL34, EL84, 6L6, 6V6, 5881, 6550 and 7591A.

Tube Amp Terms and Their Meanings

All About Vacuum Tube Guitar Amplifiers

UNDERSTANDING ELECTRONIC CONCEPTS

Tube amp electronics can be easily understood once you familiarize yourself with certain common electrical terms. There are a few absolute principles of electronics—like other scientific phenomenon such as gravity. Gravity doesn't change. You throw something up, and it always falls down. Gravity works like this all the time. So in this sense, gravity is absolute. Oh sure, there are those fun-houses where the insides are contorted to make you think down is up, but gravity doesn't care. It keeps pulling stuff down all the time. And gravity is not like those cartoons where the gravity doesn't take effect until the cartoon character notices he just ran off a cliff! To understand gravity, there are certain concepts like "up" and "down" that must be understood.

Electricity has its concepts too. An electrical circuit is when a conductor connects two or more unequal charges such that a complete pathway for electrical current exists. When you turn on a flashlight, you create an electrical circuit. Electrical current flows from the negative terminal of the battery through the flashlight bulb, through the switch and back to the positive terminal of the battery.

Current is the quantity of electrons flowing past a certain point. If electricity were water, current would be "gallons." To use the water analogy, a river would be considered high current, whereas a small spring would be low current. For example, the volume of water flowing through a drinking straw can be seen as low current, whereas the volume of water flowing from the mouth of the Mississippi River can be seen as high current. A hole in the dam is low current, but if the dam collapsed, there would be high current. The unit for measuring electrical current is "Amperes." which is usually shortened to "Amps."

Voltage is electrical pressure or you could think of it as "speed." The measurement for voltage is volts. For an analogy, think of a high-pressure car wash, this is like high voltage. If you picture a leaky faucet, that is like low voltage. For example, a 1.5 volt battery would have low pressure, err, voltage—while the 120 volt wall outlet has much higher electrical pressure. For instance, a high-pressure car wash would be like high voltage, whereas a bad shower in a $20 motel would be like low voltage (low pressure). The unit for measuring voltage is "volts."

It is possible to have high current/low voltage, high current/high voltage, low current/high voltage and low current/low voltage. Keeping with the water analogy, a slow moving river would be high current but low voltage. Niagara Falls would be high current with high voltage. A high-pressure car wash would be low current with high voltage, whereas a slow moving spring might be low current with low voltage.

In a guitar amplifier, there are different voltages to operate different circuits. Low voltages, usually 6.3 volts, operate the heaters (a.k.a. filaments or filament heaters) in the tubes. This is the part of the tube that lights up like a light bulb. The function of the heater is to warm the inside of the tube hot enough so that it will operate properly. There are also high voltages on the plate circuits of all the tubes. Some tubes are run at higher voltages than other tubes depending on the actual type of tube.

Wattage or Power is the combined effect of current and voltage. The formula for calculating wattage is simply the voltage (expressed in volts) times current (expressed in amps). Keeping with the water analogy, Niagara Falls has lots of power because it is high current (gallons) multiplied times high voltage (pressure). For example, there is much more power in Niagara Falls than a slow moving spring.

If you have 1 volt of electricity at 1 amp of current, you would have 1 watt of power. If either the voltage or the current goes up, so does the power. For example 2 volts at 1 amp is 2 watts of power. Or 2 volts at 4 amps would be 8 watts. Remember, Power = Voltage times Current. One need only know some of the information to find out the rest of it.

Resistance is any opposition to the flow of electrical current. The unit of measurement of resistance is ohms. For example, the pilot lamp in your guitar amp has resistance, and so does the speaker. But there are electronic components whose sole purpose is to resist the flow of electrons. These are obviously called resistors! But actually, anything that resists the flow of electrons has resistance. This is

All About Vacuum Tube Guitar Amplifiers

somewhat like friction, because resistance generates heat. With the water example above, resistance would be analogous to a "bottleneck" in a pipeline or a kink in a garden hose. In a guitar amplifier, resistors are used to create resistance in various parts of the circuit.

Resistors have a property that can be very useful. When an electrical current is passed through a resistance, a voltage develops across that resistance. This is easy to imagine, just think of trying to force a gallon of water quickly through a soda straw. There would be pressure building up where the water was entering the straw. This build up of "pressure" would be similar to the build up of voltage that occurs when current is passed through a resistor.

If you increase the current, that voltage that develops across the resistor will increase. If you decrease the current, the voltage will decrease. This is an extremely useful quality that can be used for a number of functions.

Resistors are used in many different circuits throughout an amplifier, but the essence of a resistor is the fact that a voltage will develop across it when electrical current passes through it. The exact amount of voltage that develops or "drops" across a resistor is known as voltage drop. A mathematical formula called Ohm's law is useful to determine exact voltage drop.

OHM'S LAW—Perhaps the most basic of absolute electronics principles is Ohm's Law. In the early part of the 1800's, Georg Simon Ohm proved by experiment that a relationship exists between current, voltage and resistance. This relationship is called Ohm's Law and is stated as follows: The current in a circuit is directly proportional to the applied voltage and inversely proportional to the circuit resistance. Ohm's Law can be expressed by the equation: $I = E/R$. The current (I) in an electrical circuit equals the voltage (E) divided by the resistance (R). Using Ohm's Law, one can deduce what will happen when an electrical current is passed through a resistance.

Using algebra to manipulate the basic equation, you could multiply both sides of the equation by R and extrapolate that Voltage (E) equals current (I) times resistance (R). $E = IR$.

Again, one could divide both sides by I and manipulate the equation to get $R = E/I$ or resistance equals voltage divided by current.

So we connect a 6 volt battery to a lantern. The lantern's bulb measures 75 ohms on an ohmmeter. The 6 volts is connected to the 75 ohm bulb and electrical current flows. The amount of current flowing through this circuit is 6 volts divided by 75 ohms which equals .08 amps. Now if we wanted to convert amps to milliamps, we would move the decimal over three places to the right converting to 80 milliamps. That wasn't so hard, was it?

But let's say we had only an ammeter, and could not measure the D.C. resistance of the light bulb. We could have put an ammeter in series with the battery and light bulb and measured the .08 amps of current. Plugging this into the equation: voltage divided by current equals resistance, we would get 6 volts divided by .08 amps = resistance; which would give you the 75 ohms! See how easy this stuff is!

But what if we had an unknown battery voltage and we connected the light bulb to the batteries? If we knew the current and resistance, we could have figured out the voltage. Current times resistance equals voltage, or, .08 amps times 75 ohms equals 6 volts.

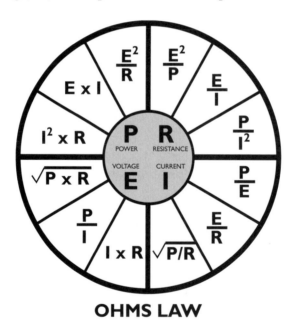

OHMS LAW

Look at the Ohm's Law Table. The center of the circle shows Power, Resistance, Voltage and Current. There are several equations expressed in terms of what is already known.

For example, let's say that we know the current and the resistance of a circuit and we want to find voltage. Looking at the center of the chart we find Voltage. We see there are three equations related to Voltage and since current and resistance are known, we choose the equation that is expressed with current and resistance. Voltage (E) equals current (I) times resistance (R). Substitute in the numbers and you have the answer.

Here is a practical application. Let's say you are playing a tweed Deluxe amplifier and you would like to know how much power it is putting out. You look at the ohm's Law Chart and find power (P). You see that voltage (E) squared over resistance (R) equals Power. So you hook a voltmeter up to the speaker leads on your amp and start playing the amp. As you are playing, you are looking at the meter to see how much AC voltage is appearing across your 8 ohm speaker. You see it staying somewhere around 12 volts. Here is the math: You square the voltage (12 volts) to get 144 and then divide by the resistance (R), which is 8 ohm for the resistance of the speaker. The result is 18 watts.

SERIES CIRCUIT—When two or more electronic components are connected in a linear fashion such that all the electrical current must pass through one component before it can go through the other component; then that type of connection is called a series circuit. With a series connection, the same current flows through every component. However, the voltage across each component is independent of the other and will vary depending on the resistance of each component. Of course, Ohm's Law can calculate the voltage. Remember that voltage (E) equals current (I) times the resistance (R). If the two or more components have different resistances, then the voltage to each component will be unique because the current flowing through each is constant. The sum total voltage of the multiple components will equal the total applied voltage. That is to say that if you connect three resistors in series, and connect the series resistor assembly to a 9-volt battery, then the total voltage developed across each resistor added together will equal the 9-volts of applied voltage.

For example, if you had two speakers and you connect the plus lead of one speaker to the minus lead of the other speaker and then used the free lead on each speaker to apply electrical current, then those speakers are said to be wired in series. In order for the current to flow, all of it must go through one speaker before it can go through the other.

In a series circuit, the resistance of each component added together equals the total resistance. Total Resistance = R1 + R2 + R3. Let's look at an example where we connect a 1,000 ohm, a 500 ohm and a 3,000 ohm resistor in series. This assembly is connected to a 9 volt battery. We know that the total current flowing through the circuit flows through each component. Since the total circuit resistance is 4,500 ohms, we can use Ohm's Law to calculate the amount of current actually flowing. I = E/R, or, current equals voltage divided by resistance. The voltage is 9 volts and the total resistance is 4,500 ohms; so the current is 9/4,500 amps. This reduces to .002 amps. Move the decimal three places to the right and convert to milliamps and you express this as 2 milliamps.

We can then use Ohm's Law again to determine how much voltage will appear across each resistor. We already know that .002 amps of current flows through each component. We know that voltage equals current times resistance and we have three resistors. The 500 ohm resistor will have 1 volt across it (.002 X 500), the 1000 ohm resistor will have 2 volts across it (.002 X 1000), and the 3,000 ohm resistor will have 6 volts across it (.002 X 3,000). Notice that the sum of the voltages add up to 9 volts!

With a series connection of resistive components such as speakers, or resistors, the overall resistance of the circuit is the sum of each resistance in that circuit. If you had two 8 ohm speakers that were series connected, then the total ohms would be 16 ohms.

With a series connection, all of the current must go through each component, so the current is always the same for each component. However the voltage will drop across each component in accordance with Ohm's law. (Remember the discussion about "voltage drop" when we talked about resistance?)

That is to say with two speakers connected in series, each will see only half the total voltage but the current through each speaker will be the same. This is exactly the opposite with parallel connected speakers because with parallel connected speakers, the voltage on each will be the total voltage but only half the current will go through each speaker. PARALLEL CIRCUIT—When two or more electronic components are connected across each other such that the electrical current has two or more possible paths it can go, then those components are said to be connected in parallel. In a parallel circuit, the voltages are the same across each component, but the current is split up among the various components in the parallel connection. In parallel connections, the current going through one component cannot go through the other, as each component is a separate electrical path.

The most obvious example of a parallel circuit is when an extra speaker cabinet is used with the regular speaker. Extra speaker jacks are almost always connected in parallel. You may already know that if you connect two 8 ohm speaker cabinets in parallel to an amp, then amp sees 4 ohms. The first time I found out about that it made no sense. It would seem that 8 ohms plus 8 ohms would equal 16 ohms. And that is correct if you were connecting them in series. But when two speakers or any two resistive components are connected in parallel, there are multiple paths for the electrons to take and it becomes much easier (less resistive) for current to flow.

Think of an hour glass. There is a small hole for the sand to flow through from top to bottom. Every grain of sand must flow through this one hole. But what if we somehow had two holes for the sand to flow through. Then every grain of sand would not have to flow through one hole. Some grains could go through one hole while others went through the other hole. By adding the extra hole, it would be like having two resistances in parallel. The actual resistance would be cut in half since there are now two paths instead of one.

But the resistance would be cut in half only because the two speaker loads are the same number of ohms. For example, two 8 ohm speakers in parallel would make 4 ohms total load.

What if you had an 8 ohm speaker in parallel with a 16 ohm speaker? Or an 82K resistor in parallel with a 47K resistor? When there are two resistances in parallel, one may find the total resistance with a formula: R1 times R2 divided by R1 plus R2.

So in the case of the 8-ohm (R1) and 16-ohm (R2) speaker parallel connected, here is the math. If you multiply 8 times 16, you get 128. Divide that by their sum (24) and you get 5⅓ ohms. That is the answer.

But what about the 82 K resistor in parallel with the 47K resistor? 82K times 47K equals 3,854 Meg. Divide that by their sum of 136K and you get 28.338K for the answer. Notice how in both examples the total ohms of a parallel circuit is always less than the lesser of the two resistances. In the first example, the smallest speaker was 8 ohms, yet the total was 5⅓ ohms. In the second, example, the smaller value was

47K and yet the parallel connection total was 28.338K. Resistance will always be less with parallel-connected components because there is always less resistance when there are more pathways. And parallel connections each provide a separate pathway for current to flow.

In a parallel circuit, although the same voltage is present across all components and will equal the applied voltage, the current flowing through each component will vary depending on the resistance of that component. Of course, the current can be calculated by using Ohm's Law. Current equals voltage divided by resistance.

For example, if we took two 8 ohm speakers and connected the plus leads together, then connected the minus leads together and then applied signal between the plus and minus leads, those speakers are said to be wired in parallel. When parallel wiring speakers of the same impedance (ohms), the total ohm value is found by dividing the ohms of one speaker by the number of speakers. With parallel connections, it is easier for the current to pass because there are multiple paths (like having an hour glass with two ways for the sand to pass instead of only one). Two 8 ohm speakers connected in parallel become 4 ohms. If you used three 8 ohm speakers wired in parallel, you would get $2\frac{2}{3}$ ohms. Of course four speakers wired in parallel would give 2 ohms. Every time you add another, there is another path the current can go and the resistance is lowered.

In a series connected circuit, there is no current division because all of the current goes through every component. However, the series connected circuit has the characteristic of voltage division.

Parallel circuits work in an opposite fashion. There is no voltage division for parallel-connected components. The voltage is the same across every component in a parallel connection. In a parallel-connected circuit, each component is a separate pathway for current to flow, so the current must divide amongst the different paths. The current will divide in proportion to the resistances involved. For example, using the same example mentioned earlier, we could look at an 8 ohm speaker in parallel with a 16 ohm speaker. We already know that the total ohms would equal $5\frac{1}{3}$ ohms. But the current does not divide evenly. The 8 ohm speaker will have twice as much current flowing through it as the 16 ohm speaker. So one third of the current would go through the 16 ohm speaker and two thirds of the current would go through the 8 ohm speaker.

You could use Ohms law to show that the 8 ohm speaker would actually get two thirds of the power and the 16 ohm speaker would only get one third. One form of Ohms law is: wattage equals voltage squared divided by resistance. Because they are parallel-connected, the voltage across both the 8 ohm and 16 ohm speaker is equal, so when you divide that constant voltage by the resistance, the 16 ohm speaker will develop ½ the wattage that the 8 ohm takes. When you look at the total wattage, the 16 ohm is getting ⅓ total wattage while the 8 ohm is getting ⅔ of the total which is another way of saying the 8 ohm's wattage is twice what the 16 ohm would be.

SERIES/PARALLEL OR PARALLEL/SERIES CIRCUIT—Sometimes an electronic circuit will have some of the components in parallel and other components in series. When both parallel circuits and series circuits are combined we know that:

1. The sum of current flowing through each component in a parallel circuit equals the applied current.
2. Parallel connected components will have the same voltage across them.
3. The sum of voltage drops across each component in a series circuit equals the applied voltage.
4. Series connected components will have the same current flowing through them.

Using the speaker example, if you had four 8 ohm speakers and wired them such that there were two pairs—the two speakers of each pair wired in parallel (this would make each set 4 ohms) and then wired the two sets in series (you would add the 4 ohms from each set to get 8 ohms); then those speakers are said to be wired in Parallel/Series.

Alternately, one could wire four speakers such that there were two pairs with two speakers in each pair wired in series with each other (this would make 16 ohms per pair) and then wire those two sets in parallel with each other (this would bring it back to 8 ohms total). Then those speakers are said to be wired in Series/Parallel.

VOLTAGE DIVIDER—If you take ten 1 ohm resistors and connect them in series—that is one lead of one resistor connected to the next, then the other end of that one connected to the next, etc., such that you would end up with a long string of 1 ohm resistors with two free ends; there are a few things we know about this:

1. Since they are all in series, you would add the sum of the resistances to get the total resistance. If you used ten 1 ohm resistors, the resistance would be 10 ohms.

2. If you applied 20 volts to one end and you connected the other end to a point of zero volts (a.k.a. ground), then each 1 ohm resistor would have 2 volts of electricity across it. That is to say, if we measured from ground, to the fifth resistor; then we would get exactly 10 volts.

3. Also with the scenario above, if you measured across any single resistor by connecting the two leads of your voltmeter to the ends of any single resistor, then the meter will read 2 volts.

To sum it up, the voltage drops evenly across an even resistance. This principle is exploited in many different ways to achieve particular results in a guitar amplifier.

For example, the volume control is basically a long carbon trace that has the property of resistance. When you turn the shaft, there is an element (called a wiper and connected to the shaft) that can be adjusted to any point on the resistive carbon trace. This will vary the resistance from the wiper to either end. Let's say we are feeding the volume control with a 10 volt signal. That 10 volt signal is connected to one end of the resistive carbon trace inside the potentiometer, the other end of the trace is connected to ground (zero volts), and the wiper can be adjusted anywhere from zero to 10 volts — depending on where the wiper is physically touching the carbon trace. When the wiper is closest to zero volts, then it too will be zero volts. If the wiper is placed half way up the trace, then the voltage on the wiper will be 5 volts.

One could take this same voltage divider (volume control) idea and combine it with frequency sensitive circuitry to get tone controls. Your treble, middle and bass controls are designed by arranging various capacitors and resistors that select certain ranges of frequencies and sending those signals to a voltage divider.

Besides almost every potentiometer on the amp operating as a voltage divider, the voltage divider concept is also used in many other circuits in the amplifier. For example the number 2 input on a Fender blackface or silverface amp is configured as a voltage divider. When you plug into the number 2 input, a switch is opened that puts two 68K resistors in series. One free end is connected to your guitar pickup and the other end goes to ground. The output signal is taken off the junction of these two resistors, thus the signal is actually cut in half. This configuration actually cuts the voltage from your pickup in half, thus reducing the overall gain by 3dB.

There are two 220K resistors on the main filter totem pole stack in the power supply of almost all blackface and silverface Fenders. By putting these resistors in series with each other, but in parallel with the two 70 uf / 350 volt filter caps, the voltage is divided exactly in half across each filter cap. This insures that there are equal voltages across each 70 uf filter capacitor.

The voltage divider concept regulates how much negative voltage the bias supply is putting out to the grids of the output tubes. Negative voltage bias supply circuits will always have two or more resistors and possibly an adjustment pot, configured as a voltage divider, to divide off just the right amount of voltage.

Sometimes, the voltage divider is used to drop the gain of a particular stage. In the 6G15 Fender reverb, for example, the 12AT7 between the input jack and dwell control uses a voltage divider on the actual plate load resistor. Two resistors are put in series such that only $\frac{1}{11}$th of the signal is divided off to go to the dwell control.

A.C. AND D.C. ELECTRICITY—Electrical current can consist of two types: Direct current (D.C.) and Alternating current (A.C.). Direct current is where the electricity flows in one direction and the direction does not change. Alternating current is where the electricity flows one direction and then changes to go in the opposite direction; hence the name: alternating current.

Your wall outlet is an example of A.C. The electricity from a wall outlet goes one direction and then back 60 times per second. Another good example of A.C. is your actual guitar signal. For example, the "A" note of a guitar will have 440 cycles of back and forth each second.

All amplifiers work on D.C. electricity, but the wall outlet only puts out A.C. electricity, so every amplifier has a power supply section that takes the wall A.C. and converts, or "rectifies," it into D.C. Usually the D.C. needs to be much higher than the 120 volts coming from the wall, so the electricity from the wall is first run into a step-up transformer (called a power transformer) that brings the voltage up to a higher level. The electricity is then put through a rectifier circuit to change it to D.C. A rectifier circuit can be made either with diodes or a tube, but it is basically an electrical check-valve that only lets the electricity flow one direction instead of both directions. This is how it takes A.C. (both directions) current and rectifies it to D.C. (all going the same direction) current.

INDUCTANCE—Inductance is the ability for a coil of wire to store energy and oppose changes in current. Sometimes an inductor, called a "choke," is used in the D.C. power supply of a guitar amp. It is used to oppose any changes in current. It works great for smoothing out the ripples in the current that occur when A.C. is rectified into D.C.

Inductors also have a special property about them that is most useful in audio circuits. An inductor likes to pass low-end frequencies but doesn't like to pass high frequencies. Inductors can sometimes be found in crossover networks used in 2-way speaker systems such as a PA speaker, for example the Altec Lansing Voice of the Theatre. Since inductors like to pass low frequencies, an inductor is put in series with the woofer. The lows go straight through while the highs are blocked. Additionally, sometimes an inductor is put across a tweeter to prevent lows from going through the tweeter. The low end will see the inductor as a short and bypass the tweeter altogether. The highs will not go through the inductor and go through the tweeter instead.

CAPACITANCE—Capacitance can be defined as how well a capacitor stores electrical charge. A capacitor is a device that stores electrical charge on conducting plates through the action of an electrostatic field between the plates. In reality, a capacitor is simply two conductors separated by a non-conductor.

Think about combing your dry hair on a winter night. Your comb will build up a static electrical charge on it. You can even use the comb to pick up small pieces of paper. The comb is an example of capacitance. The comb has an electrical charge stored on it. When I was a young child, I played with a plastic army rifle that made the hairs on my arm stand up when my arm was close to it. I was seeing the effects of capacitance.

Capacitance is measured in farads, which happens to be a great deal of capacitance, so to simplify everything; the farad is divided by a million and called a microfarad. Most of the capacitors inside an amp are rated in microfarads. If you divide a microfarad by a million you get a picofarad. A picofarad is a very small amount of capacitance. It is so small, that you could have a few picofarads of capacitance just by having two wires lay side by side! Even your shielded guitar cable has several hundred picofarads of capacitance between the center conductor and the shielding! For example, a 1000 picofarad (pf) capacitor and a .001 microfarad are the same value. To go from microfarad to picofarad you would move the decimal 6 places to the right.

In a guitar amplifier, the capacitor is used in many different ways. For example, it is useful for removing the A.C. signal component from a D.C. circuit. Sometimes a circuit has both D.C. (electrons moving the same direction) and A.C. (electrons moving in both directions.) A capacitor can be used to separate the A.C. signal voltage from the D.C. power supply voltage because it will pass A.C. signal voltage while blocking D.C. power supply voltage. Remember a capacitor is two conductors separated by a non-conductor, so A.C. cannot really pass through a capacitor; but it appears to. Actually, as a positive charge builds up on one conductor, an opposite charge will appear on the other conductor simultaneously, so the illusion is that the D.C. is passing. The capacitors in an amp used to separate A.C. from D.C. are called coupling capacitors or blocking capacitors. A typical value would be anywhere from 500 picofarads to .1 microfarads.

Capacitors are also used in tone shaping circuits. Although the capacitor can pass A.C., it does not pass all A.C. frequencies equally. The faster the frequency of the A.C. (think high-end), the easier it gets through a capacitor. The slower the frequency (think bottom-end), the harder it is for the A.C. signal to pass. Also, the more capacitance a capacitor has, the easier it is for a given frequency to get through. As the frequency goes up or as the capacitance goes up, the capacitor passes A.C. more easily. By selecting certain value capacitors, a circuit can pass more or less of certain frequencies making possible tone shaping.

Obviously, tone shaping is done in the tone controls. But tone shaping is also achieved by placing a capacitor across the resistor that feeds the cathode (place where electricity enters) of a preamp tube. These capacitors are called bypass caps because they allow A.C. to bypass the cathode resistor of a preamp tube, thus improving the amplification of those frequencies. The D.C. current (think of this as idling current) passes through the resistor and creates a small voltage drop across the resistor while the A.C. signal current goes through the capacitor instead. If we want lower frequencies to bypass the resistor, we simply use a larger cap. Typical values for a bypass cap range anywhere from .1 microfarads to 350 microfarads. The 25 microfarad value is the most common because when used in conjunction with a 1.5K cathode resistor, it bypasses all the frequencies that a guitar is capable of producing. A smaller value would not bypass all the lows and you would hear this as less bottom end.

Capacitors oppose a change in voltage, so large value capacitors are used to smooth out the ripples that occur when A.C. wall voltage is rectified into D.C. power. We want the ripples smoothed out so there is no hum present in the power supply. Typical filter cap values range from 8 microfarad to 220 microfarad with the 20 microfarad being the most common. These filter capacitors are also used to isolate or "decouple" one amplifier gain stage from the next stage. We don't want the A.C. voltage from one stage to superimpose itself on another stage. This could cause oscillations!

Also, capacitors are sometimes used in crossover networks for 2-way speaker systems. High end frequencies get through a capacitor much easier than low end frequencies, so a capacitor can be place in series with a tweeter and the highs will go straight through as if it isn't there, yet the lows cannot get through very easily. Additionally, a capacitor can be placed across a woofer so when highs approach the woofer, they go through the capacitor and bypass the woofer. Lows go through the woofer because they don't like going through the capacitor.

BIAS — A tube is a type of electrical valve. Current comes in the cathode of the tube and leaves through the plate. The electrical charge on the grid with respect to the cathode controls how much current can flow through the tube. This "grid to cathode" relationship is called "bias" (noun). On a preamp tube, the bias of the tube is such that when the tube is idling, it is idling at half way up. Preamp tubes are self-biased. In self-biasing, a resistor is used on the cathode of the tube. When current passes through the resistor, a voltage drop occurs across the resistor. Since the cathode is connected to the resistor, the cathode has the same voltage on it. The grid has no voltage on it and the cathode has a positive voltage on it, so the grid is negative with respect to the cathode.

On output tubes, sometimes but not always, a different type of biasing scheme is used. A negative voltage is constantly injected onto the control grids of the output tubes and the cathode of each output tube is grounded directly and not through a resistor. This is simply another way to make the grid negative with respect to the cathode. This is called fixed bias because a fixed amount of negative voltage is always present on the grids of the tubes. Some amps have a trim adjustment for the negative voltage supply. When one adjusts the trim adjustment of this negative voltage supply, then one is said to "bias" (verb) the amp. When it takes a very small voltage to bias a tube, such as a preamp tube or an EL84 output tube, the circuit is almost always cathode biased, as there will not be much difference sonically and a cathode bias circuit doesn't require a negative voltage supply. However, with tubes that take much more than a few volts to bias them, such as a 6L6, a much greater difference in sound can be heard when comparing cathode bias to fixed bias circuitry. Cathode biased output tubes have more compression, more overall sag, with a singing quality; and they are not as loud. Fixed biased output tubes are punchier, with tighter bottom-end and more focus. A Fender Twin or a Plexi Marshall is an example of a fixed bias amp. A Fender Champ or a Vox AC30 is an example of a cathode biased amp.

POLARITY—Probably the simplest example of polarity is the battery. One end is plus or positive and the other end is minus or negative. The battery is a D.C. electrical source. All of the electrons flow in one direction only, from negative to positive.

Some components in guitar amps must be mounted in a particular polarity. For example, all filter capacitors in a guitar amp are mounted with the minus lead going to ground with only one exception. In a negative voltage supply, the positive of the filter cap goes to ground and the negative goes to the circuit.

SINGLE-ENDED—When a single tube is used in a circuit to amplify an audio signal, then that tube is said to be operated as single-ended. All preamp tubes are single-ended. Since all preamp tubes are operated single-ended, when someone speaks of single-ended, they are usually referring to the output stage and not the preamp stage.

With amplifiers using only one output tube, the output stage is always configured as single-ended. Single-ended audio amplifiers must always be operated in class A. That means all single-ended amplifiers are class A, but that does not necessarily mean that all

Class A amplifiers are single-ended.

More on this later

VOLTAGE AMPLIFIER—Tube amplifiers circuits may be classified in a number of ways according to use, bias, frequency response or resonant quality. When classified according to use or type of service, amplifiers fall into two general groups—voltage amplifiers and power amplifiers. Voltage amplifiers are so designed so that signals of relatively small amplitude (small voltage) applied between the grid and cathode of the tube will produce large values of amplified signal voltage across the load in the plate circuit. The preamp tubes in an amp are almost always configured as a voltage amplifier, but not always. Output tubes are not considered voltage amplifiers.

A good example of a voltage amplifier is the 12AX7 preamp tube. When used as a first gain stage, it takes a small voltage from your guitar pickup, let's say ⅛ of a volt, and amplifies it a hundred times (i.e. ⅛ volt X 100 = 100/8 = 12.5 volts). It does not amplify the current, but only the voltage, so it will not have enough power to drive a speaker. This voltage amplification can be sent to a volume control, a tone circuit, a reverb circuit or many other possibilities. After the signal goes through a tone circuit for example, the 12.5 volts may drop down to ⅛ of a volt again, in which case the signal voltage is put through another stage of a 12AX7 to amplify the voltage back up to 12.5 volts. That explains why it is typical to see multiple stages of amplification in a guitar amplifier. After the signal voltage is amplified sufficiently, it is sent to a power amplifier so that current can also be amplified. It takes wattage to drive a speaker and wattage is voltage times current. The 12AX7 preamp tube gave us the voltage, now the signal goes to a power tube to amplify its current. The combined action of voltage and current will give the required wattage to drive a speaker.

POWER AMPLIFIER—When a tube delivers a large amount of power to the load in the plate circuit, it is said to be a power amplifier. Since power, in general, is equal to the voltage times the current, a power amplifier must develop across its load sufficient current to cause rated power to flow. With a power amplifier, the primary focus is on power and that is what distinguishes it from a voltage amplifier such as a 12AX7. The output tubes in a guitar amplifier are the power amplifiers. Their main purpose is to add current to the signal. This is different from a preamp tube (voltage amplifier) whose main purpose is to increase signal voltage. The preamp tubes only amplify voltage and that voltage is used to drive the power tubes, whose job it is to add current to the signal. Remember, it is the combined action of voltage and current that give us wattage; and wattage is needed to operate a loudspeaker.

GAIN—In a voltage amplifier, such as a 12AX7 or other voltage amplifier device, gain is the ratio of A.C. output voltage to A.C. input voltage. Certain preamp tubes have more gain than others. For example, with a 12AX7, if there is a change in grid voltage of 1 volt, it will produce a change in plate voltage by 100 volts. In a 12AT7, a change of 1 volt on the grid will produce a change of 60 voltages on the plate. Similarly, a 12AY7 will have a 44 volt change when the grid voltage is changed by 1 volt.

POWER AMPLIFICATION—In a power amplifier, power amplification is the ratio of output power to the input grid driving power. This is a concept similar to gain, but gain is increase in A.C. signal voltage whereas power amplification is an increase in power. You could think of it as power gain.

CATHODE FOLLOWER—The cathode follower is a single-ended class A degenerative amplifier, the output of which appears across the unbypassed cathode resistor. This circuit is used in the preamp section of some guitar amps. Though rare, an output tube could be configured as a cathode follower. The gain of a cathode follower is always less than unity. That means the voltage output is always slightly less than the voltage input however the circuit is capable of producing power gain.

So why would anyone use this type of circuit if it slightly loses gain? It is used to lower the impedance. Think of it as a buffer circuit that separates two other circuits. It gets its name because the output voltage (on the cathode) follows the input voltage. The output not only has the same waveform, but it has the same instantaneous polarity. The output is almost the same as the input, except is has more current (power), which is another way of saying it is lower impedance. Also, this circuit sounds very rich and detailed. It improves the stability of the amp because it is a degenerative amplifier. You will see it as the third stage in all early Marshalls, Fender tweed

Bassmans, most Kendrick amplifiers and many others. You can also find it in "dry signal" circuit of the 6G15 Fender stand-alone Reverb unit. Originally, Fender used it to buffer the tone circuits from the rest of the amp, but it really had great tone. I suspect that Marshall used it because they were trying to copy Fender!

PUSH-PULL—There are two popular ways that the output tubes in an amp can be configured as power amplifiers. The first way, is single-ended and we talked about that earlier. In single-ended, a single tube (or two tubes in parallel) connect to one end of an output transformer. Hence the name: single-ended. The other end of the output transformer goes to a high voltage (B+) supply. This type of configuration is always operated in Class A.

There is another way of configuring two or more tubes in an output stage and it is called PUSH-PULL. In push pull, a center-tapped primary is used on the output transformer.

The plate of one tube connects to one end of the transformer primary, the plate of the other tube connects to the other end of the primary and the high voltage supply connects to the center-tap. For this to work, the signal voltage driving these two tubes must be 180 degrees out of phase with each other. You've probably seen a picture of a sound wave where it has a hill and a valley. The top of the hill is 180 degrees out of phase with the bottom of the valley. Out of phase, simply put, means one tube is getting positive signal while the other is getting negative signal and vice versa.

What makes this work is the fact that the power tubes are driven out of phase, but the transformer "flips phase" to reinvert the signal coming off the secondary of the transformer back "in phase." How does it do this? Picture a bar of iron with a coil of wire wound around it. For example, if you situate the bar such that it is vertical and you are looking directly on top of the bar, you might notice that the turns of wire on that bar appear clockwise. Keeping the bar vertical, but changing your viewing perspective to directly underneath the bar, the turns now appear counterclockwise. So when plate current from one tube feeds the beginning of the transformer winding and moving "down" towards the center-tap, it will produce current in the secondary of the transformer that is 180 degrees out of phase with the signal produced when current is introduced from the end of the winding and flowing "up" towards the center-tap. That is why we want to drive the two tubes with out of phase signal—because the transformer will "re-invert" the signals so that they become "in phase" with each other when they appear on the secondary. Interestingly, the secondary develops signal that is the average of what the two tubes are putting out. This is why push-pull overdrive rules. When one of the tubes is pushed to cut-off (no current), the other tube is pushed to saturation. The average of those two will give a smooth overdriven waveform both at the hill and the valley of the wave.

Sometimes, to achieve greater power and more headroom, four tubes are used such that two tubes are connected in parallel to one end of the output transformer primary with the other two tubes connected in parallel to the other end. You've seen this in all high-powered amps such as the Marshall 100 watt or the Kendrick Spindletop.

There are many advantages to be gained by using the push-pull power amplifier configuration. Second harmonics and all even numbered harmonics, as well as even-order combinations of frequencies will be effectively eliminated if the tubes are balanced and if the harmonics are introduced within the output tubes themselves. Hum from B+ power supply, a big problem in single-ended designs, is substantially reduced because ripple components which are present in both halves of the primary on the output transformer, phase cancel each other. Also, filament hum which inevitably leaks from the 6.3 volt heaters to the plate circuit, will also phase cancel in the primary of the output transformer. So you end up with a quieter, more hum-free design.

Besides all that, a push-pull design can be operated in class AB which is more efficient than a single-ended Class A design. The result is greater than the sum of the parts. For example, a single-ended 6L6 might produce only 10 watts of power, so if you used two 6L6s in parallel with a single-ended design, you may end up with 20 watts. However, take two 6L6's with a push-pull configuration and you could end up with 50 watts or so. This is why you almost always see amps with two or more output tubes configured as push-pull.

Push-pull doesn't have to always be Class AB. It can also be operated in Class A. To do so, all you have to do is bias the tube where the idle plate current is half way between cutoff and saturation and limit the drive signal such that the tube never gets driven in to cutoff. This gives you class A push-pull, ala Vox AC30.

PHASE INVERTER—This is a circuit, the purpose of which, is to create two signals that are 180 degrees out of phase with each other. The phase inverter is used in a push-pull amplifier to drive the power tubes out of phase with one another. One phase inverter output supplies signal to the grid of one output tube and the other phase inverter output, which is 180 degrees out of phase with the first output, feeds the grid of the other output tube. There are several ways to achieve phase inversion. It could be done with a phase inversion transformer or with tubes. If done with tubes, it is usually done with a single triode or with two triodes.

Besides being used in push-pull power amps, you are likely to see this circuit in the preamp vibrato circuit of certain early 60s Fender amps. In these types of vibrato, an oscillator signal drives a phase inverter which modulates different frequencies. It is like having two oscillators that are exactly synchronized to modulate different frequencies in a pin-ponging effect.

TRANSFORMER PHASE INVERTER—Perhaps the simplest way to achieve phase inversion is through the use of a center-tapped inter-stage coupling transformer. The transformer-style phase inverter was used on many Gibson amps. The transformer has a centertap on the secondary that is either grounded or attached to a negative bias voltage supply—depending on whether the output stage is cathode biased or fixed biased. The two ends of the secondary each feed an output tube in the push-pull output stage. Of course, the transformer secondary must be tapped at the exact electrical center; otherwise the signals on the secondary will not be symmetrical.

This type of phase inversion has limited application because of losses and distortion inherent in transformers. For example, the loss in voltage (through leakage reactance) is greater for high frequencies than it is for lower frequencies. Also the shunting capacitance effect also increases with frequency. A really great sounding phase inversion transformer is cost prohibitive, which can explain why all the transformer phase inverters used on guitar amps lack detail and sound dull.

PARAPHASE INVERTER—The paraphase inverter uses two triodes or two sections of the same tube. We have all seen this circuit used on all "B" and "C" series tweed Fender push-pull amps. The first triode is configured as a regular amplifier. The output of this tube feeds one of the output tubes and a voltage divider. The voltage divider feeds another triode. When the signal goes through the second triode, the phase is flipped 180 degrees and the output feeds the other output tube. To keep the gain balanced, the voltage divider is used to bump the gain down slightly before the signal hits the second triode.

The inherent flaw in this design is the fact that the inverted signal will have distortion present that the uninverted signal lacks. Because of this, it is impossible to get a very clean tone. Also, because every triode is slightly different, it is nearly impossible for this circuit to ever have perfect balance. This circuit works well for blues harmonica or slide guitar

SELF-BALANCED PARAPHASE INVERTER — This circuit is nearly identical to the regular paraphase inverter, except for way the voltage divider in between the two triode sections is configured. We have all seen this circuit on "D" series tweed Fenders. The voltage divider is configured such that the second triode section's output is also put back through the same load resistor. If the second section has too much gain, it will slightly phase-cancel some of the input signal coming from the first triode section and correct itself.

Although this circuit might be more balanced than the paraphase inverter, it still has the same inherent flaw of having distortion present in one side that is not present in the other. Again, a quality clean tone is impossible with this setup.

DISTRIBUTIVE LOAD PHASE INVERTER — We've all seen this circuit on "E" series tweed Fender amps. In this type of circuit, a single tube is used to achieve phase inversion. This circuit exploits the fact that the cathode of a tube is 180 degrees out of phase with its plate. A signal is introduced to the grid of the distributive load phase inverter tube. The output of the phase inverter is taken from two different places — the cathode and the plate. These two components are 180 degrees out of phase with each other, so one output drives one power tube and the other output drives the other power tube. Because two out of phase signals are taken from the same tube, sometimes this circuit is called a phase splitter.

The quality of the distributive load phase inverter is much better than the other types covered thus far. This type of inverter uses only one triode section and that makes it different from the others. Also, with the distributive load inverter, there is always a slight loss in gain.

The other types of tube phase inverters produce gain.

We've also seen the distributive load phase inverter used on most Ampegs—including the SVT and the V4.

Long-Tailed Pair Phase Inverter—Perhaps the most commonly used phase inverter today is the long-tailed pair. We have seen this circuit on "F" series tweed Fenders; brownface, blackface and silverface Fenders; almost all Marshalls, all Boogies, all Kendricks, all Vox AC30s, all Trainwrecks and many others. It has other names such as the Schmitt Phase Inverter, Common Cathode Impedance Self-balancing Inverter, Grounded Grid inverter, or Long-tailed Pair. With this type of circuit, two triode sections are used and an exact balance may be obtained by suitable proportioning of the two load resistors.

In this type of circuit, there is no bypass capacitor on the cathode resistor and the entire cathode circuit stands on a "long-tail" resistor. The larger the value of the "long-tail" resistor, the more balance but less gain the circuit will produce. The grid of the second triode section is connected to both a grid resistor and a coupling capacitor. The capacitor is in parallel with the entire cathode circuit/long tail resistor. Any out of balance voltage across the long-tail resistor excites the grid of the second triode through the coupling capacitor.

Tone Stacks—A tone stack is a variable filter circuit in which the frequency response of an amplifier is adjustable to suit one's own taste. There are many different ways to achieve this because one can boost treble, boost bass, attenuate treble, or attenuate bass. To further complicate matters, one can have the tone stack directly in the signal path, or in a feedback path. Even though there are dozens of circuit possibilities, there are five circuits that are most common.

Single-knob Tone Stack—Certainly the most simple, a single-knob tone stack is a capacitor in series with a potentiometer. It is almost always configured as a treble attenuation circuit. For example, we see this circuit as the single-knob tone control on most guitars. A capacitor is placed in series with a pot. The cap/pot series assembly is connected between the hot lead of the circuit and ground. As the pot is turned to have less resistance, more of the highs are grounded out. Remember high frequencies go through a capacitor very easily but low frequencies do not. When the knob is turned to its least resistance, the capacitor allows high frequencies to ground out, but not lows. When the knob is turned to its highest resistance, very little of the high frequency signal can pass to ground.

There are other ways of configuring such a circuit. For example, instead of grounding the circuit, it could be connected across the grids of a push-pull output stage. As the pot's resistance is decreased, the highs would phase cancel each other. We have seen this circuit as the high-cut of a Vox AC30.

COMPOUND SINGLE-KNOB TONE STACK—We can take this treble attenuation circuit to another level by configuring it such that as the knob is turned one way the highs are attenuated, but as it is turned the other way the highs are boosted. We see this circuit across the instrument channel of a 5E3 tweed Deluxe. When the tone control is turned up, the circuit actually bypasses highs around the volume control to boost treble, yet grounds highs out when the knob is turned down. It is interesting to note that the microphone channel is not configured this way. It is configured only as a treble attenuation circuit, yet both channels use the same knob!

AMPEG TONE STACK—Almost all Ampeg amps use a high-fidelity tone stack that is also called the Baxandall tone stack. This circuit first appeared in Audio Engineering magazine in the 50s and was touted as a low-loss "active" tone control. By "active," we mean when the tone control is half way up, the circuit is "normal." Turning the control either up or down will either boost or attenuate certain frequencies. This is a clever arrangement of capacitors and pots such that there are two controls: a bass and treble. When the bass is turned up, the bass frequencies are boosted. When it is turned down, the bass frequencies are attenuated. Likewise, when the treble control is turned up, the treble frequencies are boosted. If the treble control is turned down, the treble frequencies are attenuated. Because of less loss than other types of tone circuits, this circuit was very popular for Hi Fi amplifier circuits in the 50s and 60s. It has a different sound than other types of tone circuits because there is no midrange boost or attenuation.

FENDER/MARSHALL TONE STACK—Although Fender experimented with different types of tone controls in the 40s 50s and early 60s, the tone stack used on the 5F6A Bassman eventually emerged as their favorite. It was used on all blackface and silverface amps. It is basically a passive crossover that divides the frequencies into ranges. The pots allow one to select more or less of these frequencies. All Marshall amps 50 watts and over used this same tone stack. This type of stack has a characteristic tone that is unique. We've seen the same tone stack on all Vox amps 30 watts and up.

It is interesting to note that where the tone stack occurs in the circuit makes a huge difference on how well it performs. For example, a blackface Super Reverb has the tone circuit immediately after the first gain stage where signal voltages are relatively small. Any change produced by the tone circuit has a big impact on the overall sound. Contrast this with a plexi-glass Marshall, whose similar tone stack occurs after the third stage where signal voltage is fairly high. Notice how turning the knobs on the Marshall have little effect on the overall tone of the amp. It is because the signal voltage is so high, that the tone controls have less relative effect on the signal. NEGATIVE FEEDBACK DRIVEN TONE STACK—We see this circuit over and over again on the presence circuits for Fender and Marshall amps. To understand how it works, we must first understand negative feedback. The negative feedback circuit takes a very small signal from the output of the amp and injects it back into a previous section of the amp. Care is taken to inject it into a section of the amp that is out of phase with the output. This causes slight phase cancellation. The amount of negative feedback is limited because we only want a little phase cancellation. This is done so that the amp will play more evenly at all frequencies. If a particular frequency wants to amplify more than normal, more of that frequency will be fed back, thus, phase canceling that frequency a little more than normal and correcting it to be the same level as the other frequencies. The presence control basically limits the amount of high-end that is fed back. If less highs are fed back, then there will be less phase cancellation to those frequencies and thus the highs will seem to be boosted. In the presence circuit, the frequencies controlled are above 5K. There are no notes on a guitar this high. So why would we want to boost signals above 5K? Because there are overtones and harmonics in this range and those are the frequencies give that jangly, acoustic guitar-like quality.

All About Vacuum Tube Guitar Amplifiers

VACUUM TUBES 131

Knowing some of the finer points about vacuum tubes can certainly enhance the experience of owning and playing a tube amplifier. Just as your oil in your automobile needs changing from time to time, a tube amp will eventually need new output tubes and perhaps a new preamp tube now and then. Some basic knowledge is helpful for the player to get the most from his tube guitar amplifier.

There are three types of tubes you are likely to find in your tube guitar amplifier; namely: the rectifier tube, the output tubes and the preamp tubes.

RECTIFIER TUBES

Not all amplifiers have a rectifier tube, and some amps actually have more than one rectifier tube. The job of the rectifier tube is to convert the wall A.C. to the D.C. voltages that are required to run the other tubes in the tube amplifier. This job can also be done with a solid-state rectifier, which is the reason not all tube amps have a rectifier tube. Typical rectifier tubes are the 5AR4, 5U4, 5Y3 and the 5V4. When removing a rectifier tube, remember that this tube gets the hottest of any tube in the amp and it will stay hot for a while after the amp is turned off. So you want to make sure and allow some time for the tube to cool, or use a towel or oven mitt when removing this tube, if the amp has been on for a while. A gentle circular rocking motion while pulling the tube out will make it easier to remove. Rectifier tubes rarely malfunction. There are instances where the insides make become loose, which could affect the sound, but for the most part a rectifier tube will either work or it will not. In the case of malfunction, it is likely to short out and blow a fuse. Sometimes the rectifier tube will arc intermittently, which could also cause a fuse to blow. When this occurs, there will be a miniature lightning bolt inside the tube. Once a rectifier tube arcs, whether it blows a fuse or not, it should be replaced because it will be prone to arcing in the future.

Although the rectifier tube usually goes into an 8 pin socket (called an octal socket for obvious reasons), the rectifier tubes usually have only four or five pins and a guide pin. If you look at the big guide-pin in the middle of the bottom of the tube, you will notice there is a protruding part that is called a key. This key is intentionally put on the guide pin to make absolutely certain that the tube is inserted correctly in the socket and the pin-out on the tube matches with the circuitry wired into the socket.

If you look at the middle of an octal tube socket, you will notice there is a hole where the guide pin inserts. In this hole is an indention that is also called the key. Pin #1 and pin #8 are on either side of the key. If you are looking at the bottom of the socket, pin #8 is the first pin counterclockwise from the key and pin #1 is the first pin clockwise from the key.

Looking from the top of the socket, pin #8 is clockwise and pin #1 is counterclockwise from the key. Typical five-volt rectifier tubes such as the ones previously mentioned only use the even number pins (#2, #4, #6, #8), even though there are eight pins on the socket itself.

There are also a few 6 volt rectifier tubes that use 9 pin sockets. These are found usually on small practice amps. More on 9 pin sockets later. The rectifier tube is almost always closest to the A.C. cord and farthest from the input jacks of an amp. That is because there is A.C. going to the rectifier tube and it is not a good idea to have A.C. near an input jack because we don't want A.C. hum near the input of an amp.

PREAMP TUBES

The most popular preamp tube in the world is the familiar 12AX7 twin triode. These are actually two triodes in a small glass envelope. The tube gained its popularity with designers because it is small, has very good gain, and very little hum. The reason for the lack of hum is that the filament is a 12.6 volt center-tapped filament that is almost always run in a humbucking fashion. The two ends of the filament are tied together (pin #4 and pin #5) and counted as one lead. 6.3 volts is applied between the center-tap (pin #9) and the two ends, which makes it humbucking. The hum coming from the first half of the filament is 180 degrees out of phase with the hum coming from the second half, so the hum is phase cancelled, which is another name for humbucking.

The 12AX7 has a few cousins, namely the 12AY7, 12AT7 and 12AU7. All of these tubes are 9 pin tubes with the same pin-out. The difference is basically the amplification factor. All of these tubes are interchangeable. The most gain is the 12AX7 with an amplification factor of 100. The 12AT7 has an amplification factor of 60; the 12AY7 has 44 and the 12AU7A has 20. If you wanted to lower the gain of an amp to get a cleaner sound, the 12AX7 could be replaced with a lower gain tube, for example, the 12AY7.

These type preamp tubes are 9 pin miniature tubes. On such a tube, the pins are spaced out so there could be a 10th pin; however the tenth pin is missing. This missing pin is called the key and it is there to make sure you insert the tube correctly into the socket. It is impossible for it to go in any other way. When replacing a preamp tube, look for the key on the tube and make sure it lines up with the key in the socket. If you are looking at the bottom of the tube at the key, the first pin clockwise from the key is pin #1. The next is pin #2, then pin #3, etc. to pin #9.

Some of the older vintage amps used large bottle preamp tubes that used octal sockets. These are the same sockets used by the rectifier tubes and output tubes; and of course, there is a guide pin in the middle with a key to assure proper indexing upon insertion. Common types are the 6SQ7 (amplification factor of 100), the 6SC7 (amplification factor of 70), 6SN7 (amplification factor of 20) and the 6SL7 (amplification factor of 70). Because the 6SN7 and the 6SL7 have the same pin-out, these can be substituted for each another with only a difference in gain. The other tubes mentioned cannot be substituted.

OUTPUT TUBES

Most American tube guitar amplifiers used 6V6, 6L6, 6BQ5 or 6550 output tubes. British amplifiers usually used EL34 or EL84 output tubes. All of these tubes have octal bases and use octal sockets except for the 6BQ5 and the EL84. Both the EL84 and the 6BQ5 use 9 pin miniature sockets just like the 12AX7 preamp tubes mentioned earlier. In fact, the 6BQ5 is basically the American version of the EL84 and either can be substituted with the other. There is also a military version of this tube called the 7189A. The military version is more rugged and can take higher voltages and more abuse. The first guitar amplifier I ever owned in August of 1965 was a Sears and Roebuck practice amp that used a single 7189A.

6L6 tubes can generally be substituted in a 6V6 amp, but not vice versa. Usually a 6L6 can substitute into an EL34 amp if the bias is reset. The EL34 cannot substitute in a 6L6 without some rewiring done to the socket and a larger value screen resistor installed. Generally a 6550 will substitute in an EL34 amp with only the bias reset.

There are basically two ways to configure the output stage of a tube amp; namely: cathode bias (a.k.a. self-bias) and fixed bias. When replacing output tubes in a fixed bias amp, it is very important to adjust the bias setting. Adjusting the bias is basically the idle adjustment of the tube. Just as you would adjust the idle of your automobile after replacing the carburetor, it is advisable to adjust the bias of a tube amp when replacing the output tubes. Different tubes of the same brand respond differently to a circuit and the bias voltage must be adjusted so that the tube doesn't run too hot or too cold.

On a cathode-biased amp, the bias adjustment is not necessary because there is a large resistor on the cathode of the tube (hence the name cathode-biased). When current comes through this resistor, a voltage develops on the cathode. As more current flows, the voltage becomes greater until enough voltage develops to level the flow of current and the tube reaches a state of equilibrium. You could say the tube is biasing itself, hence the nickname: self-biased.

GAIN AND OVERDRIVE

All About Vacuum Tube Guitar Amplifiers

ANATOMY OF A TRIODE PREAMP GAIN STAGE

Perhaps the most important function of a vacuum tube is the ability to amplify an input signal. Every tube guitar amp consists of two or more tubes and the associated components needed for their operation. The minute voltage produced by the electric guitar's pickup is amplified by a number of consecutive stages, such that it can become powerful enough to operate a loudspeaker. The actual amount of gain that results is mostly dependent on the number of stages used. If you consider gain as the ratio of output to input, then one may see that the more stages, the more gain.

Although amplifiers come in many different brands, shapes, sizes, colors and configurations; there are certain recurring designs that we see as common denominators among all or almost all tube guitar amplifiers. Almost every tube guitar amp uses a simple, "Western Electric style" voltage amplifier circuit in the preamp section of the amplifier. Although a gain stage can be operated with a pentode or a triode type preamp tube, besides the very few guitar amps that use a pentode tube in the preamp circuit, the overwhelming majority of preamps are configured with a triode. Many different types of triodes could be used with the most common being the 12AX7, which is a twin triode (two triodes in one glass bottle, which conserves physical space). Whether a 12AX7, 12AT7, 12AY7, or some of the "big base" preamps such as the 6SL7 or 6SN7 are employed, the basic circuit associated with this Class A voltage amplifier is identical. The actual component values may be slightly different, or even an optional component added to sculpt tone, but the basic preamp circuit remains the same.

Look at Figure 1. This is a diagram of a Resistance Capacitance Coupled Triode. It is always configured as a single-ended "Class A"

Resistance Capacitance Coupled Triode Amplifier

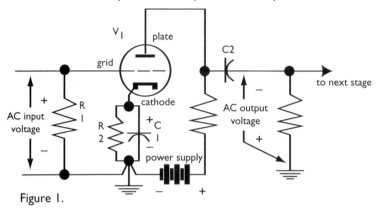

Figure 1.

voltage amplifier. You have heard this circuit time and time again on almost every Fender, Marshall, Gibson, Ampeg, National, Vox, Kendrick or almost any other tube amp. It was invented by the phone company, Western Electric.

The basic circuit consists of four resistors and two capacitors connected to a triode. I have labeled the resistors as R1, R2, R3, R4 and the capacitors as C1 and C2. The vacuum tube is called V1.

If you look at the diagram, you will see a triode vacuum tube labeled V1. Besides the normal heater filament component (not shown in diagram) that lights up inside the tube to warm up the cathode, a triode has three significant electronic components: the cathode, grid and plate. In a triode, electrons move from the cathode to the plate. Think of the cathode as the place where electrons enter the tube and think of the plate as the place where electrons exit the tube. If they enter at the cathode and leave at the plate, what is the grid?

The grid is the component in the tube that controls how much current flows from the cathode to the plate. The minute electrical charge on the grid has a huge effect on how much current flows from the cathode to the plate. To use a water analogy, if you tried to place your finger over a faucet to keep the water from flowing, it would be impossible; but if you turned the handle on the faucet, it would be easy. The grid easily controls how much current flows through the tube, so it is like the handle of the faucet.

In fact, if the grid's electrical charge is sufficiently negative enough with respect to the cathode, the tube will be completely turned "off"

All About Vacuum Tube Guitar Amplifiers

and there will be no current flowing through the tube. We call this condition "cutoff." On the other hand, if the grid becomes more and more positive with respect to the cathode, more and more current will flow. There will come a point when all the current that can flow, is flowing; and no more can possibly flow through the circuit. This condition is called "saturation." Think of a water faucet turned on full blast.

So let's review. The grid is the component in the tube whose job it is to control how much electrical current flows through the tube. The current is coming into the tube from the cathode and leaving the tube at the plate.

Notice that a power supply is connected to the cathode and plate via the cathode and plate resistors (R2 and R3). The positive side of the power supply connects to R3 which connects to the plate. The negative or ground side of the power supply connects to ground. R2 is called the cathode resistor for obvious reasons, and is connected from ground (the negative side of the power supply) to the cathode of the tube. This completes the circuit.

Now let's supposed we did not connect the grid of the tube to anything. What would happen? Current would flow from the cathode to the plate at an uncontrollable rate. The tube would quickly reach saturation because the valve would be "wide open" so to speak.

R1 is sometimes called the grid return resistor. One end of R1 connects to the grid of the triode and the other end connects to ground. This resistor establishes a ground relationship for the grid and it is the load for the input signal. If this stage were the first stage in an amp, the input jack hot lead would connect to the grid side of R1 and the sleeve on the jack would go to ground. This would essentially connect your guitar pickup across R1. When your guitar is connected across R1, as the string is plucked, the guitar pickup produces an A.C. voltage and that A.C. voltage appears across R1. Remember A.C. voltage is a flow of electrons that goes both ways. The polarity is quickly changing from plus to minus as the string vibrates. This changing polarity of A.C. voltage, since it is connected to R1, will cause more or less current from the power supply to flow from the cathode of the tube to the plate. This current flow will be a faithful reproduction of the guitar signal appearing across R1 except it will be much stronger. You could say it is the amplified version of the signal!

But why do we need a cathode resistor (R2) and a plate resistor (R3)? The cathode resistor, R2, is needed in order to set the operating level (bias) of the tube. We want the idle setting of the tube to be exactly half way between saturation and cutoff. Doing so allows a signal to be amplified without chopping off the top or the bottom of the wave. By having the valve set half way on, the grid (think faucet handle) can make more or less current flow in order to faithfully reproduce the signal to be amplified.

The more negative the grid is with respect to the cathode, the less current will flow from cathode to plate. R2 is placed between the cathode and ground. When current passes through R2, a small positive voltage develops on the cathode side of R2. A tube really doesn't know the difference between a cathode that is slightly positive with respect to the grid, or a grid that is slightly negative with respect to the cathode. It is the electrical charge relationship, relative to each other, that counts. Either way, less current flows from cathode to plate. The trick is to use a resistor value that allows the tube to idle exactly half way between saturation and cutoff; that way, the signal can be amplified such that the output signal is true to the input signal without cutting the top or the bottom of the waveform off. For a typical 12AX7 configuration, a 1.5K cathode resistor is most common. Other values can be used and I have seen them from 200 to 10,000 ohms, depending on the tube type used, and the operating voltages.

Well, that explains the cathode resistor, but what about the plate resistor? If you removed the plate resistor from the circuit, there would be no amplification. Remember that a tube circuit has both D.C. (all the electricity is going one direction) and A.C. (the electricity goes both directions). When D.C. current idles through the tube and through R3, a voltage drop occurs across R3. That is why the voltage on the plate is always less than that coming from the power supply. When the input signal is injected into R1 of the circuit, more or less current flows through R3. This causes the fluctuating signal voltage on the plate to go up and down rapidly as more or less current is run through the plate resistor (R3). This A.C. component is on top of the D.C. component (idle current) of the circuit. When the grid is at its most positive part of the input cycle and maximum current is run

All About Vacuum Tube Guitar Amplifiers

through the plate circuit, then maximum voltage drop occurs across R3 and the actual plate voltage goes down. During the most negative portion of the input cycle when minimum current flows, then a minimal voltage drop will occur across R3 and the plate voltage will go up. Try to think of the idle current as being the D.C. component and the amplified guitar signal as being the A.C. component. The D.C. component is constant and all the electricity is moving the same way while the A.C. component (guitar signal) is laying on top of the D.C. current. Typical value for R3 is usually 100K. I have seen many different values used from 50K to 250K, but 100K is the most common.

Now we need to separate the D.C. from the A.C. This is done easily with C2 which is also called a blocking capacitor or a coupling capacitor. To understand how it blocks the D.C. component and still lets the A.C. component pass, we will need to have a short discussion about capacitors. A capacitor is actually two conductors separated by a non conductor. You could easily make a capacitor by taking two sheets of metal foil and separating them with a sheet of paper. D.C. current cannot flow through a capacitor because the non conductor part of the capacitor stops the circuit. However, A.C. current can appear to flow through a capacitor.

A.C. doesn't really flow through a capacitor, but you would never know the difference because it appears to flow through the capacitor. Remember A.C. is current that flows both ways. When C2 is connected to the plate of the triode and the plate voltage fluctuates with the input signal, the instantaneous charge on one of the capacitor's plates also fluctuates with the plate of the tube. This causes an equal and opposite charge to appear on the other plate in the capacitor.

Here's another analogy to explain how a capacitor seems to pass A.C. yet blocks D.C. current: Let's say we had a pipe with water in it and in the middle of the pipe, and blocking the flow, was a water-proof membrane. As pressure (voltage) would be exerted on the water of one side of the pipe, this pressure could be felt on the other side of the pipe. If this pressure was rapidly changing, that rapid change could be felt on the other side of the pipe. The water is not really flowing through the pipe because of the membrane in the way; however, the instantaneous changes in pressure are real and verifiable just as if the membrane had not been there.

The output side of C2 is connected to R4. R4 is the output load and could be any load, including the grid resistor (R1) of another gain stage. R4 could be changed to a potentiometer to make a volume control. R4 could easily be replaced with a tone circuit or a reverb pan or any other suitable load. R4 is usually a load of 100K to 500K.

Now we have looked at everything except C1. C1 is the bypass capacitor. We know capacitors will pass A.C. but not D.C. To keep the D.C. voltage stable that appears on the top of R2, a bypass capacitor is added across R2. This is the path that the A.C. current goes. When the tube draws more or less current, the fluctuating A.C. goes through the capacitor while the idle D.C. current goes through the resistor.

The value of C1 is chosen so that its capacitive reactance (the resistive property of capacitors to A.C.) is one tenth of the resistance of R2 at the lowest operating frequency. The most common value is 25 microfarad, however, typical values range from .1 mfd to 250 mfd. A 25 mfd bypass cap bypasses all A.C. guitar frequencies when used with a typical 1.5K ohm cathode resistor (R2). By choosing a cap value whose capacitive reactance equals $\frac{1}{10}$ of the resistance of the cathode resistor, one can be assured the AC will go through the bypass capacitor and not the cathode resistor as electricity always prefers the path of least resistance. The D.C. quiescent current cannot pass through the capacitor so we are certain it will travel through the cathode resistor.

Sometimes the capacitor is left off. If you omit C1, the gain of the tube drops considerable and the tube compresses more because the A.C. goes through the cathode resistor instead of the bypass cap. Occasionally you will see a preamp circuit that omits C1, especially if the input signal feeding R1 is very strong. Eliminating the bypass cap reduces gain, but it also reduces clipping, so it is ideal for accommodating a hot input signal such as a bass guitar or a keyboard or a larger input on a later stage.

All About Vacuum Tube Guitar Amplifiers

Resistance Capacitance Coupled Pentode Amplifier

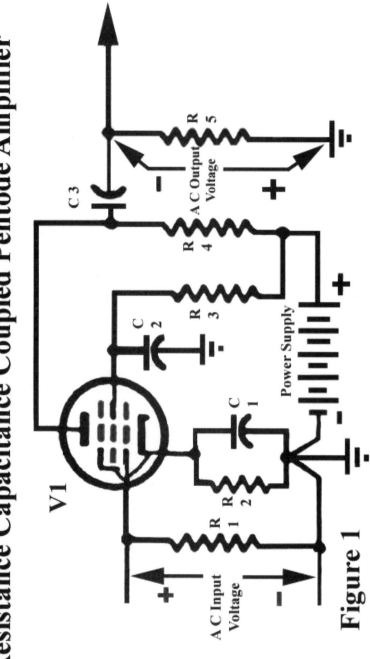

Figure 1

ANATOMY OF A PENTODE PREAMP GAIN STAGE

Although most guitar amplifier preamp circuits employ a triode in the preamp gain stage, there are some guitar amplifiers that use a pentode for the preamp gain. The pentode preamp will have more gain than a triode, and it will have more harmonic distortion.

Look at Figure 1. This is a typical configuration of a resistance capacitance coupled pentode preamp gain stage. It is always configured as a single-ended "Class A" voltage amplifier. Although not as prevalent as the triode gain stage, you have heard this circuit many times before. It was used on all lower powered British Vox amps (AC4, AC10, and the AC15). Gibson used it on several models (early Kalamazoo KEH, early Gibsonette models, BR 1, BR6F, GA 9, GA 40 Les Paul, Mastertone Special), and Fender used it on the 5C1 Champ. If you've ever heard a Matchless DC 30, you have heard a pentode preamp gain stage.

The basic circuit consists of five resistors and three capacitors connected to a pentode. I have labeled the resistors as R1, R2, R3, R4, R5 and the capacitors as C1, C2 and C3. The vacuum tube is called V1.

Besides the normal heater filament component (not shown in diagram), that lights up inside the tube to warm up the cathode, a pentode has five significant electronic components. In the diagram, these internal tube components are shown from bottom to top inside of V1 as: the cathode, grid, screen grid, suppressor grid and plate. In a pentode, as in a triode, electrons move from the cathode to the plate. Think of the cathode as the place where electrons enter the tube and think of the plate as the place where electrons exit the tube.

The grid (a.k.a. control grid) is the component in the tube that controls how much current flows from the cathode to the plate. The minute electrical charge on the grid has a huge effect on how much current flows from the cathode to the plate.

Besides the normal control grid, the pentode has two extra grids that the triode does not have. A screen grid is added between the regular control grid and the plate. This screen grid is charged with a positive high voltage and serves as an accelerator of electrons so that larger values of plate current can be drawn. However, when an electron strikes the plate, it dislodges other electrons and causes something called secondary emissions. In a triode, the grid (which is negative) repels the secondary emission back to the plate; so in the case of the triode, secondary emission is not a problem. However, in a tube with a screen grid, the positively charged screen would attract the secondary electrons thus causing reverse current to flow from plate to screen! So to overcome this effect, a third grid, called the suppressor grid is inserted between the screen and the plate. The suppressor grid is connected to the cathode so it is negative with respect to the plate. This suppressor grid acts as a shield that prevents secondary electrons from bouncing back to the screen. Most pentodes have the suppressor grid internally connected to the cathode, but there are some pentodes that have a separate pin-out for the suppressor grid. In this case, the suppressor grid is simply connected to the cathode at the tube socket.

Notice that a power supply is connected to the cathode and plate via the cathode and plate resistors (R2 and R4). The positive side of the power supply connects to R4 which connects to the plate. The negative or ground side of the power supply connects to ground. R2 is called the cathode resistor for obvious reasons, and is connected from ground (the negative side of the power supply) to the cathode of the tube. This completes the circuit. R3 connects from the positive side of the power supply to the screen grid and this is how the screen grid gets its positive charge. There is very little current flowing through R3.

Now let's supposed we did not connect the grid of the tube to anything. What would happen? Current would flow from the cathode to the plate at an uncontrollable rate. The tube would quickly reach saturation because the valve would be "wide open" so to speak.

R1 is sometimes called the grid return resistor. One end of R1 connects to the grid of the pentode and the other end connects to ground. This resistor establishes a ground relationship for the grid and is also the load for the input signal. If this stage were the first stage in an amp, the input jack hot lead would connect to the grid side of R1 and the sleeve on the jack would go to ground. This would essentially connect your guitar pickup across R1.When your guitar is connected across R1, as the string is plucked, the guitar pickup produces an AC voltage and that AC voltage appears across R1. Remember AC voltage is a flow of electrons that goes both ways. The polarity is quickly changing from plus to minus as the string vibrates. This changing polarity of AC voltage, since it is connected to R1, will cause more or less current from the power supply to flow from the cathode of the tube to the plate. This current flow will be a faithful reproduction of the guitar signal appearing across R1 except it will be much stronger. You could say it is the amplified version of the signal!

But why do we need a cathode resistor (R2), a screen resistor (R3) and a plate resistor (R4)? The cathode resistor, R2, is needed in order to set the operating level (bias) of the tube. We want the idle setting of the tube to be exactly half way between saturation and cutoff. Doing so allows a signal to be amplified without chopping off the top or the bottom of the wave. By having the valve set half way on, the grid can make more or less current flow in order to faithfully reproduce the signal to be amplified.

The more negative the grid is with respect to the cathode, the less current will flow from cathode to plate. R2 is placed between the cathode and ground. When current passes through R2, a small positive voltage develops on the cathode side of R2. A tube really doesn't know the difference between a cathode that is slightly positive with respect to the grid, or a grid that is slightly negative with respect to the cathode. It is the electrical charge relationship, relative to each other, that counts. Either way, less current flows from cathode to plate. The trick is to use a resistor value that allows the tube to idle exactly half way between saturation and cutoff. That way, the signal can be amplified such that the output signal is true to the input signal without cutting the top or the bottom of the waveform off. For a typical 6267 (EF86) tube configuration, a 1.5K cathode resistor is most common.

Well, that explains the cathode resistor, but what about the plate resistor? If you removed the plate resistor from the circuit, there would be no amplification. Remember that a tube circuit has both D.C. (all the electricity is going one direction) and A.C. (the electricity goes both directions). When D.C. current idles through the tube and through R4, a voltage drop occurs across R4. That is why the voltage on the plate is always less than that coming from the power supply. When the input signal is injected into R1 of the circuit, more or less current flows through R4. This causes the fluctuating signal voltage on the plate to go up and down rapidly as more or less current is run through the plate resistor (R4). This A.C. component is on top of the D.C. component (idle current) of the circuit. When the grid is at its most positive part of the input cycle and maximum current is run through the plate circuit, then maximum voltage drop occurs across R4 and the actual plate voltage goes down. During the most negative portion of the input cycle when minimum current flows, then a minimal voltage drop will occur across R4 and the plate voltage will go up. Try to think of the idle current as being the D.C. component and the amplified guitar signal as being the A.C. component. The D.C. component is constant and all the electricity is moving the same way while the A.C. component (guitar signal) is laying on top of the D.C. current and since it is A.C., the current is flowing both ways. Typical value for R4 can be anywhere from 100K to 500K.

The screen voltage is supplied via the screen resistor (R3). We use a resistor here to discourage electrons from flowing through the screen circuit and to drop the voltage such that it is lower than the plate voltage. The point of having a screen grid is simply to accelerate the electrons such that the tube can have more plate current. The screen has a high voltage on it. The electrons on the cathode can see this positive charge and are attracted to it. When they get close to the screen grid, they then can see the plate which: a) has more mass, and b) has a higher voltage. So they go to the plate. In a perfect world, there would be zero current flowing through the screen grid, but there will always be a few lazy electrons that are content to simply go through the screen grid circuit and not bother with going to the plate.

Now we need to separate the D.C. from the A.C. This is done easily with C3 which is also called a blocking capacitor or a coupling capacitor. To understand how it blocks the D.C. component and still lets the A.C. component pass, we will need to have a short discussion about capacitors. A capacitor is actually two conductors separated by a non-conductor. You could easily make a capacitor by taking two sheets of metal foil and separating them with a sheet of paper. D.C.

All About Vacuum Tube Guitar Amplifiers

current cannot flow through a capacitor because the non conductor part of the capacitor stops the circuit. However A.C. current can appear to flow through a capacitor.

A.C. doesn't really flow through a capacitor, but you would never know the difference because it appears to flow through the capacitor. Remember A.C. is current that flows both ways. When C3 is connected to the plate of the triode and the plate voltage fluctuates with the input signal, the instantaneous charge on one of the capacitor's plates also fluctuates with the plate of the tube. This causes an equal and opposite charge to appear on the other plate of the capacitor. That is what gives the illusion that A.C. is passing.

Here's another analogy to explain how a capacitor seems to pass A.C. yet blocks D.C. current: Let's say we had a pipe with water in it and in the middle of the pipe, and blocking the flow, was a water proof membrane. As pressure (voltage) would be exerted on the water of one side of the pipe, this pressure could be felt on the other side of the pipe. If this pressure was rapidly changing, that rapid change could be felt on the other side of the pipe. The water is not really flowing through the pipe because of the membrane in the way; however, the instantaneous changes in pressure are real and verifiable just as if the membrane had not been there.

The output side of C3 is connected to R5. R5 is the output load and could be any load, including the grid return resistor (like R1) of another gain stage. R5 could be changed to a potentiometer to make a volume control. R5 could easily be replaced with a tone circuit or a reverb pan or any other suitable load. R5 is usually a load of 500K to 1 Meg ohm.

Now we have looked at everything except C1 and C2. Both are bypass capacitors. C1 is the cathode bypass capacitor. We know capacitors will pass A.C. but not D.C. To keep the D.C. voltage stable that appears on the top of R2, a bypass capacitor is added across R2. This is the path that the A.C. current goes. When the tube draws more or less current, the fluctuating A.C. goes through the capacitor while the idle D.C. current goes through the resistor.

The value of C1 is chosen so that its capacitive reactance (the resistive property of capacitors to A.C.) is one tenth of the resistance of R2 at the lowest operating frequency. The most common value is 25 microfarad, however, typical values range from .1 mfd to 250 mfd. A 25 mfd bypass cap bypasses all A.C. guitar frequencies when used with a typical 1.5 K ohm cathode resistor (R2). By choosing a cap value whose capacitive reactance equals 1/10th of the resistance of the cathode resistor, one can be assured the A.C. will go through the bypass capacitor and not the cathode resistor as electricity always prefers the path of least resistance. The D.C. quiescent current cannot pass through the capacitor so we are certain it will travel through the cathode resistor.

The gain of the tube is affected by the screen voltage, so we do not want the screen voltage to fluctuate. C2 is the bypass capacitor for the screen grid. This capacitor opposes any change in the screen voltage which is necessary for the tube to have best performance (maximum gain). If small amounts of screen current want to fluctuate the positive voltage on the screen, C2 bypasses this A.C. variation to ground, thus stabilizing the voltage (and gain) of the pentode.

Pentodes have higher gain, so they are more prone to instability and oscillation problems. I have seen pentode gain stages that made whistling and other oscillation noises and the only way it could be stopped was by using a grounded preamp tube cover and a shielded control grid wire. Also, because of the higher gain, any imperfections in the tube such as noise or microphonics are amplified more, thus bringing out the imperfections of the tube.

All About Vacuum Tube Guitar Amplifiers

TUBE GUITAR AMPLIFIER OVERDRIVE

More than a half century ago, in the early days of electric guitar, manufacturers assumed they were designing an electrified guitar that would basically sound the same as an acoustic guitar, but much, much louder. Can you imagine that? Guitar players wanting to play loud! Because of this, all of the designs decisions were based on the presumption the designers' job was to make the amp sound as clean as possible. Nowadays, it is almost comical to read some of the early brochures from the 50s Fender tweed amps and see how they hyped their tweed amps as "distortion free." At that time, distortion and overdrive were considered very undesirable and a detriment to be avoided at all costs! Perhaps the worst slur one could offer a manufacturer back then, was to accuse their guitar amplifier of having distortion.

Of course, each manufacturer would design their amplifiers using their own electric guitars and pickups for sound testing. Connect an A.C. voltmeter to a guitar loaded with single-coil, "Fender-style," pickups and play a first position, "Mel Bay style," open "E" chord. Assuming the pickups are adjusted to the proper height, your meter will show approximately ⅛ of a volt coming out of the guitar. Now do the same test, except this time try it with a guitar loaded with humbucking pickups. You will get approximately ¼ of a volt. Now, do the same test with a P-90 soap bar pickup and you will likely get ½ of a volt. So the P-90 soap bar pickup has four times the output of the single-coil pickup and the humbucker is twice the single-coil, yet half the P-90.

A typical preamp tube, say a 12AX7 for example, has an amplification factor of 100. This means that for every 1 volt of change on the input of the tube (grid), there will be a 100 volt change on the output (plate). Even older style octal base preamp tubes such as the

6SC7 or 6SL7 have an amplification factor of 70. If you compare running a single-coil pickup guitar, such as a Stratocaster, through a 12AX7 preamp tube; versus running a humbucking pickup guitar such as a Gibson Les Paul through the same tube, there will be a huge difference on the output side. Using the 1st position "E" chord example as mentioned in the last paragraph, there will be a difference of a little more than 12 volts output; a difference of 100 percent!

So this explains why Gibson amplifiers never had much gain. The factory was testing their amps using their guitars loaded with hot pickups such as humbuckers or P-90s; and since they were going for maximum clean tone, the amps were intentionally designed with fairly low gain, by today's standards. Plug a Strat into a stock vintage Gibson amp and it will sound pretty lame. On the other hand, try a Gibson guitar through a Fender amp. That will give some distortion and if you turn it up, you could push it into overdrive.

DISTORTION AND OVERDRIVE: WHAT'S THE DIFFERENCE

Distortion could be considered as any sound that occurs in the output signal that was not present in the input signal. Overdrive is the sound of a tube when it is driven into saturation. So technically speaking, overdrive is a type of distortion. It is the type of distortion that a tube produces when it is driven into saturation. Actually, the tube will distort somewhat by having the waveform change shapes (more square) even before the tube goes into saturation. So, there are other types of distortion other than overdrive.

HOW DID OVERDRIVE COME ABOUT?

I can't prove this, but I would wager the first overdrive came along when someone plugged a humbucking guitar into a Fender amp. This type of overdrive was what is now called output tube overdrive. The guitar's pickups would have driven the preamp tube such that it would put out about double the preamp signal. This extra signal voltage would have driven the output tubes so hard that the waveform would have saturated in the output stage and thus overdrive was born.

Somewhere in the late 60s, a device came on the market that made a Fender guitar overdrive like the Gibson. It was called a LPB1 Power booster and it was made by Mike Matthews of Electro-Harmonix,

Sovtek, and New Sensor fame. I used one back then and practically everyone I knew had one. It was a simple, 9-volt battery operated, transistor gain stage controlled by a single volume knob and mounted inside a small metal box. You could plug the LPB1 into your amp's input and then plug your guitar into the LPB1. There was an on/off switch to turn it off and on but I would leave mine on. This would amplify the pickup some before it went into the amp, thus creating the same effect as having a very hot pickup.

PREAMP TUBE OVERDRIVE

It was about this same time or perhaps a year later that a brilliant designer from California, Randall Smith, came up with the idea to put an extra preamp tube gain stage in front of a Fender amp circuit. These first Boogie amps were actually Fenders that were modified. This type of distortion was different than the overdrive known up until that point. With an extra tube in front of the preamp circuit, the preamp tube was being overdriven instead of the output stage. Although this had the effect of giving nearly infinite sustain; it sounded buzzy, removed the player's dynamics and homogenized the sound.

Most of us have seen a picture of a sine wave on an oscilloscope. It looks like a hill and a valley, then a hill and a valley, etc. If we play an A-440 note, then the waveform goes up the hill and down the valley 440 times in one second. All preamp tubes are run single-ended Class A. With all single-ended designs, the "hill portion" of the sound wave is amplified by the tube simply drawing more current while the "valley portion" of the sound wave is amplified by the tube drawing less current. If you overdrive a preamp, when the "hill portion" is overdriven, there will come a point when the tube is drawing as much current as it can, regardless of any additional signal on the input. This condition is called saturation. If a tube is a valve, then saturation is when the valve is turned all the way on. On the other hand, when the wave continues and you come to the "valley portion" of the wave, the tube draws less and less current until it gets to a point where the tube stops drawing current at all. This condition is actually called "cutoff." If a tube is a valve, then "cutoff" is when the valve is turned off.

And therein lays the problem with preamp tube overdrive. When a preamp tube is overdriven, basically, the bottom half of the wave disappears. This explains why it sounds buzzy and causes listener fatigue. Actually, when you drive a Class A amplifier into cutoff, it is no longer Class A. It becomes Class AB which doesn't work with a single-ended design since half of the wave is missing.

OUTPUT TUBE OVERDRIVE

With the exception of the single output tube Class A practice amps, such as the Fender Champ, almost all output stages are configured as push-pull. A push-pull design is used because it is much more efficient than single-ended and one can get much more wattage out of a pair of output tubes when they are run as a push-pull design. Push-pull output stages are almost always run Class AB, but it is possibly to run a push-pull output stage in Class A.

A push-pull design works a little differently than the single-ended. A push-pull output stage amplifies the "hill portion" of the sound wave by drawing more current. But it amplifies the "valley portion" of the sound wave by also drawing more current. Because of this clever arrangement, the output stage can have both the top "hill portion' and the bottom "valley portion" of the wave go into saturation evenly, thus producing a very pleasing tone. Actually a push-pull configuration requires two tubes or a multiple of two. Let's look at using two output tubes in push-pull.

When the signal overdrives the "hill portion" of the wave, the first tube draws more and more current until it is drawing as much current as it can. Any additional input signal will not produce more output current. Again, this condition is called saturation. When the "valley portion" of the wave is overdriven, the second tube amplifies this valley by also drawing more current. One could increase the input signal until saturation is reached on the "valley portion" of the signal. So with output tube overdrive, both the top and bottom of the wave are saturated evenly. This is much different from preamp overdrive which simple cuts off the bottom half of each sound wave.

This is why output tube overdrive sounds so much better than preamp tube overdrive. In fact, with output tube overdrive, the overdrive is sensitive to pick attack and technique. Pick hard and the tone will overdrive more than when you pick softly. With some amplifiers, you can find that magic spot where you pick softly or roll your volume down a number or two and the tone cleans up. But lay into that pick attack with the guitar volume all the way up and you will achieve an organic, cello-like overdriven output. It is round and never buzzy.

There are stomp boxes that add distortion and there are many other effects one could use to hopefully enhance his guitar tone, but a cooking push-pull output stage with that rich and chocolatey, natural-tube overdrive has always "done it" for me.

POWER
ATTENUATORS

All About Vacuum Tube Guitar Amplifiers

ALL ABOUT POWER ATTENUATORS

How does your tube amp sound its best? This is an interesting question. Most of us like to think of our tube amplifiers as a musical instrument. When we set the amp where it sounds its best, sometimes the amplifier is too loud for the occasion. Some of these situations might include playing in a small room such as a living room or bedroom, or even a small club where a 50 or 100 watt amp is simply too much. In short, there are times when the overall volume of the amp is too loud, but if the volume control is adjusted, the amp loses its magic tone. The logical thing to do is attenuate the power after it leaves the amp and before it gets to the speaker.

WHAT IS A POWER ATTENUATOR?

When Jim Kelly came up with the idea of putting an L-pad between the amplifier and the speaker, the original intention was to have the clean volume be consistent with the lead volume during channel switching. When he overdrove the lead channel to get it sounding its best, he found it was too loud in comparison to his best clean channel sound. With his channel switching amp, an L-pad between the amp and the speaker switched into the circuit when the Lead channel was selected. Conversely, the speakers switched directly back to the amplifier's output (bypassing the L-pad) when the clean channel was selected.

WHY ATTENUATE THE POWER?

There are many instances when one would desire to attenuate the amp's power before it gets to the speaker. In a recording situation, one could attenuate the power and use close miking. This would minimize bleeding over onto other tracks.

About a year ago, I was in Los Angeles to attend a NAMM show. I was early and stumbled upon some free tickets to see Jay Leno and the Tonight Show. I was amazed at what a low volume the Kevin Eubanks band was playing. When they first began the show, the soundman didn't have them turned up through the studio monitors and their stage volume was so low, I could barely make out that the band was even playing at all. Kevin Eubanks was using a power attenuator. This allowed him to get his sound, but at such a low volume that the mixing could be done without bleed-over becoming a problem. When I see the Kevin Eubanks band playing on T.V., it seems like they are really rockin' out loud. This must be a soundman's dream—to be able to mix without bleed-over and having complete control of the mix. Think of the side benefit the band reaps by having the exact cue mix they want to hear coming from their individual monitors. No wonder that band is so tight. Everyone can hear!

Which brings us to another reason to attenuate: Maybe you want to save your hearing?

HOW DO POWER ATTENUATORS WORK?

There are many different ideas about what works best to attenuate the power. Most of the ideas break down into two basics types of attenuators. One type is the transformer driven style (i.e. Marshal Power Brake) and the other type would be the resistive type (ie. Kendrick Power Glide, THD Hot Plate).

The transformer type attenuator is basically a step down transformer to step down power. I don't like this setup. The impedances are unpredictable and I don't like the sound.

Resistive type attenuators have many variations, all in an effort to make the attenuator as transparent as possible while still attenuating the power. What goes on in an amplifier's output stage is very complex and simple. For instance, the output transformer is simply some coils of wire around an iron core. Sounds simple enough. However, as the speaker is moving a voice coil in front of a magnet, it is generating electricity at times and its inductive reactance is changing as the voice coil moves through a magnetic field. This voice coil induced electricity is hooked up to the secondary of the transformer. Normally we think of transformers in terms of the primary of that transformer. In other words, we would think of an output transformer as a step down transformer. It has high voltage on the primary, which steps down to low voltage on the secondary, which of course connects to the speaker.

Let's look at this from the speaker's point of view. From the speaker; and looking back into the amp, the secondary would become the primary and any voltage generated by the speaker would be stepped up as it moved backwards through the amp. We are talking about voltage originating from the speaker being injected backwards through the amp, stepped up and then injected to the plates of the output tubes. Besides all this affecting the sound, there are changing impedances and other complexities.

MANY VARIATIONS OF THE RESISTIVE TYPE

There was one variation of attenuator called the Tom Schultz that used ordinary 10 watt cement resistors. These are least desirable among the resistive type.

Another variation, the Rocktron Juice Extractor, used a nichrome wire resistive element. Made in the mid to late 80s, the problem was that the nichrome element would become brittle and shatter. When it shattered, the output of the amp was disconnected from the speaker causing the amplifier to see an open load. Tube amplifiers cannot be operated successfully without a load! Yikes!

An outfit from Ann Arbor Michigan made an attenuator in the late 70s and 80s that successfully used a nichrome wire. This was the Altair and was the first one on the market.

There are attenuators being made today, such as the THD Hot Plate. The THD Hot Plate is reactive and uses a light bulb with various other capacitors and inductors to compensate for frequencies normally lost during attenuation. Other features include: adjustable line output, dummy load, extra speaker jack, noise reduction, forced air-cooling.

The Kendrick Power Glide uses large wattage ceramic core wire wound resistors. Some of these resistors are in series and some are in parallel with the speaker to achieve correct impedance while attenuating the power incrementally. Other features include: adjustable line output, dummy load, and extra speaker jack. The 175-watt main load resistor is rated for 1,750 watts for 10 seconds. This unit runs very cool and doesn't need any cooling.

On most resistive type attenuators, a resistor is placed in parallel with the speaker. This will attenuate the current. It will act like a bypass, because the current can flow through the parallel resistance instead of going through the speaker. To attenuate the voltage and correct the impedance, another resistor is placed in series with the speaker. This will attenuate some of the voltage going to the speaker but will also correct the overall impedance.

Carefully chosen resistor values for the series resistance and the parallel resistance will assure correct impedance that the amplifier "sees." For Example: Let's say we have a 4 ohm amplifier connected to a 4 ohm speaker and we want to drop the power going to that speaker. We put a 4 ohm resistor in parallel with the speaker, which bypasses ½ of the current around the speaker. The current sees two paths. One path is the speaker and the other path is the resistor. But now we have a problem. When we put a 4 ohm resistor across a 4 ohm speaker, from the amplifier's point of view, it is seeing a 2 ohm load. It wants to see a 4 ohm load, so we put a 2 ohm resistor in series with the speaker and now the amplifier sees a 4 ohm load **AND** ½ of the voltage is attenuated. Wattage is voltage times current. If we have half as much current times half as much voltage, we end up with a quarter as much wattage. Now that wasn't so hard was it?

HOW MUCH ATTENUATION?

The more you attenuate, the less transparent your attenuator will become. Attenuating more than -15 dB cannot be done without some loss of tone. Lets start with a 100 watt amp that's way too loud for your application. If you cut your power in half, you have attenuated -3 dB. So a 100-watt amp attenuated -3dB becomes 50 watts.

Any of you that have ever taken two tubes out of a four tube output stage knows that -3 dB really doesn't sound like that much of a wattage drop. Cut it in half again (another -3 dB or -6 dB total) and you have a 25 watter. This would be like taking a Twin down to a Deluxe. In our example with the 2 ohm and 4 ohm resistor above, we attenuated to -6 dB.

If you dropped it to -9 dB total, you are looking at 12.5 watts, if you started by using the same 100-watt amp. Attenuated -12 dB, we would end up hearing 6.25 watts and -15 dB is like hearing 3 watts. Believe it or not, 3 watts are quite loud if you are playing in your bedroom at midnight.

All About Vacuum Tube Guitar Amplifiers

All About Vacuum Tube Guitar Amplifiers

BUILD A POWER ATTENUATOR FOR LOW WATTAGE TUBE AMP

You quit playing your 100 watt amp to avoid alienating those you love and now you play a low wattage amp. The problem is: it is still too loud. You want to get a good tone and can't get it with the volume turned almost off. What you need is a little power attenuation. A power attenuator is a device that can be placed between your amp's output and the speaker. This allows you to turn your amp up, but attenuate the power before it gets to your speaker. It differs from the master volume concept because it allows you to drive the output tubes hard. Output stage distortion is round and sweet; it is what I call the "good" distortion. Here are the plans for making a power attenuator that allows you to turn down your speaker by 6 dB without turning down your amp. The parts to build this attenuator will cost about $30. It features a toggle switch that will attenuate 6 dB in one position and go back to dead-nut stock in the other position. Your tone will not be affected in the least—only the volume will be cut to about a quarter of stock power.

THEORY OF OPERATION

The idea is to correct the load impedance while using a resistor matrix to cut the power by 6 dB. For now, let's assume we are working with an 8 ohm speaker load. First, we will put an 8 ohm resistor in parallel with the 8 ohm speaker load. This will cut the current in half, because there are now two paths for the current to go; namely through the speaker or through the resistor.

Ordinarily, this would change the load impedance from 8 ohms to 4 ohms. To correct the impedance and cut the voltage in half, we will add a 4 ohm resistor in series with the speaker/8 ohm resistor combination. This will bring the load impedance back to 8 ohms and the voltage will be divided between the speaker/8 ohm resistor and the 4 ohm series resistor.

When we cut the current in half (-3 dB) and we cut the voltage in half (another -3 dB) that the speaker "sees," we end up with a -6 dB attenuation while maintaining the proper load impedance of 8 ohms.

If you were doing this with a 16 ohm load, you would use a 16 ohm resistor in parallel (R2) and an 8 ohm resistor in series (R1). With a 4 ohm load, you would use a 4 ohm resistor in parallel (R2) and a 2 ohm resistor in series (R1).

PARTS LIST FOR 8 OHM POWER ATTENUATOR

These plans are for use with a low wattage amp, perhaps 25 watts or less. For larger wattage amps, I do not recommend making your own attenuator. You would need much larger wattage resistors and 6 db probably will not be ample attenuation on a high wattage amp.

R1—4 ohm 15 watt wirewound resistor. I recommend Mouser part # 284 – HS15 - 4

R2—8 ohm 10 watt wirewound resistor. I recommend Mouser part # 284 – HS10 - 8

SW1—DPDT Toggle switch. I recommend Mouser part # 10DS238

J1—Output jack. I recommend Mouser part# 502 - 11

P1—¼" right angle phone plug. I recommend Mouser part# 171 - 1206

CHASSIS—metal enclosure. I recommend Mouser part# 537 - 136 - P

HEAT SINK COMPOUND—I recommend Mouser part # 524 - 8109 - S

SPEAKER WIRE—about 2 feet. Use 18 gauge.

HOOK-UP WIRE—about 2 feet or less. Use 18 gauge.

Mouser can be reached at 1 - 800 - 34 MOUSE (1- 800 - 346 - 6873)

LET'S GET STARTED BUILDING

Look at all the parts and decide where they can go in the chassis and not interfere with each other. Take a minute to plan. Look over the schematic in the diagram and spend a few minutes planning— before you drill any holes. You've heard of the carpenter's rule, to measure twice and cut once. The amp builder's rule is to plan it out twice and drill the chassis holes once.

Although it is possible to build the attenuator inside the amp, I advise not to drill any holes in a vintage tube amp. Use a separate chassis and maintain the integrity of the vintage amp.

Notice that the wirewound resistors are heatsinkable. We will mount them inside the chassis, leaving enough room for the output jack, switch and a hole for the input wire. The heat sink compound is applied between the resistor and the chassis. It allows for superior heat transfer from the resistor to the chassis. You must have adequate heatsinking for the resistors to actually perform at the specified wattage rating!

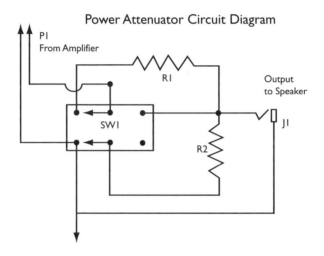

Power Attenuator Circuit Diagram

LET'S HOOK IT UP AND PLAY

The ¼" right angle phone plug (P1) is attached to the short speaker wire and will act as the input for this unit. The plug would connect to the amplifier's output jack. The ¼" female jack (J1) will act as the output and will connect to your speaker.

When the toggle switch is in the position shown on the diagram, then attenuation occurs. But when the toggle switch is switched to the other position, the circuit returns to stock.

I first designed this attenuator for one of my employees that played regularly in Austin, Texas. There is a sound ordinance in Austin and my employee complained that even his 22 watt Kendrick 2112 was too loud. He didn't need a large wattage attenuator so I designed this for him. It has worked perfectly for him and it can work perfectly for you too.

All About Vacuum Tube Guitar Amplifiers

BUILD A 2-STAGE POWER ATTENUATOR FOR HIGH POWERED TUBE AMP

When someone tells you to turn down your tube amp, they might as well tell you to stop having your amp sound its best. What makes a tube amp sound great can be found only when it is turned up to its sweet spot. Take away that and you might as well use a transistor amp. But yet most of the time, the amp's sweetest spot is far too loud for the occasion.

A master volume control will never get outstanding tone because it starves the output tubes for signal. The "good" distortion comes from the output tubes working hard. The only distortion you get from a master volume is the sound of a preamp tube with half the wave clipped off. This will only sound buzzy and over-compressed with homogenized dynamics. If you listen to it very long, you will develop listener fatigue.

What is needed is an attenuation device that could be placed in the signal path between the amplifier and the speakers so that one could turn the amp up until it sounds its best and then attenuate the signal going to the speakers so that the speakers can be turned down to the appropriate loudness for the particular room and particular situation. In a nutshell, turn the amp up and the speakers down. Not only does this allow for a great overdriven tone at the appropriate volume level, but when used in recording situations, a power attenuator will remove most, if not all, of the hiss and hum generally associated with a tube amp. When you attenuate a few db, the first thing to go is the noise!

Here are the plans for making a power attenuator that allows you to turn down your speaker by up to 12 dB without turning down your amp. The parts to build this attenuator will cost about $85.

Besides the true-bypass configuration, it features two toggle switches that will each attenuate 6 dB when switched (12 dB total attenuation with both switches on) and go back to dead-nut stock when unswitched. Your tone will not be affected in the least—but the power driving the speaker will be noticeably less. And since the load impedance is kept constant, regardless of which switches are "on," your amp will "think" it is just operating normally!

THEORY OF OPERATION

The idea is to use a resistor matrix to cut the both voltage and current and in doing so, we cut the power. Remember, power = voltage times current. We select the correct values of resistors so that we can attenuate both current and voltage at the correct proportion; thus keeping the impedance constant. This allows the amplifier to "see" the correct load. The device has two stages that operate exactly the same. For now, I will describe how one stage works because the second stage works exactly the same way. For the example, let's assume we are working with an 8 ohm speaker load. If you placed an 8 ohm resistor in parallel with an 8 ohm speaker, half of the current would go through the speaker and half would go through the resistor because there are now two paths for the current to go; namely through the speaker or through the resistor.

Adding this 8 ohm resistor in parallel with the speaker would change the total load impedance that the amp "sees" from 8 ohms to 4 ohms. We need for the load impedance to be 8 ohms, so to correct the 4 ohm load and to get it back up to 8 ohms; we simply add a 4 ohm resistor in series with the existing parallel connected speaker/8 ohm resistor. By adding this 4 ohm series resistor, we are also cutting the voltage going to the speaker in half. So to summarize; the 8 ohm resistor in parallel with the speaker cuts the current in half and the 4 ohm resistor in series with the existing speaker/8 ohm resistor cuts the voltage in half. By using both resistors simultaneously, we are keeping the total load impedance at 8 ohms.

When we cut the current in half and the voltage in half; we end up with a –6 dB attenuation while maintaining the proper load impedance of 8 ohms.

If you were doing this with a 16 ohm load, you would use a 16 ohm resistor in parallel (R2) and an 8 ohm resistor in series (R1). With a 4 ohm load, you would use a 4 ohm resistor in parallel (R2) and a 2 ohm resistor in series (R1).

Two-Stage Power Attenuator Circuit

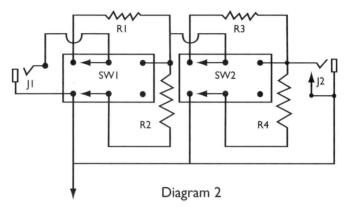

Diagram 2

I have explained how one stage works. The other stage works exactly the same way. Look at Diagram 2. R3 and R4 are the same values and function respectively as R1 and R2. Each stage has a toggle switch and each toggle switch attenuates the power by 6 dB. It makes no difference which switch is flipped on first. You can use either one or both to go from true bypass to -6 dB to -12 dB. I would not recommend making a third stage because when you attenuate more than 15 dB, attenuation will begin to affect the tone. Attenuation with this device, however, does not degrade tone and will be unnoticeable except for the loudness from the speaker.

PARTS LIST FOR 2-STAGE POWER ATTENUATOR
SW1 and SW2—DPDT Toggle switch. Mouser part # 691-2BL62-73
J1—Input jack. Mouser part# 502–11
J2—Output jack. Mouser part# 502–12A
Chassis—metal enclosure. Mouser part# 537–138-P
Heat sink compound—Mouser part #5878-CT40-5
Hook-up wire—about 2 feet or less. Use 18 gauge.
Mouser can be reached at 1 – 800 – 34 MOUSE (1-800-346-6873)
FOR 4 OHM VERSION:
R1 and R3—2 ohm 100 watt wirewound resistor Mouser part# 284-HS100-2.0F

R2 and R4—4 ohm 50 watt wirewound resistor Mouser part# 284-HS50-4.0F

FOR 8 OHM VERSION:

R1 and R3— 4 ohm 100 watt wirewound resistor Mouser part# 284-HS100-4.0F

R2 and R4—8 ohm 50 watt wirewound resistor Mouser part# 284-HS50-8.0F

FOR 16 OHM VERSION:

R1 and R3—8 ohm 100 watt wirewound resistor Mouser part# 284-HS100-8.0F

R2 and R4—16 ohm 50 watt wirewound resistor Mouser part# 71-RH50-16

Power Attenuator Layout and Wiring

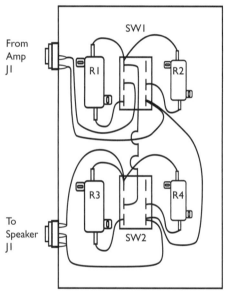

Diagram 1

LET'S GET STARTED BUILDING

Look at all the parts and decide where they can go in the chassis. Refer the diagram 1 to look at a suggested layout. Look over the schematic. Refer to diagram 2. Look at the parts and invest a few minutes planning—before you drill any holes. You've heard of the carpenter's rule: measure twice and cut once. The amp-builder's rule is to plan it twice and drill the chassis holes once.

Although it is possible to build the attenuator inside the amp, I advise not to drill any holes in a vintage tube amp. Use a separate chassis and maintain the integrity of the vintage amp.

Notice that the wirewound resistors are heatsinkable. We will mount them inside the chassis, leaving enough room for the jacks and switches. The heat sink compound is applied between the resistor and the chassis. It allows for superior heat transfer from the resistor to the chassis. You must have adequate heatsinking for the resistors to actually perform at the specified wattage rating! The component values shown will work for amps up to 100 watts and still have a margin of safety. If you would like to use the power attenuator only with a 50-watt amp, you may safely cut the wattage values of all the components to half the listed values. This would save you a little money and the device could be made smaller.

Also, notice that the output jack has a shorting switch and the input jack does not. The switch is a safety device in the event you happen to turn on the amp without a speaker plugged in to the attenuator. The switch lead on the jack simply connects to the ground lead on the same jack.

LET'S HOOK IT UP AND PLAY

When the toggle switches are switched in the position shown on diagram 2, then maximum attenuation occurs. But when both toggle switches are switched to the bypass position, the circuit returns to stock.

You can place the attenuator in the back of your amp and leave it there.

Or you could also mount the chassis box on the wall of your amp or the inside back panel for easy access to the toggle switches. If you needed to change the attenuation level for different songs, one could use DPDT footswitches instead of toggle switches and it would be possible to change attenuation level while playing. Another usage would be for changing attenuation while channel switching. For example: turn up the volume on the overdrive channel and use the attenuation for best lead tone, but turn off the attenuation when you switch to the clean channel whose volume is already turned down for maximum loud clean headroom.

SERVICING
YOUR AMP

All About Vacuum Tube Guitar Amplifiers

WORKING ON AMPS: 10 KEYS TO SAFETY

I will always remember the first time I saw the inside of a vacuum tube guitar amplifier. I was in awe. With all those wires and electronic components; I couldn't help but wonder what genius level was required for someone to build one. It seemed anyone capable of doing repairs and service on a vacuum tube amp must have been a direct descendent from Einstein or a member in good standing of the Mensa society.

As complicated as the inside of an amp may appear, it is still only a preamp, a power amp and a power supply. Each section is fairly simple and when you break it down, it is even simpler. Maybe that is why more and more people are becoming interested in doing their own servicing and repairs to their amps—in much the same spirit as a sports car driver may wish to set the timing or adjust the valves on his iron.

Just as you want to allow the engine of your car to cool before attempting to perform maintenance, there are certain precautions to take when servicing your amp. Perhaps the biggest hazard associated with servicing a vacuum tube amp is electrical shock. All tube amps use very high D.C. voltages. At best, getting shocked will annoy you while getting your immediate attention and at worst it could cause electrocution!

Regardless of how careful, if you service enough amps, you are going to shock yourself. Most of the time, the shock is more of an annoyance than anything else. Lord knows I have shocked myself hundreds of times, but there are precautions that can be taken to minimize the risk of getting shocked and also minimize the effect of getting shocked.

Besides risking a shocking experience, you are also risking blowing the amp up. Install a new filter capacitor in backwards and it will explode in your face upon powerup. If you accidentally directly ground a main B+ connection, there will be a mushroom cloud of smoke as the guts of your power transformer incinerate rather rapidly! Here are some guidelines to help prevent electrical shock, help prevent loss of property and minimize the consequences if you should get shocked.

1. CHOOSE A GOOD WORK AREA. I would choose a wooden table or some other nonconductive surface. Stay away from any area with a wet or damp floor. If you were working on a metal table in a damp basement, for example, any minor accident could turn into something major.

2. UNPLUG THE AMP UNLESS YOU ARE TESTING IT. This seems pretty basic doesn't it? It is not good enough to simply turn the switch off, because there is still 120 VAC going from the wall to the switch/fuse/etc. If you accidentally bump up against one of those live wires, your body will be the ground connection between earth and the wall voltage coming from the utility line. Ouch!

3. PUT THE STANDBY SWITCH IN THE PLAY MODE. Most amps have a bleeder circuit that discharges the caps, but it will not discharge all of them unless the standby switch is in the play mode. Of course, if the amp lacks a standby switch, this rule would not apply. If the amp doesn't have a bleeder circuit, you will need to discharge the capacitors manually, but all will not discharge unless the amp is in the play mode, so it is a good practice to always put the standby switch in the play mode.

4. DISCHARGE THE FILTER CAPACITORS IN AN AMP BEFORE WORKING ON IT. Don't go the macho approach and short them with a screwdriver to ground! That looks cool if you want to show someone a large spark/small lightning bolt. Here are two excellent ways to discharge the capacitors in an amp.

The first way involves making a bleeder device. It will consist of a 100K ohm ½ watt resistor soldered to a short piece of hookup wire. After the wire is added to the resistor, you will need two alligator clips. One gets soldered to the free end of the resistor and the other goes to the free end of the wire.

To discharge the capacitors, you would clip one end of the device to ground and the other end to the "plus" end of any electrolytic filter capacitor in the amp. Assuming you have the standby switch in the play mode as described earlier, the capacitors will all be drained in about a minute.

The other way to discharge filter caps is the Gerald Weber quickie method. I simply use an ordinary jumper wire and short pin 1 of any 12AX7 preamp tube to ground. If there are no 12AX7s in the amp, use the plate of any preamp tube. This also takes about a minute and the capacitors will be drained. As in the first method, if the amp has a standby switch, it must be in the "play" mode. The reason this works so well is because the plate of the preamp tubes connects indirectly to the filter caps through the plate load resistor. The resistor, which is usually a 100K ohm value, limits the current flow to ground the same as a bleeder resistor.

5. WORK WITH ONE HAND AND KEEP THE OTHER ONE IN YOUR POCKET OR BEHIND YOUR BACK. If you have both hands near the chassis, it is likely that one will be touching the chassis. If you were to shock yourself by touching a "hot spot" with the other hand, the electricity would go from one arm, through your heart, and through the other arm to ground (chassis). It is not a good idea to shock yourself in the heart. If you have one hand behind you or in a pocket and you get shocked, the consequences are less.

6. USE INSULATED TOOLS. And remember to keep your hands on the insulation. The insulation does no good if you are touching the metal portion of the tool.

7. BLOW OUT YOUR AMP WITH COMPRESSED AIR. Little bits of wire, trimmings from leads and other debris can become lodged in an amp's innards. Blowing the amp out with compressed air can remove all this litter quickly. One little piece of wire clipping accidentally going from the heater circuit to ground and you will become familiar with what a burned up transformer smells like. Worse yet, if it tests fine and later someone takes it on a gig and a lead trimming or a bit of solder debris shorts something out, you are done. If you don't have compressed air, turn the chassis over to get gravity on your side. Shake the amp chassis and do a visual inspection to make sure there is no debris that could cause a malfunction.

8. DOUBLE-CHECK YOUR WORK. Before plugging an amp back into 120 VAC wall power, take a minute or two to go back over everything you did. Are the new components installed with correct polarity? Did you remove the bleeder wire shorting the filter caps to chassis ground from when you discharged the filter capacitors? Double check yourself and you will be surprised how many times you save yourself. The ABC's of good amp servicing is **Always Be Checking**.

9. Use a current limiter when applying power for the first time to a recently serviced amp. If there are any paths to ground and you power up without a current limiter, you will burn something up. It is that simple. If you use a current limiter, you will never burn anything up.

For example, suppose there is an accidental short across the heater winding. The current limiter will save you from instantly destroying the power transformer. There are too many other things that could go wrong to risk blowing something up by not using a current limiter.

Directions to make a current limiter are on page 328 of my second book, *Tube Amp Talk for the Guitarist and Tech*. It is basically an extension cord with a light bulb in series. When an amp is plugged in, all the current going to the amp must also go through the light bulb. Since you are using a 100-watt bulb, the current is limited to .83 amps. That little current is not enough to allow any serious melt-downs. When a short circuit occurs, the circuit limiter circuit will want to draw more current, so the light bulb lights brightly, indicating a malfunction in the amp.

10. Never play guitar while troubleshooting the inside of a live amp. Most electric guitars have the strings grounded to the sleeve of the output jack. This connects to the shielded part of the guitar cable and that shielding ultimately connects to the chassis of the amplifier. If you are holding a guitar or otherwise touching the strings with one hand while touching parts of the amplifier with the other hand, the potential is there to shock yourself badly. Have someone else play the guitar while you troubleshoot or if you just need a signal, lay the guitar down and hit the open strings. It won't sound that good, but it will enable a signal into the live amp while you are working on it.

Working on amps can be a lot of fun, but it is even more fun when you practice safe servicing. Most of us aren't in the habit of servicing amps daily, so it is best to take every precaution and take as much time as needed. Use you brain and re-think everything. Servicing your own amp will bring the pride and satisfaction of knowing the job is done to your liking.

All About Vacuum Tube Guitar Amplifiers

HOW TO RE-GRILL A BAFFLEBOARD

If you have an amplifier that needs a new grill cloth you may want to replace it yourself. Here are some tricks that could save you time when working with the grill cloth.

Some grill cloths are textile fabric, such as the Vox, but most are non-textile—such as Fender, Gibson, Ampeg, etc. (The textile fabric grillcloth, such as Vox requires an entirely different technique and we will cover it last.) The non-textile type is what we will concern ourselves with for now.

GRILL CLOTH IS HEAT SENSITIVE. That means if you recover a grill and the grill cloth is too loose, you may tighten this type grill cloth by applying mild heat. The heat must be slow and even. I use an electric quartz heater that is about 30 inches tall and about 4 inches wide. I place an amp whose grill is loose, near the heater, such that the heat is applied evenly. This causes the threads to shrink, which tightens the entire grill cloth! Do not get the heat source too close to the grill cloth as it will cause the grill cloth to melt. When this stuff melts, it disintegrates, so you want to keep an adequate distance from the heat source to prevent destroying the grill cloth.

Never point a heat gun at a piece of grill cloth unless you intend to melt it or catch it on fire. A heat gun gets too hot, but a hair dryer can work wonders with stretching out and removing creases in the grill. I have my grill cloth on large rolls, which stay perfectly flat; however when an end user buys a small quantity of grill cloth, it may be folded with some creases. Mounting the grill cloth properly and then heat stretching the grill cloth may remove the creases. If the heat stretching doesn't remove all the creases, I would help the heat process along with a hand-held hair dryer. The proper amount of even heat can may the difference between a perfect re-grilling or a sloppy looking end result.

GRILL CLOTH HAS AN ORIENTATION. Look at both sides of the grill cloth. Which side goes out? You should be able to tell by looking. There are exceptions. For example, the late 50s tweed amps such as the 1959 Bassman (whose grill cloth had the yellow stripes) comes to mind. With this grill cloth, there is a horizontal yellow stripe when viewed from one side but a yellow dotted line when viewed from the other side. I have seen original amps with the cloth mounted either way. Maybe they didn't look before they applied. I think the yellow solid stripe out is correct and that's the way I like that particular grill cloth.

Even after you find the right side, there is still a top and bottom on certain grill cloths. Especially, you will notice this on grill cloth that has tinsel. For example, the Kendrick Black Gold amps use a black grill cloth with gold tinsel running horizontally. If you put the tinsel out and orient the grill one way, the gold tinsel seems to disappear. It simply is not visible. On the other hand, if you rotate the cloth 180 degrees, the gold tinsel comes out and is very sparkly. Look at your grill cloth and whichever way looks the best is usually correct.

GRILL CLOTH MUST BE STRAIGHT. This is the hard part. You begin by confirming the correct orientation of the grill cloth and then cutting a piece to size. Be sure and cut it with a 3 inch excess on all four sides. (This means the actual piece of grill cloth will be 6 inches longer and 6 inches wider than the baffleboard you will be covering!) When you are cutting the grill cloth, use the natural lines and/or patterns in the cloth to make sure you are cutting perfectly straight.

Once the grill cloth is cut to correct size, you would start by fastening the grill cloth in one corner. I use staples to fasten about three inches on each leg of one corner. While stapling this down, stretch the grill cloth simultaneously. You want the cloth to be fairly tight. After the first corner is completed, work down a short side. Being careful to make sure the lines and/or pattern lines up with the short side, and stretching enough to keep it tight, staple the next corner. Again staple about 3 inches of each leg of that second corner.

In a similar fashion, work down to the third corner. Keeping the lines in the grill perfectly straight, stretch the cloth and staple the third corner. Again, staple the cloth about 3 inches on each leg of this corner.

The first three corners were fairly easy. This last corner is the hardest part. It is hard, because you have to line up two edges simultaneously before stapling. If you are having a hard time seeing the lines or pattern, you may use a piece of masking tape to help your eye see the line. Taking care to make sure the lines are perfectly straight on both legs of the unstapled corner, line it up and staple this last corner. TAKE YOUR TIME AND DOUBLE CHECK. Yes, it is important to double-check your work at this point. Make sure that all the lines on these corners are perfectly straight. If you have to remove a staple or two and redo a section, now is the time to do it. Getting the corners right will make the rest go easier. Sometimes, you can staple the grill correctly and because of the looseness of the grill's weave, the cloth will "slip." This is the time to correct such slippage before finishing the sides.

FILL IN THE SIDES. Looking at the pattern, work down one side and staple the cloth. I would use a staple every inch or so. Rotate the baffleboard and do the next side, taking care to keep the lines perfectly straight.

The third and fourth side may require a little pulling and tugging while stapling. These sides are not as easy as the first two sides because the tension is beginning to mount on the grill itself.

DOUBLE CHECK AGAIN. Once you are done with the grill, look at it closely. Are there any lines that are wavey. Perhaps those can be straightened by adding more staples. It doesn't hurt to look at your job with a critical eye.

HOW TO REPLACE A TEXTILE GRILL CLOTH.

On certain amps, such as Vox, the grill cloth is made from actual cloth. Imagine that! These grill cloth types cannot be stretched very tightly and require a different technique to replace.

There is a twine mesh that is sold where needlepoint supplies are sold. This stuff looks like a screen made from twine. Get a piece big enough to cover the baffleboard.

After you remove the old grill cloth, you take a piece of this mesh and cover the front of the baffleboard with it. Staple it down very tight.

Paint the mesh and baffleboard with flat black paint. Take care to mask the cabinet properly so that you don't get overspray on the cabinet. Once the paint has dried, you get some Elmer's Glue and a small paint roller. The idea is to get a light but even coat of Elmer's Glue on the mesh. You want to use a small amount of glue because you don't want it to bleed through the grill cloth, but at the same time, you have to have enough to stick to the grill. Once you are satisfied you have an even coat of glue on the mesh, carefully place the textile grill cloth over the mesh.

LINE UP THE PATTERN AND LINES. If it is a Vox, you want to make sure the diamonds are symmetrical. This is your small window of opportunity to fine tune how the lines and patterns "line up."

LET IT DRY. Once the glue has dried overnight, replace any string trim or molding around the grill.

All About Vacuum Tube Guitar Amplifiers

RECOVERING YOUR VINTAGE AMP

Do you own a vintage amp in cosmetic disrepair that you wish looked better, but at the same time wondered, "How will recovering affect the value of my vintage amp?" I would never recommend recovering a vintage amplifier that is original and in fair condition or better. Sometimes minor defects add character to a vintage piece. At the same time, an amplifier in poor to ugly physical condition can be an eyesore.

Collectors talk of keeping something original, but originally that piece wasn't ripped, wasn't rotted nor did it have tears, cigarette burns and other aberrations. Originally it was new. Let's use a vintage car analogy. Would you rather have a vintage automobile with a ripped up original interior, rusted and rotted original paint with lots of rust and peeling on the bumper? Or would you rather have a vintage automobile that had a re-upholstered interior done right, using original style materials and color, a fabulous new paint job with correct colors and a perfectly re-chromed bumper? I'm sure you will agree that a perfect restoration can enhance the value of an otherwise ugly piece. On the other hand, use the wrong materials and colors and you could actually take away some of the value. I have had clients that have bought vintage amps, had them restored to original spec and sold them for hundreds and, in certain cases, thousands of dollars more than their buying price. On the other hand, there are people that tried to recover their own amp, screwed it up a little and when they eventually sold it, lost more money than what they thought they saved by doing it themselves.

Unless you are absolutely certain that you know what you are doing, don't attempt to recover a vintage amplifier; unless of course you are willing to take a loss when you sell it. A recovering job will only enhance the value if it is done right.

If you have the type of personality that you like to rush things and hurry everything up, don't attempt to recover a cabinet. Recovering requires patience and planning. If you are the type of person that doesn't have an eye for detail, again, you should not attempt to recover your own amp. Having said all this, here are the steps to re-covering a vintage amplifier cabinet.

PLAN YOUR RECOVERING JOB

Take some close up pictures of the amp from various angles for starters. A digital camera works great for this. I recommend you make drawings of distinguishing characteristics, such as which way the tweed stripes are going, what the corners look like, which pieces were applied first, what size each piece of covering material must be, where will the handle and other hardware mount. What is the spacing of the feet, etc? Document everything first and you won't be asking yourself later, "How was that mounted?" Or "Where do those feet go?"

After you have documented everything with appropriate photos and drawings; remove hardware, speakers, chassis, etc., and look at how the inside of the cabinet should look. Make additional drawings or photos if needed.

Before you order your covering, you will need to determine how much material you will need. After figuring the size of each cut piece of covering material and considering the way each piece is wrapped around internally, use graph paper to cut out miniature pieces using a scale of one square of the graph paper equals one inch of covering material. When using tweed, remember to draw the direction of the stripe on each graph paper scale model so that the stripes will be running in the correct direction when they are actually cut out.

After finding out how wide the covering material comes, make a scale of that width on another piece of graph paper. Now take your scale model pieces that have been cut out and lay them on top of this graph paper. Place them where you will end up with the least amount of waste and each piece is positioned so the tweed stripes are correctly oriented. From this, you will be able to determine how each piece will be cut out and how many yards of covering material you will need.

You will also need to purchase some glue. I use animal glue, but that is not practical for most people because it requires a hot glue spreader which costs thousands of dollars and it requires an experienced operator. This only leaves a few options. Contact cement is not recommended because it is very difficult to work with and the vapors are toxic. Aerosol spray glues are very expensive and they never work well. I would recommend Elmer's glue because it is easy to work with, easy to clean up and it will stick well.

You will also need a few small paint rollers for applying the glue. The 3" size works well. A heat gun or hair dryer is essential as well as a utility knife with extra blades and a few lint-free white rags.

PREPARE THE WOODEN CABINET

Just like body work prepares a car for its paint job, cabinet preparation is essential before any covering can be applied. All previous covering and glue should be removed. Most vintage amps used animal glue originally. It can be removed with heat and it is water based (water dissolves it). A heat gun can be very useful in removing the outer covering. If you don't have a heat gun, use a hair dryer set on high. If you are removing tweed, sometimes a wet towel placed on the tweed for an hour will pre-dissolve the animal glue and make it come off very easily. To get animal glue off completely, place a wet towel on it and let it set. The glue will puff up and can be easily scraped up with a paint scraper.

All wallowed or stripped mounting holes should be filled in with epoxy (Bond-O) and re-drilled. This would include both the chassis mount holes, back panel screw holes and sometimes, speaker screw holes. You will want to check the correct drill bit size BEFORE you fill in the holes with epoxy.

Finger joints that are loose should be re-glued with wood glue or super glue. Any cracks should be repaired. Look for cracks around the chassis mount holes. These can be repaired with epoxy as well. Fill dings and dents then sand smooth. Sand the entire bare cabinet until it feels smooth.

CUT OUT THE COVERING

You already determined which way the pieces are to be on the covering material such to minimize waste. The next step it to actually cut the covering material into the correct size pieces you will need to cover the cabinet. You want all the pieces cut to size before you begin gluing.

When cutting out the material, I like to use a straight-edge and a utility knife. You will need to use some sort of backing to use a utility knife. I have a large glass table that works great, but if you are careful, you could use a board positioned underneath or a thick cardboard. Any time you pick up a utility knife, you should be very careful.

APPLY THE COVERING MATERIAL

When you are ready to apply a piece of material to the cabinet, sparingly apply glue to both facing surfaces with a small paint roller. You should not use too much glue, but every square inch should have glue and the glue must be even. When you have the glue spread properly on both surfaces, apply some heat to pre-dry the glues. A heat gun or hair dryer is used for this. It is important to pre-dry the glue but don't get it completely dry. Both the covering surface and the wooden cabinet surface should be tacky and nearly dry. Now apply the piece and stretch out any wrinkles. When using tweed, if you cut the pieces with the stripes oriented correctly, you will have no problem duplicating the positioning.

After a piece is applied, squeeze out any and all air bubbles that may occur. You want every square inch of covering material to be glued to the cabinet. You do not want any air bubbles preventing any of the covering from becoming glued to the cabinet. When using Tolex, a straight-pin can be used to poke a tiny hole in the covering to allow air to escape and thus eliminate air bubbles. Use your lint-free rag to wipe up any excess glue.

When you are cutting the corners, use a utility knife with a brand new blade. Always cut the bottom corners first as this will give you a chance to practice before you cut the more visible top corners. And be careful with the utility blade. Point the blade away from you and pay attention.

When you finish covering the cabinet, leave it alone for a couple of days to dry. If it is Tolex, it may take several days for it to dry, depending on how much glue you used and how well it was pre-dried during the initial application.

After it is dried, you need to spend a little time gong back over your work and touching up. If you are using Tolex, the Tolex will not stick very well to itself. So the seams will not be glued well. Go back over the seams and re-glue them with super glue. I position the cabinet so that gravity will work with me. I don't want any super glue running down the Tolex. Start by separating the seam slightly and applying the super glue inside the seam. Push the seam together using a paper towel to blot up any excess glue around the edges. Use a clean paper towel to hold down the edge until the super glue sets. Using paper towels will avoid the possibility of gluing your skin to the cabinet!

If you are using Tolex, after the Super glue has dried, replace the hardware, reassemble the speaker and chassis and you are done. If you are doing tweed, don't replace the hardware until you do one more step.

SEAL AND AGE THE TWEED

Tweed is a porous fabric that will look awful if it is not sealed. Any little bit of dust or dirt can become permanently embedded in the tweed and water can penetrate it easily, thus causing irreparable damage. Therefore, sealing the tweed will make it look original as well as protecting the tweed by making it non-porous and thus water resistant.

There are many types of lacquers and polyurethane products that would work for sealing tweed. I think the best looking is the Polyshades by Minwax. It is available at any Sherwin Williams, Walmart or Home Depot. It comes in different colors and I recommend the Honey-Pine Matte finish to get a 50s tweed look. You can apply it with a paint brush or a lint-free rag. I use a rag. You will let each coat dry thoroughly (at least 24 hours) and then sand lightly in between coats with steel wool or Scotch-brite. Perhaps a 220 grit sand paper will work nicely. You don't want to sand into the fabric; you are just leveling everything out so that it feels smooth. You will need about 3 or 4 coats total. When you sand it, touch it and feel for smoothness. When you have enough coats, the finish will feel smooth—just like a 50s tweed. Reassemble the hardware and you are done.

All About Vacuum Tube Guitar Amplifiers

HOW TO RESURRECT
A DORMANT TUBE
GUITAR AMP

If you are into tube tone but unwilling or unable to part with the thousands of dollars it costs to purchase a collectible tube amp or purchase a new reproduction amp, the alternative could be to obtain an old tube amp for a few coins and simply resurrect it. There are many opportunities available, but usually only to the person that is actively looking. These opportunities could present themselves at flea markets, garage sales, estate sales, newspaper want ads, green sheet ads, auctions, eBay and through personal acquaintances.

I am sure you will find some very cool pieces that have potential for sounding great. Perhaps you will find something as simple and great sounding as a 100 watt Bogen Tube PA Head (as used by the late and great Terry Kath of Chicago Transit Authority), or something tragically hip such as a 1930's Gibson EH 150 from someone's Grandpa's estate sale.

DON'T TURN HER ON YET!

The first thing to do: Give the amp a thorough visual inspection. You want to look at everything very closely and observe the little things. Take your time. You are looking to see if everything looks safe before attempting to turn it on. Look at the A.C. cord. Is it frayed or in disrepair? How is the A.C. wall plug on the end? Does it look like you could plug it in without shocking yourself?

You also need to look to see that the amp is connected to a speaker load. Tube amps can be damaged very badly if you attempt to operate them without a speaker connected. You want to look to see if there are wires going from the amp to the speaker. Also, if you have a meter, you could temporarily disconnect the speaker from the amp and check to see if the speaker has continuity. Note: You will get a shorted reading if you try to measure the D.C. resistance of the speaker without disconnecting it from the amp first. Although the D.C. resistance in ohms will measure slightly less than the actual A.C. impedance, you should get D.C. resistance readings anywhere from slightly less than 2 ohms to as much as a 14 ohm reading—depending on the speaker type and configuration. Once you determine that the speaker load has continuity and proper resistance, hook it back up to the amp as it was before. Still looking at the speaker; is the cone torn? The speaker cone could be repaired or the speaker could even be reconed if necessary. If the speaker cone is torn, but there is proper impedance to the speaker, it may work well enough for now and you could get the speaker reconed later.

Look at the fuse. Is it the correct value? Is it blown? If you have a blown fuse, this is an indication the amp was malfunctioning when it last was operated. Or if the fuse is replaced with the wrong value of fuse, it could mean there was a problem that may still exist with the amp. If you look at the fuse and it is covered in foil, this may be a tip off that the power transformer may be burnt up.

If you suspect a problem with the power transformer, smell it. If you have a burnt power transformer, it will smell burnt; it may look burnt, too. But if it smells burnt, guess what? It probably is.

If the 120 volt A.C. cord needs replacing, now would be the best time to do it. I would recommend a grounded A.C. cord as a replacement. On a 3-prong, grounded A.C. cord, the third wire (green) will simply attach to the metal chassis of the amp and the other two leads (black and white), will go to the same points in the circuit as the old A.C. cord.

If the speaker load is not proper, you will need to correct that before attempting a "power up." Find the output of the amp and hook it to a speaker. If you don't know where the output wires are, here is where to find them: They are coming out of the output transformer. The output transformer is situated in the circuit between the speaker load and the output tubes. The primary wires (usually brown, blue and red) go to the output tubes and the high voltage supply. The other wires, which are usually thicker wires than the primary wires, would go to the speaker. If there are multiple wires for the secondary of the output transformer and you don't know which two to use, take a meter and determine which pair of secondary wires read the most resistance. There will not be much difference in the different taps, only a fraction of an ohm, so you will need a digital ohmmeter. Whichever secondary pair has the most D.C. resistance between them should go to the speaker load for now. Almost any speaker load will work for testing, any handy 4 ohm or 8 ohm speaker will generally do the trick for simple testing.

There is a technique to determine the impedance output of the transformer leads, but it is beyond the scope of this chapter. (Complete instructions on how to determine impedances of the leads of an unknown transformer can be seen in my instructional DVD, *Tube Guitar Amplifier Servicing and Overhaul*.)

Once you determine the amp has a proper A.C. cord, a proper speaker load and the fuse is not previously blown or abnormal, then it is time to look to see that all the tubes are installed and in place. If there is a tube missing, it is pointless to attempt a "power on." You need to have all the tubes in place before trying your luck at plugging it into the wall A.C. outlet and seeing what you've got.

If you have a current limiter, you should plug the amp into a current limiter. If the fuse has been burnt, remove the output tubes and rectifier before plugging the A.C. cord into the current limiter. If the fuse was intact, just plug the amp into a current limiter and turn it on. To power up an amp that has been dormant, one must use a current limiter. The current limiter acts as a bottleneck in the power supply A.C. circuit. You could say it bottlenecks current such that the current is limited to less than 1 amp.

If you don't have a current limiter, you can make one very easily. Complete instructions are in my second book, *Tube Amp Talk* on page 328. But basically, a current limiter can be as simple as a 100 watt light bulb in series with the amp. You can put one together with an extension cord and a utility 100 watt lamp. You wire the extension cord so the lamp is in series with one of its two legs. If the amp becomes shorted, the only current that can pass is whatever the light bulb will allow to pass. For example, a 100 watt bulb would pass less than 1 amp of current maximum. (100 watts divided by 120 volts = .83 amps). So even if the amplifier is shorted out, only .83 amps of current may flow through the amplifier at any given time. This will save you burning anything up if there is a short or some other malfunction.

Also, if the filter capacitors are all dried up, and they probably will be if the amp has been dormant; you want to limit the current so the D.C. across the caps will help to reform them somewhat. When the amp is in a current limiter, very little current is allowed to pass through the A.C. supply circuit. By the time the limited current gets divided up between the filament circuit, the bias circuit and the B+ circuit, there is very little current available to leak through the filter caps. Running the amp on the limiter for a few hours allows for the DC to be placed across the cap to help reformed it some—without fear of too much current running through the filter caps and making them worse.

Plug the amp into 120 A.C. and turned it on, keeping it in the standby mode. Or, if it doesn't have a standby switch, you can just turn it on and let it warm up.

Look at the tubes while the amp is warming up. Does it look like they are all lighting up? If the tubes are all lighting up, that is a good sign. Look for any arcing in any of the tubes. If you see arcing in any of the tubes, turn the amp off immediately. If no arcing is present, leave the amp in the play mode and plugged into the current limiter for a few hours before attempting a full current "power up."

A word of **CAUTION** when removing tubes: Some vintage amps use obscure tubes. Also, the tube configuration may not be documented. To make matters worse, the tubes may have their labels worn off. So you want to pay attention about "which tubes" go in "which sockets"—**BEFORE** any tubes are removed from their sockets. Different tube types have different pin-outs and to put a tube with one pin-out into a socket with a different pin-out is a recipe for disaster.

Continuing your inspection of the outside of the amp, look at the cabinet. Does it need repair? If the wood joints are coming apart, a little super glue and a couple of clamps can bring it closer to stock without flattening your wallet. Does the covering need attention? Maybe it can be cleaned and touched up as opposed to recovering it. How is the grill cloth? If the grill cloth is loose, it could actually rattle against the baffleboard and cause an ugly sound. Most non-textile grill cloth materials will shrink up and get really tight with a little heat. I am not talking about a heat gun as that would easily burn a hole in almost any grill cloth. I am talking about the radiance, let's say, from a portable electric heater. Some carefully and evenly distributed heat from one of those type heaters can make the grill cloth tight enough to bounce a quarter.

If you have a dirty Tolex amp, here is a trick I use. This is only to be done on Tolex. If you don't know what Tolex is, don't try it. Specifically, do not try this with tweed or bookbinding material as it will not work. Take the Tolex covered amp cabinet and remove the speakers and amp chassis. Cover and protect the tube label or anything else on the cabinet that you would not want to get wet. You can remove the handle and feet if it is convenient to do so. Leave the chassis and speakers at home, but drive the cabinet and grill to the local coin operated car wash. Select the whitewall tire spray mode on the spray nozzle. Spray this on the Tolex and let it soak. After a little soaking, rinse it with the spray wash and the spray rinse. Since Tolex is made from rubber, this cleaner will make a Tolex amp look like brand new. It deep cleans all the little crevices. You can also use this cleaning technique on a non-textile grill cloth with excellent results.

Let's start by unplugging the amp and removing the amp chassis from the cabinet. After the chassis is removed, take your time and do a very slow and detailed visual inspection on the inside of the amp. You want to look at every component individually and look at every solder connection individually. Don't try to rush the visual inspection. You've got to use a little common sense here because you are looking to "see" if anything looks amiss. If you see a wire dangling into the air, then a light goes off in your head; "That shouldn't be like that." Or if a resistor is charred looking, perhaps it has been overheated severely and should be changed. Or what if the electrolytic capacitors have ruptures in the ends and chemicals are obviously leaking out of them? You want to take note of everything you see as it will help you in deciding what all needs to be replaced during the overhaul.

SERVICING THE POTS, JACKS, SOCKETS AND SWITCHES.

We will inspect and clean the tube sockets, pots, switches and jacks. To do this you will need a burnishing tool, a dental pick (or paper clip or straight pin), and some Tech spray. The burnishing tool is a very small, double-sided file that can be bought at an electronics store. I bought mine at Mouser (1-800-34 MOUSE). Their part number is #524-9338. I use De-Oxit brand Tech Spray, but almost any good technician's spray cleaner will work fine. The spray can will usually have a small piece of plastic tubing that fits into the nozzle of the aerosol spray can. This tubing helps to direct the actual spray and is especially handy for spraying inside switches and in specific small areas.

Every volume knob, tone control, vibrato knob, etc., is controlled by a potentiometer—"pot" for short. You want to spray the Tech Spray inside the pot and then turn the pot's control shaft quickly in both directions. This will "scrub" the innards of the pot. It will make the pots "feel" new again, and it will also get rid of noise in the pot. If you clean the pots and later notice they are still noisy, then you may need to replace that particular offending pot or there could be something else wrong with the amp.

Once all the pots are cleaned, let's clean and re-tension the sockets. Remove one tube only and spray the pins on the tube. Reinsert the tube, observing correct orientation in the socket, and take it out, and reinsert. You are trying to "scrub" any corrosion off of the metal contacts which are inside the tube socket. You may need to spray the pins on the tube a second time and repeat. The tube is supposed to "float" in the socket and the pins will all move within the socket itself, however the pins coming out of the bottom of the tube should seat tightly within the female connectors inside the sockets themselves. If it seems too easy for the pins to go in the socket, then the female connectors inside the socket may need to be re-tensioned.

Re-tensioning the sockets should be done after the amp is drained of electricity. If you don't know how to drain the amp, the idiot method is to place or clip one end of a jumper wire on the metal chassis and then touch the other end of the wire to every pin inside the socket. Hold it for 20 seconds on each pin and that will do it. Basically you are shorting each pin to ground. Or you can simply short a jumper from pin #1 of any 12AX7 type tube to ground for about 20 seconds and it will drain all the electricity of the entire amp. Either way, make sure the standby switch is in the "play" position and the A.C. cord is unplugged.

You would simply take the dental pick or other small tool and manipulate the female part of the connector so that it will be tighter on the pins of the tube. Once you have finished cleaning and re-tensioning one tube socket, move to the next one and continue, until all sockets have been cleaned and all have adequate tension.

Next, you will want to clean and burnish the jacks and spray out the switches. On most guitar amps, a jack with a shorting switch is used for the input jack and sometimes for the output jack. Have ever heard the noise that occurs when you plug a guitar cord into and amp's input, but without terminating the other end of the cord? To prevent that noise from occurring, most amps have a shorting jack that shorts the amp's input to ground when a guitar cord is not plugged into the jack. Sometimes the switching part of the jack gets dirty and doesn't make good contact. You want to spray the parts of the jack and plug a guitar cord into the jack while observing the switching part of the jack being engaged by the plug on the guitar cord. Take your burnishing tool, which is simply a small double-sided fine grain file, and place it between the two contacting surfaces. With the burnishing tool between the two contacting surfaces of the switch, remove the guitar cord plug. There should now be a slight pressure on the burnishing tool. Move the tool in a sawing motion and this will clean and resurface the contacts on the switching part of the jack. Do this on every jack. Sometimes you will notice the jack will have a sprung switch. That is to say that the contacts are not touching or they are not touching tightly against each other. Sometimes a sprung jack can be re-tensioned by inserting a guitar cord plug into the jack, and this will open the switch. With the switch remaining open, use a pair of needle nose pliers to carefully adjust the contacts closer to each other. Test it by removing the guitar cord plug and watching to see if the contacts appear to be working correctly. Sometimes a jack will not re-tension well, in which case you will need to replace the jack with an identical part.

And finally you will want to spray some Tech spray in the actually toggle switches on the amp. As soon as you spray a switch, rock the bat of the toggle switch back and forth several times. This will "scrub" the inside of the switch and help the contacts to seat properly.

FILTER CAPS

More than likely, your amp will require new filter caps. Filter caps are supposed to be changed every six to ten years so it is unlikely that a dormant amp would have fresh filter caps. Basically, you want to make a list of every electrolytic cap in the amp and replace it with the same value. Pay attention to the voltage ratings on a cap. You must use at least the voltage rating that was used before or you may go higher. For example, there may have been a 16 uf at 450 volt capacitor and you could use a standard value 16 uf at 475 volt value as a replacement.

Sometimes the amp will use a multi-section filter cap that looks like a metal tube. In such a case, if you cannot find the correct replacement part, you can simply replace the multi-section can cap with individual capacitors. For example, you may have a 40 uf / 450 volt, 20 uf / 450 volt, 20 uf / 450 volt multi-section cap. In such a case, you could replace that one component with a 40 uf / 500 volt cap and two 20 uf / 500 volt individual caps. Before removing any of the wires from the can cap, make a drawing of how every thing was wired as stock. Perhaps you have two red wires going to the 40 uf terminal, perhaps a blue wire going to one of the 20 uf terminals and a green wire going to the other 20 uf terminal, and perhaps three black wires going to the grounded lead on the multi-section cap. You would make this drawing and then take the red wires and move them to the plus terminal of a 40 uf / 500 volt cap. Likewise, the blue wire would go to the plus terminal of a 20 uf / 500 volt cap and the green wire would go to the plus lead of the other 20 uf cap. Now all the minus terminals from the individual caps will connect together and connect to the three black wires. You will also want to ground the minus terminals to a chassis ground. Sometime there is an exception to this. On some Marshalls and certain other amps (Ampeg comes to mind), the multi-section can caps may be ganged together in series such that the minus of one can does not go to ground, but to the plus side of another cap. Just look at everything really good before you take anything apart and make sure you put it back like you found it, only with new filter caps. Actually, all electrolytic capacitors in the amp should be replaced. There are usually a few that are used as bypass capacitors on the preamp tubes and there may be one or two in the negative voltage bias supply (if the output stage is set up as fixed biased) or on the cathode of the output tubes (if the output stage is set up as self biased).

REPLACE THE TUBES

Tubes are like guitar strings: Different brands sound different, they sound best when they are new; the don't sound good when they are old, and if you don't change them, eventually all will break.

Replace the tubes with exact replacements. There are some very good tubes being sold today thanks to the good people at Magic Parts, New Sensor, Antique Electronic Supply and Groove Tubes. Before you remove the tubes from the amp, you want to make sure you know which tube goes where. Some amps have a tube configuration label. If yours doesn't have one, you could make one up on the computer and print it out for your own use. Or, another technique is to clean the chassis really clean near the sockets and use a Sharpie felt tip marker to write directly onto the chassis the type of tube that goes into each socket. If your amp is cathode biased (aka self biased), you will not need to adjust anything after you replace the tubes. You will know if it is cathode biased as there will be a big resistor going from the cathode of the output tubes to ground. The cathode is pin #8 on most guitar amp output tubes (6V6, 6L6, 5881, 6550, KT66, EL34, etc.)

On the other hand, if the cathode is grounded directly to the chassis without a cathode resistor, the amp will be fixed bias. If the amp is fixed bias, you will need to set the bias. There are many bias probe devices being sold on the market today. I use the transformer shunt method which can be done with a simple multimeter and the method is outlined in my first book, *A Desktop Reference of Hip Vintage Guitar Amps*. Just as one would have to set idle when changing a carburetor on their car's engine; on a fixed bias amp, you will also need to set the idle, errr.... bias of the output tubes.

Now for the fun part: Start the amp off by using a current limiter. The current limiter will prevent anything from blowing up should you have something hooked up wrong. Once you have ascertained everything has been done correctly, take it off of current limit, set the bias, and wail. Always remember: better musicianship begins with louder amps.

All About Vacuum Tube Guitar Amplifiers

CAP JOBS DONE RIGHT

Everyone wants their vintage tube guitar amplifier to sound its best. In the same spirit a Porsche 911 Turbo owner would want his mosheen "tuned and timed," a "cap job" could be the difference between an amp that performs with amazement and one that merely functions. Just as you are sure to eventually need a new set of tires or a new battery for your car; sooner or later, every vacuum tube guitar amp will eventually need a cap job to replace deteriorated filter capacitors. It doesn't matter how well you take care of your amp, filter capacitors will all eventually need replacement for your amp to sound its best.

WHAT DO FILTER CAPACITORS DO?

Every vacuum tube guitar amp requires high voltage D.C. power and yet our wall electricity is relatively lower voltage A.C. power. To fulfill this need, guitar amps have a power supply that converts low voltage A.C. (electrons flow both directions) to high voltage D.C. (electrons all flow in the same direction.)

There are only a few components needed to achieve this. First, we need a power transformer to step the wall voltage up higher. Second, we need a rectifier such as a rectifier tube or diode which acts as a check valve to allow the electrons to move in only one direction. And finally, we need the filter capacitors to smooth out the voltage. You see, when the electricity is initially converted from A.C. to D.C., it is first pulsating D.C. So it is all going the same direction, it is just that the electricity is pulsing. Filter caps are used because they will oppose any change in voltage. The filter capacitor stores energy and releases energy such that the voltage remains constant. When the D.C. begins to pulse an abundance of electrons, the filter cap stores the extra electrons, thus reducing the actual pulse and opposing any change in voltage. When the D.C. is in between pulses, the filter cap will discharge just enough to oppose any change in voltage. Think of the filter cap as a big battery that keeps the voltage constant. The rectifier circuit is simply keeping the battery, err, filter cap charged.

WHAT IS A TUBE AMP?

It could be argued that a vacuum tube guitar amplifier is nothing more than a modulated power supply. The filter capacitors are an integral part of the power supply. Their ability to store energy and oppose changes in supply voltage affects how the amp sounds. With almost all tube guitar amps, the 60 Hertz A.C. wall pulse becomes 120 Hertz pulsating D.C. after it is rectified to D.C. When the filter caps are not performing properly, this 120 Hertz pulse is not filtered out of the supply voltage which feeds the preamp and output tubes in the amp. The result is an ugly, somewhat out-of-tune sub tone as the 120 Hertz power supply ripple current modulates the notes you are playing. Think ring modulator! The resultant sound is both annoying and non-musical. You can hear this unnatural sound even on brand new amplifiers, if the manufacturer used Taiwanese filter capacitors. The Taiwanese filter capacitors will test fine on a capacitance meter, but remember the capacitance meters typically test at 9 volts or less. In an actual tube guitar amplifier, the main power supply voltage, that is connected to the filter caps, could be somewhere between 400 and 500 volts D.C. These foreign-made filter capacitors will not filter out the ripple current and your amp tone will be severely affected. It will sound like it needs a cap job!

THE FIRST DILEMMA—THE RIGHT PARTS ARE NOT AVAILABLE

Six hundred volt electrolytic capacitors have fallen by the path of the 8-track cartridge and are no longer available. Worse yet, many technicians are replacing six hundred volt electrolytic capacitors with five hundred volt capacitors. In some instances, the five hundred volt replacements will work fine, but in other sections of the circuit, particularly the main filter section, and a modification must be done if five hundred volt capacitors are to be used.

To compound matters, certain Chinese and Taiwanese electrolytic capacitors have become very popular at amp repair shops because of their low prices. These capacitors are made with the internal conductors "etched" to create more surface area in the plates. The extra surface area of etching allows the capacitor to have a larger microfarad value, but in a smaller physical size. You can easily spot these because the physical size of these foreign capacitors is much smaller than a good quality capacitor of the same microfarad value. With capacitors, bigger really is better. These smaller type electrolytic capacitors will test fine on a capacitor tester because a capacitor tester is using a 9 volt test circuit. Even though they would probably work fine in a low voltage circuit, under the high voltages in a vacuum tube guitar amp, they will not do their job. Part of their job is to filter out "ripple current"—a byproduct that occurs when the wall A.C. voltage is rectified to D.C. voltage. And of course, D.C. voltage is necessary for a tube to amplify. If the cap fails to filter out the ripple current, then the ripple current modulates the notes you are trying to amplify and the result is an "out of tune" sub-tone that is not a harmonic of the note being played. The extra note moves around and never in a harmonic way. We hear this as an ugly "ghost note," competing with the note you are playing. Another job of the filter cap is to store electric energy. When the tubes need this energy, the cap supplies it. If the cap is not storing the energy properly, then the amp will lack dynamics and punch. The attack will be mushy and in some cases the front of the note will be compressed off.

VINTAGE AMPLIFIERS NEED MODIFICATION

The amps of yesteryear sometimes used parts that are no longer available. To get performance from an amp, a simple modification is necessary. For example, let's take a look at a typical Fender tweed Bassman. It uses five filter capacitors. There are four 20 uf 600 volt filters caps in the cap pan and one 10 uf 450 volt cap on the board. The two filter caps that connect directly to the rectifier tube (via the standby switch) are called the main filter caps. The problem on the Bassman and many other tube amplifiers is that they used 600 volt caps which are not available in a size that will fit in the cap pan. Some amp technicians replace all the caps with the original microfarad value, but they use 500 volt caps. It would seem that the 500 volt cap would work because the plate voltage, at idle, is much less than 500 volts. Looks can be deceiving. The 500 volt capacitors simply will not work for the long term.

Because of the lack of voltage drop across the internal resistance of the power transformer B+ winding, during moments when the power tubes are not drawing current, the voltage could jump up as much as 100 volts. You can see exactly how much plate voltage your particular amp will jump by doing an experiment. Take a voltmeter and with the amp on and in the play mode, measure the plate voltage on the plate of any output tube (pin #3 for a 6V6, 6L6, EL34, 5881). Now remove the output tubes from the amp (to simulate cutoff conditions) and with the amp on and in the play mode, take a plate voltage reading on the same pin of the socket. Notice how the voltage goes up considerably when the output tubes are not drawing current.

So if the instantaneous plate voltage goes up over 500 volts when you simulate cutoff, then you need to use a filter cap circuit with more than enough voltage rating to cover it. If you use 500 volt caps in an amp whose cutoff condition plate voltage goes to 520 volts, then those caps will become internally ruptured by the excess voltage and they will quit working after a few hours of use.

WHAT DID LEO DO?

Actually, Leo Fender later modified the power supply in his blackface amps to improve the main filter's voltage rating up to 700 volts. He did this by using two 350 volt caps in series. If you look at a blackface amplifier, such as a Super Reverb or a Twin Reverb, you will notice that the main filter caps (that's the ones that connect to the rectifier tube) are wired differently from the tweed or Brown Amps. Look at the 5F6A Bassman schematic and notice that the two main filters are in parallel with each other and both connect to pin #8 of the rectifier tube (via the standby switch). Two 20 uf 600 volt capacitors were used for the main filters. When two capacitors are placed in parallel, the capacitance is added to compute how much capacitance the circuit actually "sees." So the circuit is seeing 40 uf of capacitance at a 600 volt rating. However, when you look at the AB763 Super Reverb schematic, you see that two 70 uf capacitors at 350 volts are used. This is called a "totem pole stack." It looks like a "totem pole" when you see it on the schematic. When two capacitors of the same microfarad value are placed in series, as is the case with the "totem pole stack," the capacitance that the circuit "sees" will be half of the microfarad value of one capacitor; but the voltages are added together to compute the voltage rating.

Also, notice that the AB763's cap configuration shows a 220K resistor across each of the 70 uf filter caps. This is important as these two resistors assure a full 700 volt rating because they will force the voltage to divide evenly across the caps. Since each cap is 350 volts, and since the voltage is dividing evenly, the main filters look like a 700 volt cap at 35 microfarads to the rest of the circuit. If the resistors are left off, due to inherent differences in caps, the voltage will not drop evenly across the caps and the voltage rating will not be the full 700 volts. That is why those two resistors are important to the circuit.

MAKING THE AMP PLAY IN TUNE

Actually, when vintage amps were originally designed, no one ever thought the amps would be played very loudly. People just didn't crank the amps back then. PA systems were not like the ones of today and many bands used high impedance microphones and plugged them into the normal channel of their amps. In fact, some bands only used one amplifier and two or more players would plug into the same amp!

When amps were tested, they never really turned them up much. Everyone was going for clean. So the stock microfarad value of filter caps in vintage amplifiers are almost always too small to do their job when the amp is cranked.

The more current an amp is drawing, the harder it is for the filter caps to filter out the 120 Hertz ripple current. As the microfarad value is increased, it becomes easier for the filter capacitor to filter out that ripple current. If the volume control is about a third of the way up, the amp isn't drawing very much current and the stock filters will be adequate. If, on the other hand, the amp is dimed, then the stock value filter caps will not be enough microfarad value to filter out the ripple.

When an amp is overdriven it uses more current. The microfarad value needed to filter ripple current is directly proportional to the amount of current the circuit is using. That is to say that 40 uf of capacitance works fine in a Bassman whose volume is set to three. 35 uf works fine in a Super Reverb turned ⅓ of the way up. But in the case of an amp turned up to overdrive, more capacitance is needed to tame down that ugly ripple current and hold everything together and "in tune."

I did a sound experiment and kept increasing the value of the main filter until I could dime a two 6L6-style amp and not hear any "ghost notes." Through trial and error, I found that two 220 uf capacitors, placed in series, would filter out the ripple—no matter how loudly I played the amp. And by using two 350 volt 220 uf caps with a 220K ohm resistor across each one, my circuit would actually "see" 110 uf at a 700 volt rating. This is about triple the stock microfarad value. I might also add that the two 220 uf capacitors had to be quality capacitors (not those little foreign caps) or the ripple current would remain. The extra filtering also helps the bottom-end of the amp. It takes more energy to produce a low note than a high note. The higher microfarad value capacitors store more energy, so when the tubes demand a lot of power to reproduce a low note, the higher rated filter will be able to provide it.

On a single-ended amp such as a Champ, or a Gibson GA-5, I still like to use 100 uf of capacitance for the main filter, but usually the supply voltage is such that a single 100 uf 500 volt capacitor will work fine. This is really even more important to increase the main filtering in a single-ended amp, because there is not hum cancellation in the output transformer as is the case in a push-pull amp such as a Twin Reverb.

Vintage Marshall amps were designed using multi-section can caps. My advice here is to avoid LCR brand British caps. These sound like they were made in Taiwan. I would recommend F&T brand, German caps for replacement in a Marshall.

THE TUBE MANUALS SAY, "DON'T DO IT."

If you read a vacuum tube manual, it will tell you that 110 uf is too much capacitance to use with a standard rectifier tube such as a 5AR4 or a 5U4. Tube manuals are written by companies that sell tubes and not by experts in tube guitar tone. The tube manual will also tell you the design maximum of a 6V6 is 350 volts, and it should never be run higher than that, yet almost all 6V6 guitar amps exceed that voltage. So we can't always go with the manual, sometimes we must go with our ears and what works.

Actually there is a trick to using that much capacitance on a rectifier tube. The trick is to move the main filter (the totem pole stack) from the cold side of the standby switch to the hot side. The reason is simple. When a filter is not charged and a large voltage is placed on it, as is the case when the main filter is connected to the cold side of the standby switch, the cap will want to absorb that charge. For a fraction of a second the cap will look like a short circuit to the rectifier tube, while the cap is charging. If you move the main filter "totem pole stack" to pin #8 of the rectifier tube, then as the tube begins to warm up, and begins to deliver some voltage, the main filter caps will trickle charge. By the time the amp is warmed up, the main filters will have already become fully charged. When the standby switch is flipped, the oversized main filters actually "help" the rectifier to charge the other caps in the circuit. Actually, if you look at the Super Reverb schematic, you will see that Leo moved the main filter "totem pole stack" to pin #8 of the rectifier tube just as we are suggesting here. When you move the main filter "totem pole stack" to pin #8 you will notice a loud 'pop,' while flipping the standby switch to play. That is the sound of the main filter cap "helping" the rectifier to charge all the other caps in the amp. The pop is normal and not to be concerned with. It is much better to have the main filter helping the rectifier than to have the main filters on the cold side of the standby switch and look like a short circuit to the rectifier. In fact, you will blow a lot of rectifiers if you use higher valued caps and have the caps on the cold side of the standby switch. So if you recap your Bassman or any other amp whose main filters are connected to the cold side of the standby switch, remember to move the main filters to the rectifier tube so they will be fully charged by the time the amp warms up.

BIAS-SUPPLY FILTER CAPS

With regards to the type of output stage in an amp, there are basically two possible types of tube amp designs; namely: cathode-bias output stage or fixed-bias output stage. Either type uses one or more electrolytic capacitors that must be replaced when performing a cap job. Let's start with the fixed-bias type of amp first. In this design, output stage bias is achieved by injecting a fixed negative voltage to the grids of the output tubes. So an amp of this design must have a negative voltage power supply which will use one or more electrolytic filter capacitors.

On a fixed bias blackface Fender amp such as a Twin, Bandmaster, Showman, Pro or a Super Reverb; the negative voltage supply is on a small, separate circuit board near the pilot lamp. This type of circuit always uses a small diode and one or more electrolytic capacitors. The common value for this capacitor is somewhere between 10 and 100 uf at anywhere from 50 to 160 volts. In a tweed Bassman, for example, two 10 uf 160 volt capacitors are used. In a blackface Super Reverb, a single 50 uf at 50 volt is used.

The bias filter capacitor or capacitors should be replaced with at least the stock value or better. If you use a higher microfarad value or voltage value, it will work just fine. Increasing the microfarad value can only make the amp hum less and increasing the voltage value will only make the amp more dependable. There are some amps that need the voltage value of the bias filter caps increased because the original design left little if any safety margin. For example, there are certain blackface Fender amps that use a 50 volt cap and if you check the actual voltage in the circuit, it runs at about 55 volts. Check the negative D.C. voltage coming off the diode. The voltage rating of the cap should be high enough such that there is at least a 20% safety margin. I know there are amps that have 55 volts coming off the bias supply diode and the design calls for a 50 volt cap. On such an amp, I would recommend changing to a higher voltage capacitor. Perhaps a 75 or 100 volt rating would give you better reliability.

Since the negative voltage bias supply is negative, the filter capacitors associated with it are configured with opposite polarity to the main power supply filter capacitors. That is to say that the "plus" end of the bias filters will go to ground and the "minus" end will go to the circuit. Please note that this is exactly the opposite of the orientation of all the other filter capacitors in the power supply whose orientation is such that the "minus" end connects to ground and the "plus" end to the high voltage circuit. If any of the electrolytic capacitors in the amp are installed with incorrect polarity, they will let you know about it when you power the amp up. The odor is horrific when they blow up like a firecracker. Plus the mini-mushroom cloud could be photographed for future amusement. Get them installed in the correct polarity and you will never know that smell.

CATHODE BYPASS CAPACITOR

Not all amplifiers are configured with a fixed-bias output stage. Some amplifiers, such as a Vox AC30 or a Fender Champ, are config-

All About Vacuum Tube Guitar Amplifiers

ured as a cathode-bias output stage. In a cathode-bias design, one or more resistors are connected between the cathode of the output tubes and ground. When current passes through the resistor, a bias voltage is developed. The cathode resistor will also have an electrolytic capacitor across it. This will bypass A.C. voltage around the resistor. So what you end up with is A.C. signal voltage going through the capacitor and D.C. idle current going through the resistor. The bypass capacitor will always be an electrolytic capacitor and like other electrolytics will need periodic replacement. The microfarad value and voltage value of this capacitor will affect the tone of the amp. Generally, a larger physical size cap will sound better because this particular cap will need to pass quite a bit of A.C. current. That is why a much higher voltage rating cap is generally used. For example, there may be only 40 volts on the cap, but a 100 volt cap or higher will be used. The bigger physical size of the higher voltage cap will accommodate the high A.C. current much better. The microfarad value will affect the bottom-end response of the amp. A larger than stock microfarad value will improve the bottom-end as compared to the stock value.

When changing the bypass capacitor in a cathode-biased amp, one must consider the best spot to mount the capacitor. Electrolytic capacitors are heat sensitive and output tubes are generally very hot. Remember too, that hot air rises. So you would want to mount the bypass cap or caps away from the output tubes themselves. The bypass cap will be across the cathode resistor electrically, but it doesn't have to be there physically. That cathode resistor will get very hot and if you mount them right next to each other without cooling space in between, you are likely to experience that horrific odor we spoke of earlier. I like to move the bypass cap away from the tube socket and away from the cathode resistor for coolest operation. The bypass capacitor will mount with the same polarity as the filter capacitors and that is: "minus" to ground and "plus" to the circuit.

OTHER BYPASS CAPACITORS

Regardless of whether your particular amp has a cathode-bias or fixed-bias output stage, all amps will have a preamp section. All the tubes in the preamp section are cathode- biased and will need to have their bypass capacitors changed. If the amp uses dual triode 12AX7,

12AU7, 12AT7, or 12AY7 style tubes, the cathodes will be pin #3 and pin #8. These will go to a cathode resistor and this resistor will almost always have an electrolytic bypass capacitor across it. These capacitors will be different from the bypass capacitors in the output stage because the preamp tubes do not get very hot and do not draw much current or much voltage. A typical preamp tube will develop only about 1 or 2 volts on the cathode and 1 or 2 milliamps of current. That being the case, the bypass caps in the preamp do not get the punishment or the heat associated with output stage bypass caps. In other words, almost any voltage rating will do. Anywhere from 6 to 25 volt ratings are typical. The value of this capacitor will affect the bottom-end response of the amp, so unless you are deliberately voicing the amp to be different from stock, I recommend staying with the stock value.

As is the case with other electrolytic capacitors, you would want to avoid foreign-made components. I like to use American-made Sprague capacitors for the bypass capacitors in the preamp. Also, there are certain vintage amps that use a dual capacitor to bypass two resistors. You are likely to encounter these types of capacitors in blackface fender amps.

These type capacitors have three leads: namely one "minus" lead and two "plus" leads. Although these dual capacitors are no longer available, you can make one easily by taking two capacitors of the correct value and twisting the "minus" leads together, then treating it as a single lead. As with all bypass capacitors, the "minus" goes to ground and the "plus" goes towards the tube.

OTHER CONSIDERATIONS WITH COUPLING CAPS

In most cases, changing all the electrolytic capacitors in an amp will complete the cap job, however in some cases; one or more of the non-electrolytic coupling caps in the amp will need changing. These types of capacitors are usually made from tubular foil and rarely if ever malfunction. These coupling caps are used to block D.C. plate voltage while still passing A.C. signal voltage. In rare cases, they will leak a small amount of D.C. voltage which will adversely affect the tone of the amp. Let's say a coupling cap is leaking even a half a volt. Since the preamp tubes bias in at 1 or 1.5 volts, if a coupling cap is leaking even a half a volt onto the next preamp stage, the biasing of the next stage will be WAY off and the amp will lack power, dynamics and in general will sound like garbage.

These coupling caps always attach to the plate circuit of a preamp tube (pin #1 and pin #6 on a dual triode such as a 12AX7, 12AU7, 12AY7, 12AT7, etc.). In order to check these capacitors, one must leave the plate side connected to the circuit, but unsolder the non-plate side. With the capacitor unsoldered on the one side, you would power up the amp and connect a good quality digital voltmeter between the unattached side of the capacitor and ground. If there is a reading of 2.5 volts D.C. or more, the coupling cap is leaking and should be replaced. The brand of capacitor will make a huge difference. As with other electronic capacitors, never use cheap Radio Shack components or anything made in Taiwan in a guitar amp. The coupling cap should be either a Mallory 150 series or a Sprague Orange drop capacitor. There are other good ones to use, but stay away from any high-end audio type capacitors. High-end audio circuits are going for transparent sound whereas a guitar amp uses components to enrich and fatten the tone. If you ever plugged a guitar into a home stereo, you already know how cold the high-end audio designs sound for electric guitar.

EPILOG

So to sum up, let us say that to perform a cap job on an amp, all electrolytic capacitors should be changed with the same value caps as the original except for the main filters. Correct polarity should be observed. The minus end of the cap will always go to ground except when used in a negative voltage bias supply, in which case the plus end goes to ground. Care should be taken when replacing 600 volt caps with 500 volt replacements. If the applied voltage (the voltage measured from your rectifier cathode with the other tubes removed) is over 500 volts, parallel-configured 600 volt caps should be replaced with a series-configured pair of 350 volt caps. To accommodate louder playing styles, the main filter caps should have their value increased. I use a single 100 uf at 500 volt for the main filter in a single-ended (Champ style) amp, but with other amps that require higher voltage ratings, I recommend using a "totem pole stack" made with two 220uf at 350 volt capacitors and two 220K ohm resistors. I do not recommend increasing the microfarad value of any of the filter caps that supply voltage to the preamp tubes. To do so will cause a loss of bottom end. It is important to also check the voltage across the negative bias voltage filter cap and the output tube bypass cap to make sure the voltage rating is higher than the actual voltage used in the circuit. With capacitors, you want to allow a margin of safety and make sure the cap is rated for more voltage than it will ever "see" in the circuit.

With a little TLC, a vintage amp can be made to perform as good as and in some cases even better than new. And always remember: a great sounding amp will inspire the player to perform on a level not available when playing a bad sounding amp rig. Inspiration occurs when the amp is performing on the same level as the player!

All About Vacuum Tube Guitar Amplifiers

SQUEEZING AMP PERFORMANCE FROM LAME TO INSANE

Besides the highly collectible early Marshall, early Fender and early Vox amps, there are many cool vintage tube amps that were never quite as popular as the early Fenders or Marshalls and as a result, could be bought for a few hundred dollars. I am talking specifically about the less expensive vintage tube amps such as the Magnatone, Airline, Silvertone, Dan Electro, Ampeg, National, Supro, Gretsch, Traynor and Harmony. Most of these amps were not very aggressive, but could be made to sound outstanding with a little tweaking.

Because early amp manufacturers thought they were going for clean, undistorted sound; many of the design decisions were made in favor of low gain, low distortion. Not only that, but since players didn't play as loudly in the 50s and 60s, softer power supply designs were used and sometimes cheaper speakers. Since every component in a tube amp makes a difference in tone, one could see that making small improvements in different areas of an amp's design could change it from an endothermic device into a firebreather. By beefing up the power supply, perhaps getting a little more gain from the preamp, using a more efficient speaker, little bit better tubes, etc. one could take a lame sounding amp that was not suitable for gigging and change it into a tone monster that could be played with a pounding drummer.

FIRST THINGS FIRST

Before any circuit modifications can be considered, the amp must first be in top condition electronically. By this, I mean you need to have good tubes, good filter caps and everything in good working condition. In most cases, you will need to do a cap job and overhaul at the very least. The cap job is part of the power supply. Perhaps when you are doing the overhaul it would be advised to increase the capacitor value of the main filter cap.

POWER SUPPLY

Let's look at the power supply. Is it configured to give the output stage the most it can offer? Sometimes, power supplies are underfiltered, sometimes they are hooked up to have less power than is available and sometimes the power resistors are too large and drop the preamp voltages too much. So let's look at it. If the rectifier uses a pi filter before the center-tap of the output transformer, you need to move the center-tap of the output transformer to the cathode of the rectifier tube. You have seen this circuit on those "E" series Fenders. Moving the output transformer center-tap to the rectifier cathode will increase the B+ voltage going to the output tubes and therefore improve wattage and headroom.

Look at the filter capacitor connected to the cathode of the rectifier tube. If it's value is less than 100 uf, I would recommend bringing it up to at least 100 uf. If your amp's B+ voltage is higher than 400 volts, you may need to use two 220 uf caps at 350 volts and arrange them in series so that the circuit will see 110 uf at 700 volts. If the B+ voltage is 400 volts or less on the rectifier output; I recommend replacing the filter cap with a 100 uf at 500 volt value. You need to have 100 uf on this part of the circuit to get the voltages up and supply the output tubes with the power they need.

SPEAKER

Of course, a critical component in any amplifier is the speaker. After all, the speaker is what is vibrating the air to make sound. You can test the speaker for cone excursion. Disconnect the speaker from the amp and touch the terminals from a fresh 9-volt battery to the speaker. The cone should move significantly and you should be able to easily "see" it move. You should also hear a loud, "thump." If the speaker is barely moving or if the "thump" is barely there, it may be time to replace the stock speaker with a good sounding and efficient speaker.

I remember taking an Ampeg amp and changing the speaker to an EVM 15" speaker. It sounded like the amp's volume had doubled—just by using the better speaker. I could have gigged with that amp. Of course, there were a few other things that were modified—besides the speaker.

OUTPUT STAGE

An output stage cannot sound its best without good sounding output tubes that are biased for best tone. Tubes wear out a little from day to day and you can have a set of old tubes that produce sound, but just don't perform like a new set. When you replace the tubes with good, fresh ones; you are bringing out the best performance of which the amp is capable. And you don't need NOS tubes to get great tone. There are some very good sounding tubes being made today and we are blessed with many quality choices.

PREAMP

Power tubes need gain in order to be driven and preamp tubes take the small signal from your guitar pickup and amplify it strong enough to drive the power tubes. There are many factors that control the gain of a preamp tube. Most preamps are configured with multiple stages, so if you can squeeze a little more gain from each stage, the result is much more gain. When you multiply 100 times 100 times 100, you get a much bigger number than 80 times 80 times 80. The gain improvement works this way when you increase the gain of multiple stages.

Some gain killers that are common in these amps in question could include low plate voltages, too high cathode resistor values, unbypassed cathode resistors, and voltage divider coupling of stages. Let's take these one at a time.

PLATE VOLTAGES

Most preamps perform best at around 200 volts. The supply voltage comes from a dropping resistor that is connected to the power supply. I am not talking about the plate resistor which is usually a 100K resistor connected to the plate of a preamp tube. I am talking about the resistor that feeds the plate resistor. It will be connected to both the plate resistor and a filter cap. If you make this resistor smaller, you will increase the amount of voltage feeding the tube. Let's say you check the plate voltage and see 140 volts. You check the other preamp tubes plates and see 150 volts or so. You want to improve every stage's voltage supply and there is almost always a resistor that will increase all the preamp tube plate voltages simultaneously.

In almost every case, the supply voltage for all the preamps come from a dropping resistor that is connected to the phase inverter supply. If you can find that resistor, reducing its value will raise all the preamp tube's supply voltages. The resistor will connect to both the plate load resistor of the phase inverter at one end and a plate load resistor from one or more preamp plate load resistors at the other end. Each end will have a filter cap going to it. To reduce the value of this resistor, you may simply add another resistor in parallel to it. That will decrease the amount of voltage the resistor drops. You could place a decade box over the stock resistor and keep experimenting with different resistance values until you get the plate voltage of the preamp stage up to around 200 volts or so. You will probably only need a ½ watt resistor, but to figure the wattage rating needed for any resistor, you can square the voltage across it and divide by the resistance. This will give you the wattage in the circuit, but you will double it for design safety.

All preamp tubes will have more gain if they get more B+ voltage. For pentode preamps such as the 6267, 6SC7 or the EF86, the gain difference by increasing the supply voltage is truly dramatic. A small increase in voltage can mean a large increase in overall power and touch sensitivity.

CATHODE RESISTOR VALUE

The cathode resistor is what is biasing the preamp tube. It is common for the preamp tube to be biased on the cold side as most of these designs were going for clean. Let's say you look at the cathode resistor on a 12AX7 style tube. If it is feeding only one cathode, the correct optimum value will be around 1.5K. With a plate voltage of 200 volts, the 1.5K will bias the tube for optimum gain and efficiency. Often you will see a cathode resistor on a 12AX7 style tube whose value is 3K or 4.7K. This will work, but it won't have the gain of a 1.5K cathode resistor. Check the cathode resistor value for each preamp stage and change it to a 1.5K if needed.

Sometimes two cathodes will be tied together and only one cathode resistor is used. In this case, the optimum cathode resistor value will be an 820 ohm.

UNBYPASSED CATHODE RESISTORS (IN THE PREAMP)

Check every preamp tube cathode resistor and see if it is bypassed by a capacitor. If the cathode resistor is missing a cathode bypass capacitor, you are not getting full gain from the stage. You can add a bypass capacitor and its value will affect tone. A typical value is 25 uf at 25 volt. You can use a larger one, say a 50 uf, 100 uf, or 200 uf, if you need more bottom-end or use a smaller one if you need less bottom and more bite. A 10 uf, 5 uf, 1 uf or .68 uf are all typical values depending on what you want it to sound like. This is where listening to the amp will usually let you know what it needs. The smaller value caps accentuate the highs more and the larger values accentuate the lows. Don't be afraid to experiment with several different values to see what sounds best for your particular amp.

VOLTAGE DIVIDER COUPLING OF STAGES

Another common gain killer design was to couple one stage to the next with a voltage divider. Perhaps the signal fed a pair of 470K ohm resistors in series to ground. The junction of the two resistors fed the grid of the next stage. This circuit would drop the gain by 3 dB because it cuts the signal in half. If you find a voltage divider coupling circuit, you can usually change it by shorting the first of the two series resistors with a straight wire. This will insure full signal transfer.

EPILOG

Remember that successful amp voicing is measured by the inch. If you get a little more power from new output tubes and that new speaker has 3 dB more efficiency and you get rid of a voltage divider interstage coupling for another 3 dB increase and improve preamp voltage and thus get another 3 dB gain and then bypass an unbypassed cathode resistor for another 5 dB gain, it all starts to add up. With a few dollars worth of parts, a little know how, some listening tests and a lot of love, you can squeeze an amp's performance from lame to insane.

All About Vacuum Tube Guitar Amplifiers

FINE TUNING REVERB AND VIBRATO/TREMOLO

Perhaps the most popular effects ever put on guitar amplifiers are reverb and vibrato/tremolo. But how many really sound great? Most vibrato circuits lack intensity or oscillate too fast and reverb circuits can be weak or boingey. Here are some tips for fine tuning your reverb and vibrato/tremolo.

FINE TUNING YOUR REVERB

There are basically two types of reverbs used on guitar amps: transformer coupled and capacitance coupled. Each type uses its own type of reverb pan which are not interchangeable. For example, Ampegs and most Gibson amplifiers used a capacitance coupled high impedance reverb pan. In such a design, the transducer is coupled to the reverb driver tube with a big capacitor. This design is different from the Fender-style reverb which used an output transformer to couple the reverb driver tube to a low impedance reverb pan.

TROUBLESHOOTING A REVERB

If your reverb doesn't work at all, it is almost always a problem with either the cables that connect the reverb pan to the amp or the reverb pan. Check the RCA connectors and make sure the reverb cables make good contact. Sometimes the RCA connector needs to be re-tensioned slightly just to make sure the connection is a solid, metal-on-metal connection. Inspect the springs in the bottom of the pan. There are two-spring pans (like Fender) and three-spring pans (like Boogie). Are the springs connected at each end to a transducer? If the pan is in a reverb bag, check to make sure there is a bottom on the pan. Usually bottoms are made with cardboard and wooden reinforcements. This is to keep the reverb bag from bunching up under the pan and deadening the springs. If the springs are deadened, the reverb will not function.

IMPROVING THE SOUND OF THE REVERB PAN

Each reverb pan has its own sound and some sound better than others. If you look at the bottom of the reverb pan, you will notice the springs connect on each end to a transducer. There are two transducers that are different. One transducer acts like a speaker on one end and the other transducer acts as a microphone on the other end. The side that functions like a speaker vibrates the springs while the transducer that functions as a microphone picks up the springs' reverberated sound which is amplified to make reverb. Each transducer uses a coil of wire wrapped around a small plastic bobbin that is situated on a laminated iron core. Most of the time, the coil on each transducer is loose. A loose coil robs the reverb of efficiency.

Here is a trick I learned about how to improve the function of the transducers. You will need a bamboo chopstick and a razor utility knife. Using the utility knife, slice a very thin shim of bamboo from the chopstick. Look at one of the transducers and notice that there is a coil wrapped around a bobbin and the bobbin is situated on an iron laminated core. You want to insert the shim between the laminated iron core and the bobbin. This will tighten up the coil and prevent it from moving on the core. Do this to the coils on both transducers. This will improve efficiency because when the coil is tight against the core, more energy goes into the spring instead of going into moving the coil. And on the microphone side, more energy is induced into the coil rather than moving the coil. The shims used to tighten the coils just make everything work better.

If you modify your pan and it still sounds bad, you may need to get a new pan. Sometimes a pan will sound boingey no matter what you do. A new pan is the solution.

If your reverb footswitch hums, it is likely a shielding or grounding problem with the footswitch and/or the footswitch cable. The footswitch cable should be shielded and the footswitch housing should be connected to this shielding.

FINE TUNING YOUR VIBRATO/TREMOLO

There is a difference between vibrato and tremolo. The tremolo effect is actually a pulsating volume such as what you would find on a blackface Super Reverb, although it is inaccurately labeled as vibrato on that particular amp. True vibrato involves a slightly change in pitch rather than a change in volume.

Whether an amp has vibrato, tremolo or both (as with a Vox AC30), there are always two parts to these effects. All have a low frequency oscillator and all have some type of modulator circuit.

Let's look at the oscillator first. In almost every case, the oscillator is a phase shift type oscillator made from a preamp tube that has three small capacitors going from the plate of the tube back to the input. The capacitors each shift the phase by approximately 60 degrees so that the output is in phase with the input and that makes the tube oscillate. There are resistors going to ground at the junction of each of the feedback capacitors. One of these resistors is actually a variable resistance potentiometer and changing its resistance value will make the oscillator go faster or slower.

If one wants to slow down the overall speed of the oscillator, there are a couple of easy ways to do this. Increasing the value of one of the feedback capacitors by a factor of two or increasing the value of one of the resistors going to ground by a factor of two will cause the overall speed to slow down. You may use the footswitch to turn the oscillator off and on and check to make sure it starts up properly. You don't want to set the speed so slow that the oscillator doesn't want to start oscillating.

There are different types of modulators. The most common type is the optocoupler, as found on most Fender blackface amps. The optocoupler is actually a neon bulb that lights in sync with the oscillator. Part of the optocoupler is a light-sensitive resistor that grounds out the guitar signal when the neon bulb lights up.

Another type of modulator circuit connects the oscillator output to the grid resistors of the output tubes. This fluctuating voltage modulates the bias on the tubes such that the amplification of the output tubes is more or less as the oscillator oscillates. This is my personal favorite type of tremolo, but when the tube is putting out the most, it can sometimes be a little distorted. An example of this type of modulation would be a Princeton Reverb amp. The effect is greater if the idle bias current is set lower. This makes the tremolo more intense.

Another type of modulator circuit as found on the Vibro Champ and certain Alamo amps, is the preamp tube bias modulator. The output of the oscillator is sent to the cathode resistor of a preamp tube. This circuit is just another way of changing the bias of the preamp tube such that there is more or less gain in sync with the oscillator.

Another type of modulator is found on Magnatone amps. These amps use a varistor to make the phase shift and in doing so affect the pitch. This is true vibrato.

The Vox amplifiers used phase shifting to get true vibrato. The circuitry was very complicated.

The brown Fender Concert, Pro, Super and Bandmaster amps ran an oscillator through a phase inverter which altered the grid bias of two preamp tube sections. A crossover separated signals into highs, mids and lows. The highs and lows were modulated out of phase while the mids were grounded out. This gave a swirling pitch shift effect. These were very complicated circuits.

TROUBLESHOOTING A VIBRATO/TREMOLO

When troubleshooting a vibrato or tremolo, the first question should be, "Is the oscillator oscillating?" In almost every case, when a vibrato/tremolo is malfunctioning, the problem is with the oscillator. Some amps require the footswitch to be in the "on" position in order for the oscillation to start. Other amps do not require the footswitch. An analog A.C. voltmeter connected from the plate of the oscillator tube to ground will show if the tube is oscillating.

If the oscillator isn't oscillating, it is because either one of the three feedback capacitors is leaking a small amount of D.C., or the bypass capacitor connecting to the cathode of the oscillator tube is weak, or the tube itself is weak. Rather than spend time trying to figure it out, it is much easier to use the Texas shotgun approach and just change all three feedback capacitors and the cathode bypass capacitor. At this point, the oscillator should be working—assuming you have a good strong tube.

IMPROVING THE SOUND OF THE VIBRATO/TREMOLO

Sometimes the vibrato or tremolo effect lacks intensity. In an amp whose oscillator modulates the output tubes, setting the output tube bias such that there is very little idle current will make the effect seem more intense.

With amplifiers that use optocouplers, light leakage into the opto-coupler can ruin intensity. Sometimes the black shrink tubing covering the light-sensitive resistor can crack, thus letting stray light into the device. In this case, wrapping the optocoupler in black electrical tape will block light from entering the device and make the effect more intense.

Sometimes the optocoupler itself is malfunctioning. The light-dependent resistor has leads that are crimped and not soldered. These leads can become loose, compromising the connection. In such a case, a new optocoupler is the only way to go. Don't try soldering the light dependent resistor as the heat will destroy the device.

In certain blackface amps, the oscillator can cause annoying ticking in sync with the oscillator. Locate the 10 megohm resistor (color code = brown, black, blue) that connects to the vibrato circuit optocoupler. Solder a .022 uf 400 volt capacitor across this resistor. In other words, the capacitor will be connected in parallel with the 10 megohm resistor. It doesn't really matter what kind of capacitor you use as this component is not in the signal path. Any .022 uf at 400 or higher volts will work. This modification will quieten the ticking without altering anything else about the amp.

All About Vacuum Tube Guitar Amplifiers

TUBE GUITAR AMP IDIOSYNCRASIES

Study the various different models of vintage tube guitar amps made over the last several decades, and you are certain to begin noticing peculiarities with certain models that are unique to that particular model. When a tube guitar amp is designed, there are certain to be either imperfection in the layout, unintentional circuit interactivity, mistakes in the grounding, parasitic oscillations, circuit instabilities or other idiosyncrasies of that design. Sometimes the designs are modified before the amp goes to market, but other times these idiosyncrasies don't show up until years later. If you are building a replica amp you may not want to copy everything. Or, if you own one of these amps, you may choose to modify the amp design so the idiosyncrasy disappears.

THE TWEED TWIN 5D8—Like most other amps, the number 2 inputs on both channels of this amp have less gain than the number 1 inputs. But they do it by putting the input section of each channel out of phase! There are no other amps that use this circuitry to my knowledge. It has a much different sound than merely padding the signal.

Some of the signal is tapped off the wiper of the volume control. This signal is injected into the triode section that is normally used for the number 1 input, and since it is going through another triode, the signal is 180 degrees out of phase. This is layered on top of the first signal so there are two signals out of phase with each other. This results in phase canceling, which gives you less overall signal. This was done intentionally and is a cool idea. It is not like other amps where the number 2 input is lower in volume, but in phase. This one is lower in volume because it is out of phase. You don't have to use this feature. If you don't like it, simply plug the guitar into the number 1 input.

THE TWEED TWIN 5F8A—When Leo Fender began offering the 5F8 Twin Amp, this was the first amp ever made with four 6L6 style tubes. Because four output tubes draw twice as much current of two output tubes, the B+ high voltage winding on the power transformer should have been wound with larger diameter wire. To use thicker wire would have increased the current handling capabilities.

I noticed this when I reverse engineered an original 5F8 Twin power transformer and made an exact copy of the original. In my tests, I set up an adjustable fixed bias (the original bias circuit is not easily adjustable), and noticed that the output tubes could not be biased very hot or the plate voltage would drop dramatically. The wire is so small that if any significant current goes through it, the smaller diameter wire "bottlenecks" the current and the plate voltage drops quite a bit.

Think of it this way. In the B+ winding of a transformer, there is some resistance in the actual copper wire itself. You could use an ohmmeter and measure exactly how many ohms of resistance. Everything else being equal, the smaller diameter wire will have more resistance than a thicker wire. Smaller wire resists additional current flow. With four output tubes and double the normal current flow, the smaller resistance makes a larger voltage drop. This can be proven by Ohm's law, where voltage drop equals current times resistance. Yes, the higher current will cause a larger voltage drop across the internal winding of the transformer and the plate voltage will go down considerably as the transformer overheats.

Consequently, the high-powered tweed Twins are all biased cold from the factory. Oddly enough, that circuit design sounds best with the output tubes biased towards the cold side of the spectrum. I think 50 mA per side (25 mA per output tube) of plate current sounds best anyway. My guess is this problem was probably not discovered until many were shipped and by that time, they were already working on the next year's model, the 6G8 Twin, which does not have that problem.

THE 5E3 TWEED DELUXE—We've talked about this amp before. This amp has many idiosyncrasies. When I started building replicas of this amp nearly two decades ago, I noticed that one or two amps from every ten or so built had a funny parasitic oscillation. The oscillation sounded like a bad speaker and only showed up on certain low notes (Open E mostly). If the speaker was replaced, the amp did the same thing. It took a while to figure this one out, but the problem was the proximity of the .1 uf coupling cap (this is the cap that connects to pin 6 of the 2nd preamp tube) in relationship to the instrument channel input jack. Take this cap and rotate it 90 degrees counterclockwise, and the problem will vanish. You may want to do away with the wire under the board that connects the end of the coupling cap to the grid of the output tube. I would disconnect that wire under the board and run a shorter, separate wire on top of the board. There are a few companies making replica 5E3 amps, that never figured this out. If you are buying a replica amp or kit of this type of amp, you need to know about moving that capacitor's location.

Another peculiarity of the 5E3 Deluxe is the fact that it is one of the few amps ever made that doesn't use a voltage divider for the volume control. That's right. I've heard people talk about how they need a slower taper pot for the volume control because it comes up too fast in volume. Get the slowest taper pot you like, it won't help — because the volume pot is not a voltage divider like every other amp ever made. The volume control works by loading down the signal. This has a unique quality to it that can be desirable if you know how to use the interactive volume controls of the 5E3.

To get a fabulous clean tone, plug your guitar into the instrument channel and put the volume about half way up on that channel. Turn the mike channel volume control (the channel you are NOT using) almost all the way up. Notice when you turn the volume of the mike channel up, how it scoops the mids from the channel you *are* using. Adjust the volume of the channel you are not using to get the best clean tone. Usually this ends up with the mike channel volume control between 10 and 11. This interactive volume control feature enables one to get that grand piano clean tone that is available from the 5E3 Deluxe.

THE BLACKFACE AND SILVERFACE REVERB AMPS — These amps are prone to parasitic oscillations. There are numerous mistakes in the layout that cause instabilities in the circuit. There are a few companies that offer kits of these amps. The kits are copies of the original and therefore all the mistakes are copied too.

The first flaw in the layout I noticed was the filter cap ground for the screen supply circuit. It was grounded with the preamp filter caps! The screen supply filter cap should be grounded near the cathodes of the output tubes. Also, the main filter and the B+ centertap should be grounded there. If you correct the screen supply filter ground and put it where it is supposed to be, then you might notice a parasitic oscillation coming from the reverb circuit. This drove me crazy until I figured out the other problem. The cathode resistor ground from the reverb driver tube (12AT7 pin 3 and 8) should be moved to the same place as the screen supply filter ground and the oscillations will go away. I theorize that when the amp was designed, the amp had a parasitic oscillation in the reverb. Instead of moving the cathode resistor ground to where it should be, they simply moved the screen filter ground to the preamp. This stopped the oscillation, so they went with it.

There are other layout design flaws. For example, the tone caps were put on the board and long wires were run to the associated pots for those controls. The tone caps should have been mounted on the pots themselves. This will reduce the amount of grid wire by about 2 feet! You wonder why the CBS engineers put those parasitic oscillation suppression circuits in the silverface amps? They could have just corrected the layout and not needed the remedial circuits.

THE PLEXIGLASS MARSHALL—Take any plexiglass front Marshall and turn up the treble, volume and presence at the same time and you will get a "mosquito" on top of the note. This is a parasitic oscillation that was built right into every amp.

Of course, if you turn down the treble, the volume and the presence, you can get the oscillation to go away. Parasitic oscillations are affected by volume (same as gain) and high-end response (think presence and treble).

If you go through the amp and shorten every grid wire—particularly the grid wires going to the output tubes you might be able to get rid of this "noise." If that doesn't work, move the tone caps from the board to the pots. This will reduce the grid circuit by a foot or more and should stabilize the amp.

THE EARLY MESA BOOGIE—Because of the tremendous gain used in the early Mesa Boogie amps, stability was a major problem. Instead of correcting the layout, a remedial circuit was used in which a wire was soldered to one of the output tube's plate circuit and then the wire was routed back towards the input jack and wrapped around the input wire. The input wire is the one connecting the input jack to the grid (pin 2) of the first tube. The wire coming from the output tube plate was not physically connected to the input wire; it was simply wrapped around it. By wrapping it, a very small capacitance was created which acted as a negative feedback circuit that only fed back the high frequency oscillations. This had the effect of phase canceling those oscillations. The characteristic "nasally" sound of the early Mesa Boogie amp is a result of this remedial circuitry. If you remove this "feedback wire" the amp will oscillate so badly that it cannot be used.

FIVE MODS TO PRESERVE YOUR VINTAGE AMP

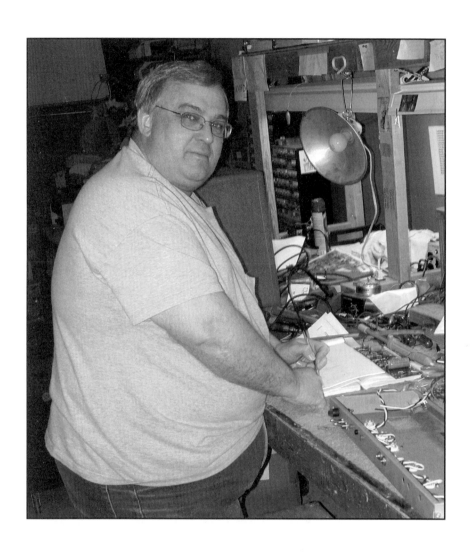

All About Vacuum Tube Guitar Amplifiers

FIVE MODS TO PRESERVE YOUR VINTAGE AMP

Vintage tube guitar amps were made to make beautiful music and at the same time, they are collectible and should be kept as pristine as possible. Ironically, it is in the interest of keeping them pristine that I recommend these simple, reversible mods. If you own a vintage amplifier and don't ever play it, read no further. But if you are a player that collects fine vintage amplifiers or a collector that desires to enjoy owning them by playing them, read on.

Replace the leather handle. If you don't replace it, the old leather will break and you will no longer own an original non-broken handle. There are many manufacturers that make exact replacement leather handles. Take the original handle off, put it in a zip-lock baggy and store it in the back of the amp. When you replace the handle with a new one, you will be able to pick up the amp by the new handle without fear of the original handle breaking. In fact, if you haven't replaced the handle, I suggest you never pick the amp up by the handle because 40-50 year old leather will break. A new leather handle may cost you $15, but if you don't replace it and you break the original handle, it will devalue your vintage amp investment by $100 or more and you will still have to spend the $15 to replace it later! If you ever want to sell your vintage amp, remove the new handle and replace it with the original handle. This will give you an extra handle for later use, all the while protecting your vintage amp investment.

Replace the speaker. If the speaker is non-original, you need not replace it; but if the speaker is original, replace it. If you don't replace it, and you play it, you will blow it. Almost all vintage speakers of yesteryear used paper bobbin voice-coil formers. Now if you've ever seen a piece of paper that was 40 or 50 years old, you can imagine the condition of that voice coil former. To compound natural aging, remember that a voice coil gets hot when the speaker is being played. Paper is made from wood. What happens to wood when it gets hot and cold, hot and cold over a period of decades? It deteriorates. The word "brittle" comes to mind. If you don't replace your speaker now, while it is still working, and you play the speaker; you will blow it. The paper bobbin will crumble and the voice-coil will literally fall to pieces.

By replacing the speaker before you blow it, you can always offer the amp with the original speaker later, should you ever decide to sell the amp. This will protect your investment and could make a difference of several hundred dollars later. And when you put it back to stock, you will be left with a good speaker to use for another project.

I remember a story of which I was the offender of not taking my own advice. I purchased a 5F6A Bassman in pristine condition from a pawn shop in Austin, Texas back in 1989. The amp had the original Tung Sol output tubes and all the speakers were pristine examples of P10R Jensens. I knew to replace the speakers, but I told myself I wouldn't play them very loud and I would be careful. Well, one day I was playing the amp at less than half volume and I noticed a rattling sound coming from the amp. To make a long story short, one of the speakers blew which caused a chain reaction blowing two more.

Replace all electrolytic capacitors. I have heard collector's protest this advice because they wanted to keep their amp original. They want to keep their "paper" caps. Well I've got news for you. There is no such thing as a "paper" electrolytic capacitor. All vintage electrolytic capacitors are aluminum can type. However, shrink wrap technology didn't exist in the 50s, so manufacturers labeled the aluminum can electrolytic caps with a cardboard tubing. If you were to remove the cardboard tubing, underneath you would find—an aluminum can! Today, the shrink wrap technology has replaced the cardboard tubing, but underneath is the same style of construction—aluminum can.

There are many reasons you should replace the electrolytic capacitors. For one thing, according to their manufacturers, electrolytic capacitors do not last but 6 to 10 years. All capacitors are supposed to block D.C. current, but when the electrolytic cap gets old, it allows D.C. electrical current to leak. When D.C. current leaks through the capacitor, it is basically shorting your power transformer to some degree. This electrical leakage is additive by how many capacitors are in the circuit and how much each is leaking. If there are five capacitors and each leak 15 milliamps, then you have 75 milliamps of current that is leaking through them and constantly stressing the power transformer anytime the amp is on. As an example, let's say you have a typical 6V6 vintage amp whose power transformer is capable of handling 125 milliamps without overheating. If the caps are leaking 75 milliamps all the time, then the transformer's rating has been diminished to only 50 milliamps. With such a low rating, when the amp is played, the transformer will overheat. In extreme cases the transformer will get so hot it blows. Worried about the filter caps not being original? How would you like a blown power transformer?

And besides leaking electrical current, they almost always rupture and leak chemicals. I don't know if these chemicals are toxic, but I do know they corrode the amplifier chassis. And I have one more argument for those that insist on keeping the original caps. The caps in your vintage amp are not original now. Originally they didn't leak current and they didn't leak chemicals.

If you are really anal about this electrolytic cap thang, here is a technique we have used to satisfy collectors that want to open their vintage amp and have it look dead nut original. Remove the old capacitor from the amp. Next, carefully remove the aluminum can from the center of the cardboard tubing. This is a tricky procedure involving a drill press. You would drill out the center of the cap using a small drill bit and gradually use slightly larger bits until the center of the cap has been completely removed. Sometimes when you are drilling, the cardboard tubing loosens from the can and the can just spins around inside the tubing. In this case, one could carefully uncrimp one end of the cardboard tubing and remove the entire aluminum can. It should simply slide out. Once the can is removed from the tubing, take an American-made cap and slide it into the empty cardboard tubing. Remember to insert the new capacitor such that the polarity matches the polarity marked on the cardboard tubing! We wouldn't want anyone to think the cap was in backwards. A little beeswax around the end will seal it up and you will have a cap that looks original and performs like an original. And best of all, you don't have to worry about burning up your power transformer.

If the new American-made cap happens to be a larger diameter than the original cap, take a sharp razor (we use a box-cutter type utility knife) and slice a longitudinal cut along the underbelly of the original cap. I like to make this cut completely opposite the writing on the cardboard tubing. This will allow a larger diameter cap to be inserted. Remember to mount the new cap assembly with the underbelly slice facing down so no one will ever see the slice when you are done.

I want to emphasize to use American-made capacitors and not Taiwanese capacitors that have an American name. Specifically you should avoid Illinois Capacitors as these are made in Taiwan and sound like blown caps when they are new! To get original performance, use Sprague Atom capacitors or Tech Cap brands. Ruby capacitors, made in China, are also very good.

Replace the 2-prong A.C. cord with a 3-prong A.C. cord. Most vintage amplifiers have a brittle and decaying 2-prong A.C. cord. If you continue to use it, expect for it to deteriorate to the point of shorting out and either catching on fire or blowing a house breaker or both. There is another reason you should replace it; a 2-Prong system is not grounded and you run the risk of shocking yourself. When you remove the old A.C. cord, put it in the zip-lock baggy with your removed leather handle. If you keep the 2-prong original A.C. cord, you can easily put it back or offer it to the new owner in the event you ever sell the amp.

When installing the 3-prong A.C. cord, there will be three wires: a black, a white and a green wire. Look at the 2-prong wiring. It will have a black and a white wire. To convert to 3-prong, the black from the 3-prong goes where the black of the 2-prong was, white goes to where the white of the 2-prong was, and the extra green wire gets grounded. The best way to ground it is by soldering it to a grounding lug and fastening the grounding lug to one of the transformer mounting screws. This can be done without altering the chassis or drilling any holes and it is easily reversible.

When you complete this mod, you can play the amp without fear of getting shocked and without fear of catching anything on fire. And you are not devaluing your investment.

Re-tube the amp with new tubes and set the bias. Any tube amp will sound its best with fresh tubes. And anytime you replace output tubes the bias should be adjusted properly unless the amp is selfbaised (a.k.a as cathode biased). There are some very good tubes available today. You don't need to spend hundreds of dollars on new old stock tubes when there are so many great choices with new manufactured tubes. I absolutely love the JJ brand E34L tubes which are made in the old Tesla factory in Slovakia. I have ABed these with original Mullard EL34s and I like them better. Russian tubes have gotten much better over the years and there are even some Chinese tubes that could pass for NOS in blind sound tests. Groove Tubes manufactures some great sounding tubes made in the USA.

Old Stock tubes haven't been manufactured in 25 years or more. That means whatever is remaining of this Old Stock has been culled over and over again for the last 25 years. You would be hard pressed to find NOS preamp tubes that were not noisy or microphonic. And the best output tubes have been selected by the first four owners of the batch and what is left now are the culls. John buys 200 tubes in 1990 and keeps the best 50 for himself and sells the remainder to Bill in 1993. Bill selects through them and takes the best 50 for himself before selling the batch to Paul in 1996. Paul can't seem to find any good ones, but keeps the best 25 of the lot and then you see the rest on ebay as a lot of 75 NOS tubes. And because these are old stock, they are sold as is!

Save yourself the aggravation. Get some new tubes from a reputable source and you will be much happier. I have a story for you. A friend of mine knows that I love the Mullard EL37 output tubes. These are great in a 6L6 circuit. He purchased a factory matched set of the military version of this tube (the CV586) on the internet. They came in factory matched boxes and were wrapped in the original paper cardboard packaging. He gave them to me for my fiftieth birthday. Because they were so nice and pristine looking, I saved them for about a year before putting them into an amp. When I put them in, one of the tubes immediately started ticking—thus making the pair unusable. There is no recourse here because the sale happened long ago. My friend paid over $300 for those unusable tubes.

That is another reason for using new manufactured tubes. When the tube eventually malfunctions, you are not out much moola. If you blow an expensive new Old Stock tube it is going to be a bigger loss.

People ask me if they must set the bias when changing output tubes. Some amps are self-biased, in which case, they will adjust themselves. But with a fixed biased amp, it is best to adjust the bias when changing output tubes. Would you replace the tires on your car and not check the air pressure on each tire? Do you think it is a good idea to change the carbeurator on your engine and not set the idle? Or how about doing a valve job on your vintage car without setting the valve clearance? You don't have to set bias, but if you don't set bias, don't expect your amp to perform its best.

All About Vacuum Tube Guitar Amplifiers

HOW TO MAKE YOUR AMP LOUD

Certain amps have always been more popular than others, and it seems the ones that were least popular were the brands that were not very loud or responsive. These cool vintage tube amps usually sell for much less, since they were never the most popular. Some of these brands would include; Danelectro, Silvertone, National, Gretsch, Airline, Supro, Alamo, Magnatone, Guild, and many others. The good news is that with a little bit of tweaking, one could take an amp with an un-thrilling design and make it into something gigworthy and fun to play. I'm talking about a tube amp that will jump up and do some tricks instead of rolling over and playing dead.

I had an Airline amp in my shop the other day that was mediocre at best. The design was very conservative. The sound was what I would characterize as low-gain homogenous, clean and compressed. The cathodes were unbypassed, the plate voltages were low and the amp just didn't feel like it wanted to please. With a little bit of tweaking, that amp became a thriller and a killer. The owner didn't pay but a few hundred (not a few thousand) for that amp and it performs now like an amp costing five times the money.

To get more volume from a tube amp, one must consider first the speaker. Speakers are rated for SPL (acronym for Sound Pressure Level). The Actual SPL is measured in dB. For example, a low efficiency speaker might be rated at 87 dB SPL. That means if you put one watt into that speaker and listened from 1 meter away, the sound pressure level would measure 87 dB of loudness. A higher efficiency speaker can make your amp much louder. Remember, there is only 3 dB difference between a 50 watt and a 100 watt amp. And there is only 3 dB difference between a 25 watt and a 50 watt amp. A speaker whose SPL is 3 dB more than another speaker will sound much louder. For example, if you connect a speaker whose SPL is 95 dB to a 100 watt amp, it will be as loud as a 50 watt amp that is connected to a speaker whose SPL is 98 dB.

JBL speakers are very high SPL—near 100 dB or more. That is why they sound so loud. I remember doing a gig with a 50 watt amp going through a high efficiency ElectroVoice 15" speaker that was much louder than the other guitar player's 100 watt amp going through a 2x12" cabinet.

But after speakers are considered, the next item to address is tubes. Tubes don't last forever. There is a difference between them functioning and performing. The insides of the tubes have tiny elements such as the cathode, grid and plate. These metals, when new sound much better than after they have been played for 1,000 hours. The problem is that you don't notice the ever-so-minor deterioration from day to day. I remember several years ago when I was going to do an important recording. I did a scratch track and was not impressed with the sound. At first, I tried moving the mike around and trying that. My drummer asked me when was the last time I changed tubes. It had been about a year and played that amp really hard a couple of times a week. I popped a fresh pair of output tubes and a new phase inverter preamp tube and couldn't believe the dramatic difference in the tone. The tubes I pulled out were Mullard EL37s and I remembered when I had installed them a year earlier, how impressed I was with those tubes when new. Of course, you will obtain optimum performance when changing output tubes if you set the bias; assuming you have a fixed bias amp.

Fresh electrolytic capacitors can also impact the loudness of an amp. If the electrolytic filter capacitors are leaking very much current, they could bog down the power transformer. An amp will always sound its best with new American made caps. If you think you are going to save money by replacing the caps with cheap Taiwanese-made electrolytic caps, don't bother. The Taiwanese caps will make the amp play out-of-tune when it is cranked and they produce a non-musical, and very annoying distortion.

Besides speakers, tubes and filter caps; some amps just have a lame design and need help. The plate voltages on the preamp tubes make a huge difference in preamp gain. As the plate voltage goes up, so does the gain. Some designs use very low voltage on the preamp tubes. This voltage can be raised by making the power resistor value smaller. I am not talking about the 100K plate resistor that goes to the tube's plate. I am talking about the power resistor (coming from the direction of the rectifier) that feeds the 100K plate resistor and the associated electrolytic filter cap. If this power resistor is made smaller, there will be more voltage on the plate. Although a 12AX7, for example, has a design maximum of 330 volts; I think they sound harsh when run that high. Typical vintage amp values might be 120 to 150 volts. You could get much more gain if you bumped the voltage up to 200 to 220 volts. This voltage has enough headroom and substantial gain. The power resistors in almost all amps are run in a string from the rectifier to the screen supply, to the phase inverter to the last gain stage, etc, to the first gain stage. If you make the first power resistor in the string (that's usually the one in between the screen supply and phase inverter supply) smaller, then all the voltages in the string will go up. This is a good way to raise all the preamp voltages a little bit by only changing one resistor. If each stage has more gain, by the time the signal gets to the output tubes, it is noticeably louder than stock.

The omission of bypass capacitors constitute another lame design seen with Silvertone, National, Supro, Magnatone, and many other amps. Remember, when those amps were designed, they thought they were designing for the best clean tone. If you leave off the bypass cap across the cathode resistor on a preamp tube, then the tube has much less gain and when hit with a big signal, it doesn't clip, it just compresses instead. The result is a cleaner, lower gain sound. That's why that circuit works well for bass guitar. I like to experiment with these preamp circuits by placing a 25 ufd at 25 volt cap across the cathode resistor on a preamp tube. With a 12AX7, the cathodes are pins 3 and 8. If you look at the socket, those pins will go to a cathode resistor whose value could range anywhere from 470 ohms to 4.7K. If that resistor does not have a bypass cap across it, then it is said to be unbypassed and the tube will not supply full gain. The size of the cap will affect the frequency response of the gain. A smaller capacitor will have more highs. A generic value that is fairly even is the 25ufd at 25 volt. You will see this as the most common bypass cap. If the amp is boomy sounding and has (in general) too much bass, try a 10 uf or a 5 uf or even a 1uf. The smaller the value, the more it will tame the bottom-end and bring out more highs. If there are multiple gain stages, check to see if the other stages are bypassed. Just adding a couple of bypass capacitors could improve gain by 10 dB or more!

All About Vacuum Tube Guitar Amplifiers

IMPROVED REVERB RELIABILITY FOR THE FENDER VIBRO KING

Anytime a new unique design is introduced, the first year of manufacturing will sometimes reveal design flaws perhaps that were not evident in the prototype testing. In 1984, there was a radical design change in the Chevrolet Corvette. It was the first year with a new body style and the first year with a digital dashboard. Of course, I had to have one so I ordered one from the factory. There were many bugs, including the body cracking above the rear axle. Chevy made over two hundred changes when the new 1985 models were introduced. There is nothing like experience to reveal design flaws.

Fast forward to the early half of the 90s. Fender had just introduced their new boutique amp, the Vibro King. Designed by my good friend and a brilliant amp designer, Bruce Zinky, it sported a 100 watt transformer on a 60 watt design. From a design point of view, this is like putting a turbo-charged V8 in a Honda Civic. It had the beefy output transformers and a built-in 3-knob reverb. It was Fender's first new point-to-point construction in two decades and sported three 10" AlNiCo speakers. Yes, this was something coming from Fender that was worthy of attention.

As with any radically new design, there were certain bugs which did not show up in the prototype testing phase. The main niggle was the reverb circuit. Fender wanted Bruce to design it with parts they already were stocking, so he chose the EL84 output tube to drive the reverb. This could have been very successful, after all, Leslie and Hammond both used 6BQ5 tubes (American version of the EL84) in their reverb designs. And if you remember, Hammond actually invented reverb; so the EL84 seemed like a good choice.

The problem was the high plate voltage coming off of that 100 watt transformer onto the EL84. Those tubes don't like high voltage and the voltage was over 440 volts on the plate. This caused the tubes to sometimes fail. Just as that Corvette didn't like being over-revved; the EL84 would have preferred a lower operating voltage.

After many of these amps were made, Fender noticed a large failure rate in the circuit and later changed the design to a 6V6 output tube to drive the reverb output transformer. This worked better, but there was not a good supply of good sounding 6V6s at that time.

If you have a first generation Vibro King, there are some modifications that could be done to make the amp's EL84 reverb circuit more reliable.

If money is of no concern, the easiest fix is to change the EL84 reverb driver tube to a 7189A. Of course, since the 7189A has not been manufactured in 20 years or more, the only ones available that will work will be New Old Stock (a.k.a. NOS). New Old Stock is simply a new tube that is decades years old, and never used. Since the inside of a vacuum tube is a vacuum, the internal components of a tube are essentially inside a "time capsule." There are many wonderful sounding Neumann U47 tube microphones in operation today that use a V14 tube. The last year the V14 was made was 1945. It was used for sonar in German U boats during World War II! So you can see a tube can be very old and work as new.

Expect to pay $100 or so for a 7189A output tube. I would only accept American made in this tube as a foreign made one could be a misbranded EL84. The 7189A is actually a military grade EL84 that can take higher voltage and consequently has better ratings. This makes for very good reliability. The tube can also be used in any EL84 or 6BQ5 circuit. They work especially well in a Vox AC30, but again, they are expensive, if you can find one.

Perhaps a better way to go is simply modify the amp such that the voltage is less on the existing EL84. I like this option better, because when it is time to change tubes, you will not have to go on a scavenger hunt looking for a 7189A. And who knows how much that tube will sell for in 5 years.

I perform a modification on the power supply of a VibroKing to drop the plate and screen voltage of the reverb drive tube to an acceptable level. Like Fender, I designed it using parts that are common to most service shops. To do this mod, you will need:

1. A filter capacitor. I used a 22 mf 500 volt Tech Cap but a 20 uf 500 volt Sprague ATOM cap will work equally well. The Tech Cap was physically smaller, so I chose it.

2. Three 1K five watt resistors. These are very common as they are used as screen resistors in all Marshall amps and most amps with EL34s.

3. Two 56K one watt carbon composition resistors.

4. A small piece of perfboard or other suitable circuit board material. I used Garolite (a.k.a. Epoxy glass laminate), but even Radio Shack perfboard will do fine.

5. A terminal strip with at least one mounting lead.

LET'S START THE MODIFICATION

Remove the amp chassis from the cabinet and remove the small cap pan cover from the chassis. Compare what you see on your cap board with the picture labeled Figure 1. Notice we removed the solid black 25K three watt resistor from the capacitor board. Notice where the orange and the yellow wires are soldered. We are going to route the power through three 1K resistors before the reverb driver transformer. This will drop the voltage considerably. To do this, we will use the

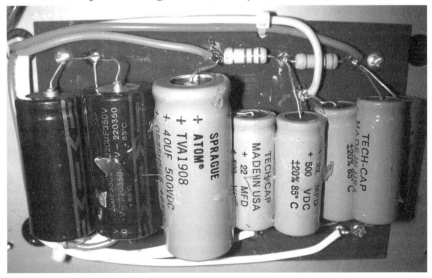

Figure 1.

Figure 2.

existing filter cap under the pan with the yellow wire as our main filter that connects to the EL84s reverb driver transformer. It will also connect to a pair of parallel dropping resistors that will feed an added filter cap and a 100 ohm screen resistor. More on this later.

Now look at Figure 2. Notice the three, 1K five watt resistors placed in series. This will actually be seen as a single 3K 15 watt resistor by the circuit. That is because when three 1K resistors are in series, the total resistance will equal 3K. I mounted these resistors on a piece of circuit board material and I used the existing screw (that holds on the back panel) to fasten it to the back inside wall of the chassis. You only need one screw because the circuit board is long and cannot move even with only one screw.

Now look at Figure 3. Notice we added a terminal strip to remount the 100 ohm screen resistor and an additional filter cap. The 22 uf filter cap has its plus end going to the circuit, and its minus end going to a convenient ground. We now have two 56K one watt resistors in parallel (green, blue, orange, silver) with each other to make a 28K two watt resistor. It is hard to see both resistors in the picture because one is directly beneath the other, but they are both there. These two parallel resistors are located between the reverb driver transformer and the EL84s screen resistor (black, brown, black).

     All About Vacuum Tube Guitar Amplifiers

Figure 3.

Notice that we added the filter cap to the screen resistor and we moved the wire off the original screen resistor terminal.

So to trace the circuit and recap, we removed the 25K three watt resistor from the cap pan.

Instead of hooking the EL84 directly to the 6L6 screen supply, we dropped it first through the three 1K five watt resistors. After the resistors, the circuit sees the primary of the reverb driver transformer and the two 56K one watt parallel resistors. These drop the voltage feeding the 100 ohm (EL84) screen resistor. A 22 uf filter cap is at the junction of the two 56Ks and the 100 ohm screen resistor.

Now it is time to double check your work against the pictures and test it out. The voltage on the plate of the EL84 should be around 350 volts and the screen supply of the EL84 should be around 300 volts or less. These voltages are not chiseled in stone and will actually vary somewhat—depending on the particular EL84.

You will probably notice the reverb tone will be smoother and of course the EL84 will last much longer and will not be prone to malfunction.

All About Vacuum Tube Guitar Amplifiers

TEN WAYS TO VOICE YOUR TUBE AMP

Most serious guitarists prefer tube amps for their obvious better tone than their transistor counterparts. But besides their tonal superiority, tube guitar amps can be easily voiced to really "zero in" on the exact tone you've been wanting. If you have an amp that sounds pretty good, but you would like to "kick it up a notch" to sounding great, here are ten methods of voicing your tube amp that could be the ticket to tonal nirvana.

For starters, you need to know exactly what it is you are missing. The only way to do this is to play the amp and listen carefully. How is the attack? Does the amp compress off the front of the note, or does it have sufficient front-end "sting" to the attack. How is the gain? Is it dirty enough? Is it clean enough? What about the frequency voicing? Is it too boomy or is the bass clear and focused? Are the mids thick enough for your taste? Is there enough shine on the top-end? These are the kind of questions you need to ask yourself while playing the amp so you will have an idea of what is missing. To get the voicing perfect, you need to have an idea of what will satisfy you. It is like a chef cooking a huge pot of chili. He tastes it and asks himself what does it need? More salt? More pepper? More jalapenos? More garlic? You get the picture. After you have done some critical listening, refer to the section below to see where you can voice the amp to get whatever has been missing.

Tube amp voicing is done by one of the following methods:

Let's take these one at a time and see what we can do with each.

1. PREAMP TUBE SUBSTITUTION — This is done by taking tubes that are similar but with different gain characteristics. Obviously we must use tubes with the same pin-out. For example, the 12AX7 tube has many different versions, each with different amounts of gain:

12AX7—amplification factor of 100

5751—amplification factor of 80

12AT7—amplification factor of 60

12AY7—amplification factor of 44`

12AU7A—amplification factor of 20

Substituting a lower gain tube for an existing higher gain tube is generally done to improve clean headroom. The lesser gain tube will not overdrive the rest of the amp as much—thus resulting in a cleaner overall sound. Conversely, a higher gain tube could be substituted for a lower gain tube to increase gain.

Hint: To find similar tubes, use a tube manual such as the *General Electric Tube Characteristics* manual. Look up the tube you are seeking substitutions for and consult the pin-out drawing in the back of the manual. All of the tubes with the same pin-out will be listed. You need only note these tubes and check their listings in the front of the tube manual to verify they use the same filament voltage and the plate voltage ratings are within spec. After that, it is simply play and listen.

When substituting preamp tubes, take into consideration that almost every brand of guitar amp ever made has the signal inputs on one side of the amp chassis and the signal is amplified as it makes its journey from the input side and moving towards the output side. So the preamp tube that is closest to the input jack will make more difference than the next one. And the second gain tube will make more difference than the one later in the circuit. It is because any change made in the first gain stage tube will be amplified by all the other tubes downline.

2. OUTPUT TUBE SUBSTITUTION—This is done to alter the overall sound of an amp. For example, the 5881, 6L6, KT66, 7581A are all similar tubes that can be substituted in a 6L6 circuit. Because the 7581A and the KT66 are rated for more plate dissipation, they will be cleaner sounding than the 5881 or the 6L6. If you are going for more clean headroom, you could try using a KT66 in a 6L6 or a 5881 circuit, for example. Using a KT66 or a 6L6 in an EL34 circuit can also increase clean headroom.

When substituting output tubes, it is imperative that the amp be re-biased to adjust the idle level of the new tube. If you are not sure you know what you are doing, bring your amp to a competent tech and ask to A/B some different output tubes.

Similarly, you could substitute a KT77 for an EL34 or a KT88 for a 6550. Sometimes, if more headroom is required, a 6L6 can be substituted for a 6V6.

If you want more grunge, an EL34 can be substituted for either a 6L6, 6550 or and of the KT style tubes. Be careful because the EL34 requires that pin #1 be jumpered to pin #8. On 6L6 type amps, since a 6L6 tube doesn't even have a pin #1, the builder sometimes will use pin #1 of the socket as a mounting post. One must be careful to remove the wires on pin #1 and tie them elsewhere and then jumper pin #1 to pin #8 when substituting to EL34 style tubes. There is also a military version EL34 called an E34L. This tube is more robust and has a better clean tone.

3. SPEAKER SUBSTITUTION—Sometimes changing the speaker can make all the difference in the world to bring the amp to life. Older speakers can lose their efficiency and an inefficient speaker will lack liveliness of attack. If you want a sportscar feel, you need a speaker that wants to work hard. Efficiency is rated in SPL which is the acronym for Sound Pressure Level. The SPL is how many decibels the speaker puts out at one meter away from the speaker when it is driven with only one watt of power. An SPL of 98 or higher will give you the liveliness needed for a punchy sound.

Maybe your existing speaker lacks highs, in which case the top end will sound muffled and boomy and lack definition. Or perhaps it lacks the satisfying bottom-end that you crave.

Speakers have holes and humps in different frequencies. You are not looking for a home stereo speaker that is ideally flat at all frequencies from 20 to 20K Hz. You want a speaker that rolls off somewhere around 6K to 8K or less. If the speaker reproduces high-end too well, you will hear abnormally high "hiss" and fret noise, which can become very annoying.

Hint: Bring your amp to a music store that sells speakers and ask to test drive some various different speakers. You want to play and listen critically until you find the speaker that does it for you.

4. ALTERING COMPONENT VALUES IN THE PREAMP SECTION—With a vintage amplifier, you must remember the designers thought they were going for clean. Many designers would leave off bypass capac-

itors that would attach from the cathode of a preamp tube to ground. This would make the amp play much cleaner, but it would also make it compress more and not have much gain. Some examples of these designs would include: all Gibson amps, Harmony, Montgomery Wards, Silvertone, and many others. By adding a bypass capacitor, the gain will increase noticeably. You need only find the cathode resistor and place the bypass cap across the resistor with the minus side of the cap facing ground.

Or you may wish to alter the actual value of the existing bypass cap. A 25 uf at 25 volt would be a typical bypass cap value and would give a fairly flat frequency response. But when voicing an amp, you may not wish a flat response. You may need more cut and less boom, in which case a smaller microfarad value would be the ticket. You will have to at least double or half a microfarad value in order to hear any difference. For example, if you used a 25 uf at 25 volt and thought it needed to have more cut and less boom, you would try a 10 uf at 25 volts (which is the nearest standard value that is at least a factor of two) and do a listening test.

If the amp lacked bottom-end, you may wish to go with a larger value cap. This will make the bass response hold together better and will also help gain.

Besides altering or adding bypass capacitors, you may wish to scrutinize the cathode resistor. This is the resistor that goes from the cathode of the preamp tube to ground and it is in parallel with the bypass cap. Cathode resistors are easy to spot because they will typically be from 470 ohm to 10K ohms or somewhere in between. The most common value is 1.5K ohm. This resistor sets the preamp tube biasing. If you make the resistor smaller, the tube is biased hotter and will break quicker. It will also have more gain. If the resistor value is increased, the bias of the tube will idle colder, which may be what you need if you are going for a better clean tone. Be careful because if you make the cathode resistor value too small, the amp will hiccup when you stop playing. It is funny to hear this as it is almost like an echo.

Or, you could go the other way, depending on what kind of tone you are going with. You could actually remove a bypass cap and increase the value of the bypass resistor and the amp would lose gain and get much cleaner. Perhaps you want the amp to be really clean.

This will clean it up.

5. CHANGING THE OUTPUT TRANSFORMER—The output transformer makes more difference to the tone than any other single component. Stay away from pricey plastic bobbin transformers. You want to go with paper bobbin transformers because they sound better than plastic. Also, with output transformers, bigger is usually better. The more iron it has, generally speaking, the better it sounds. You may want to get rid of your tiny Pro Reverb transformer and replace it with a 4 ohm Bassman output transformer. The Bassman transformer is huge by comparison.

6. CHANGING THE POWER TRANSFORMER—This is usually done when an amp has too much sag or lacks efficiency caused by the original transformer having rust. Transformer cores are made with thin laminates of silicon grain oriented steel. Each laminate or "lam," as it is called, is insulated electrically from the others' and for good reason! If the lams were not insulated from each other, since steel is a good conductor of electricity, the transformer would "think" there was just another secondary winding—that was shorted! This phantom "secondary winding" would draw lots of current and drag everything else down. Sometimes, when a transformer gets rusty, the adjacent lams can become conductive and there goes your power. The transformer will heat up because the energy, which should have been going to the real secondaries, is being drawn into the big, shorted, single-turn "phantom" secondary. Changing to a new power transformer can really change the feel of an amp.

For example, if your amplifier has too much sag, and even if it isn't rusty, you may wish to use a replacement transformer that has a higher current rating on the B+ winding. Actually, the higher current B+ winding simply uses larger diameter wire and that is what gives it the higher current rating. For example, you may have a little practice amp with a 125mA rated B+ winding. Changing it to 250mA would make a noticeable difference in the feel. Likewise, if you had an amp such as Fender Super Reverb or Pro Reverb Amp, which uses a 250 mA B+ winding, you can get a stronger bottom-end and faster response by using a 400 or 500 mA power transformer, such as the one found in a Fender Twin.

Installing a beefier, four-output tube style, power transformer in a two-output tube style amplifier is a neat trick to get that quicker punch and bigger bottom. This type of voicing and was used successfully on: Park Amps which used a 100 watt Marshall power transformer on two EL34 output tubes; the first handwired Bedrock amps which used a one Amp B+ winding on a single pair of EL34s; Fender Vibro King which used the Fender Twin power transformer on two 6L6 output tubes; and the Kendrick Specialty Jazz Amp which uses a massive 500 mA B+ winding with only a pair of KT66 tubes. Modifying an amp's performance by using a heftier power transformer would be like installing a bigger engine in a smaller race car. The performance will be faster with a higher performance feel.

7. REBIASING THE OUTPUT TUBES—To bias a set of output tubes is to set how high or low the tubes are idling. A tube that is idling low will have more headroom, won't be as loud, and it will sound somewhat cold if it is too low. On the other hand, a tube that is idling too high will break up too quickly and will lack a good clean tone. It is generally louder and can even sound mushy depending on how much too high it is idling.

There is no one, perfect, bias setting on an amp. Any setting that doesn't blow up the amp and sounds good to the player is valid. In general, we want to set the bias in an amp such that the tubes are idling low enough to get a good clean sound and high enough to break where it feels right.

If you have a tech or a tube amp friend that knows how to set the bias of an amp, you could always have him adjust the bias to different settings so you can listen to different settings while playing. Try a few different settings to see how those settings feel. Have your tech make a note of how much idle plate current you like best with that particular amp. Once you know what you like, you can easily duplicate it later with a new set of output tubes.

8. ALTERING COMPONENT VALUES IN THE OUTPUT STAGE—Regardless of what type of output stage you use—it could be single-ended style which doesn't require a phase inverter, or push-pull style which does require the phase inverter—coupling capacitors are used to couple the pre-amp or phase inverter to the output tubes. These capacitors feed the grid circuit of the output tubes. If the amp has a phase inverter, the coupling cap will couple the outputs of the phase inverter

to the inputs (grids) of the output tubes. This coupling capacitor is a filter whose value can drastically change the tone and balance of highs to lows.

For example, if you have an amp that is really boomy and it lacks clarity, making this coupling cap value smaller will take out the boom and put in some bite with clarity. You will have to change the value by at least ? to hear any difference in tone. A standard Fender value might be .1 uf. You could try listening to a .047 uf and if the sound is still too boomy, the next value to try would be a .022 uf. But on the other hand, let's say you have a bass guitar and you want to really get every last ounce of bottom-end out of that output stage. You would want to try larger values such as a .2 uf or a .5 uf. The bigger values let more bottom-end through, whereas the smaller numbers trim the bottom back.

9. ALTERING ELECTROLYTIC CAPACITOR VALUES IN THE POWER SUPPLY— All power supplies use electrolytic capacitors to smooth out the ripple current that occurs when A.C. wall voltage is rectified to D.C. voltage. Some amplifiers use electrolytic capacitors that are just too small to get rid of the ripple current at high, overdriven volumes. In such a case, the 120 Hz ripple current will modulate whatever note you happen to be playing and you will end up with an out-of-tune undertone that is not a natural harmonic of any note you play! Your notes will all sound out of tune! Not only that, but bigger caps, used in the main filter supply, pack more punch. This is particularly noticeable when playing the low E string. The bigger filter holds the note together. The smaller filter causes the front of the note to be compressed off. Usually doubling or tripling the stock filtering is what works to get rid of the ripple current and improve low end punch.

I once took a Super Reverb amp and started adding more and more microfarads of filtering to the main filter supply while I was sound testing the amp at loud volumes. I kept adding more and more filtering until I could hear the out-of-tune ripple current go away. At that point, I was using two 220 uf at 350 volt capacitors. Stock value is 70 uf at 350 volt—so basically I was about tripling the stock value. A tube manual will tell you not to do this, but I have done this for years without incident. The secret is to hook the main filter supply up to the rectifier tube such that it "trickle charges" as the rectifier tube is warming up. This is what Leo Fender did with the blackface Super Reverbs anyway. On a tweed amp, such a 5F6A Bassman, one would need only move the main filter supply from the cold side of the standby switch to the hot side, which is basically the same thing as connecting the main filter caps directly to the rectifier's cathode (pin #8). When this is done, the main filter actually "helps" the rectifier tube to charge all the other caps when the standby switch is switched from standby to play.

10. SUBSTITUTING RECTIFIER TUBES OR TYPES — Different rectifier tubes and also solid-state rectifiers, produce unique envelopes of attack. If we substitute rectifier tubes or types, we can alter the way an amp performs. This is most noticeable in the initial attack of the note and the way the note compresses. For example, a solid-state rectifier recovers very quickly, so it is the fastest attack. We would say the solid-state rectifier has the least sag, and it will also be louder than the tube type rectifiers.

The 5AR4 aka GZ34 rectifier tube is not as quick as the solid state rectifier, but is quicker and louder than a 5U4GB.

The 5U4 would be much spongier with a softer attack when compared to a 5AR4 or a solid-state rectifier.

On lower wattage amps, the 5Y3 is the spongiest with the 5V4 having a much quicker response time and less sag. Solid-state can also be used on lower wattage amps and it will always attack faster than a tube.

There are solid-state rectifier adapters that look like the male part of a rectifier tube, but with a solid-state device so that you can plug it into your rectifier tube socket and convert it to solid-state without having to rewire anything.

Remember when changing rectifier tubes or types that the output tubes of the amp will usually need to be re-biased because changes in the rectifier will alter the high voltage supply, thus affecting the performance and biasing of the output tubes.

"AB763" SCHEMATIC

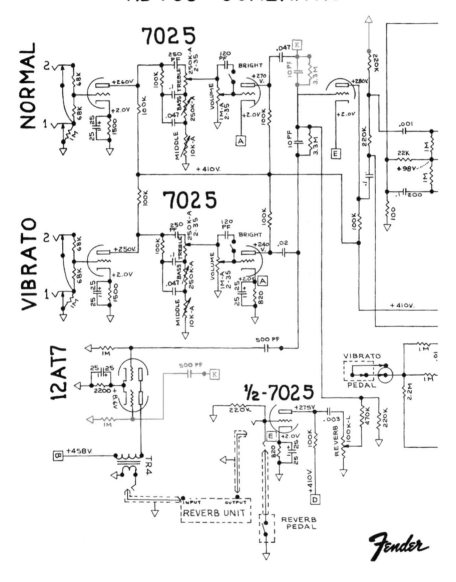

All About Vacuum Tube Guitar Amplifiers

REVERB AND VIBRATO ON BOTH CHANNELS OF YOUR BLACKFACE OR SILVERFACE FENDER

Installing reverb and vibrato on both channels is an easily reversible modification that can be done in a short amount of time with only a few dollars worth of parts. One problem with stock Fender two-channel amplifiers is that the channels are out of phase with each other, so bridging the channels doesn't work well. On a stock Fender, instead of the channels working together in unison, when bridged, they fight each other and cancel each other — resulting in low volume and thin tone. When you modify the amp to have reverb and vibrato on both channels, an extra stage of gain is used by the normal channel, which inverts phase to match the vibrato channel. Now the channels can be externally bridged with a short patchcord so that both channels can be run in parallel. To bridge the channels, the guitar is plugged into the number 1 input of either channel. A short guitar patchord is plugged into the number 2 input of that same channel and the other end of the patchcord goes into the number 1 input of the other channel. Now the channels are bridged and the tone knobs of both channels are operative.

Beside bridging the channels together, one could also use an A/B switch to switch between channels and each channel could have reverb and vibrato, but with independent tone and volume controls.

This modification works on almost all Blackface and Silverface two-channel amplifiers. All of these use essentially the same preamp circuitry so I am using the AB763 diagram to show the modification.

What parts are needed?

Only a piece of shielded wire, a 4" piece of hookup wire, two resistors and two capacitors are needed. These parts are available at any electronics shop for a few dollars. You will need:

1. One 10 picofarad ceramic disc capacitor at 500-volt rating
2. One 500 picofarad ceramic disc capacitor at 500-volt rating
3. One 1 Megohm half-watt resistor
4. One 3.3 Megohm half-watt resistor
5. One foot and a half of coaxial shielded wire
6. A four inch piece of 20 gauge solid hookup wire.

THEORY OF OPERATION

In a Fender two-channel amplifier, the reverb driver tube, which is a twin triode style 12AT7, has both sections tied together in parallel. What we will do is separate the two grids of the reverb driver tube so that each channel has its own reverb driver tube grid. We will disconnect the normal channel preamp output from the power amplifier section and connect it to a 500 pf cap that feeds the reverb driver tube grid. A 1 Meg grid return resistor is added from the reverb drive grid to ground. Also, using a 3.3 Meg resistor in parallel with a 10 pf capacitor, we will route the normal channel to the reverb mix tube. This will have both channels sharing the reverb driver tube and reverb mix tube. Since the vibrato is connected to the output of the reverb mix tube, both channels will also have vibrato.

LET'S GET STARTED

Unplug the amp and remove the chassis from the cabinet and discharge the filter capacitors. On most Fenders, if you put the standby switch in the PLAY mode and wait about 30 seconds, the filter caps will discharge. Or you can run a jumper wire from pin #1 of any preamp tube to ground and wait about a minute. In either case, the standby switch is placed in the PLAY mode.

Take a good look at the inside of the amp and compare it to our AB763 Layout drawing. The modification appears on the drawing in red ink. There are two disconnections that must be made first. Starting at pin #6 of the first preamp tube, follow the wire back to the board and notice the .047 uf cap. On the end of the .047 uf cap nearest the face of the amp, there is a wire going from there, all the way across the component board to a 220K resistor. You will begin

All About Vacuum Tube Guitar Amplifiers

by disconnecting the wire from the 220K resistor. Once the wire has been disconnected, you will need to solder a wire from that same end of the resistor to chassis ground. This ground connection is shown in the AB763 Layout diagram.

Note: If you aren't concerned about reversing the mod later, you could just leave the wire soldered to the 220K resistor and cut the wire about 4 inches from the solder connection at the 220K. Strip the insulation off the end of the stock wire and ground it. This will save you having to use extra hookup wire and you will save making one solder connection; but you will no longer have the original wire if you decide to put it back to dead-nut stock later.

Next, we will locate the 12AT7 reverb driver tube, which is the third preamp tube. Notice that there are jumpers connecting pin #1 with pin #6, pin #2 with pin #7 and pin #3 with pin #8. On this tube you will only disconnect the wire going from pin #2 to pin #7. Pin #7 will also have a wire coming from the board. You want to leave that wire hooked up to pin #7. You are only disconnecting the one going from pin #7 to pin #2. We will use the pin #2 to add reverb to the normal channel.

Connect one end of a 500 pf cap and one end of a 1 Meg resistor to pin #2 of the reverb driver tube. The free end of the resistor gets grounded to the chassis and you are going to stand the 500 pf cap up in the air for now; this will be a point "K."

Next we are going to connect a 10 pf capacitor across a 3.3 Meg resistor. This is a parallel connection. Twist the ends on both sides but don't solder just yet. We want to desolder pin #7 of the 4th preamp tube and connect one side of the twisted ends to pin #7. There will be a stock wire connected to pin #7; leave it there. Now solder this connection. Take the free ends of the 10 pf cap and 3.3 Meg resistor that are twisted together and stand them up in the air for now, this will be a point "K."

We are now ready to get out the shielded wire and connect the signal path. These connections are not shown on the AB763 layout drawing, but they are labeled with the letter "K." Remember the .047 capacitor that connects to pin #6 of the first preamp tube? The end of that cap closest to the face of the amp is also a point "K." We are going to connect a wire from that point "K" to the point "K" that is the free end of the 500 pf cap. Then we will connect a shielding wire from there to the point "K" that is the free end of the 10 pf 3.3M as-

AB763 "LAYOUT"

All About Vacuum Tube Guitar Amplifiers

sembly. When using shielded wire in an amplifier, you only ground one end of the shielding unless you want the amp to hum more than it should. So you will need two lengths of wire and each wire has only one end of the shielding grounded. The other end's shielding will be cut very short and the plastic jacket of the wire will be stretched over the shielding to prevent accidental shorting. If you wanted to get fancy, you could put some shrink tubing on the end that will not have the shielding grounded.

When you are finished with the above procedures, take five minutes to inspect your work carefully. Double check everything. When you are certain everything was done correctly, hook up a speaker and plug in a reverb pan and test it. Assuming the amp is working properly, put the chassis back in the cabinet and retest the amp.

Although this is a simple modification, it is not intended for the novice. If you are a novice and would like to try the mod anyway, find a friend that is handy with soldering and get him to help, or bring a copy of this article to your local technician for his help.

All About Vacuum Tube Guitar Amplifiers

WORLD CLASS VINTAGE TUBE AMP TONE FOR UNDER A GRAND

When you sound better, your playing improves and you love it even more. You want and need that great vintage tube tone, but you don't have an extra uncommitted ten thousand dollars laying around to buy an excellent condition 18-watt Marshall Bluesbreaker. And what's worse, your wife has taken a stand against you dropping five big ones on an original 5F6A Bassman. Is there any solution to get great vintage tube guitar amp tone for under a G?

There are a few promising options that can deliver the tone you crave without flattening the wallet. Specifically, there are two different types of Sears and Roebuck, Silvertone amps that could be bought for a few hundred dollars and then overhauled to be better than stock and modified slightly to correct the design deficiencies. The cost—less than a thousand dollars for everything.

My fave would be either the Silvertone twin twelve amp with two output tubes or the Silvertone Six ten-inch speaker amp with four output tubes. Neither of these amplifiers uses a rectifier tube. These amps were actually quite popular when I was growing up in the mid 60s because they were loud enough to gig with, and sold for a fraction of the price of a Fender. It was a lot of amplifier for the money back then and it still is today. Perusing eBay, I see these models plentiful and selling for the $300 to $400 range, which is about twice the new price they sold for in the 60s. So there is your "significant other" justification argument," Dear, they are going up in value! They have doubled in the last 40 years."

There were a couple of different versions of the Twin Twelve. You want the two output tube model called a Model 1484 and you do not want the four output tube model called a Model 1474. The reason you don't want the four output tube model is because it features a cathode-biased output stage that strings the filaments of a couple of preamp tubes together and connects them to the cathode circuit of the output tubes in lieu of a cathode resistor! As clever as this may seem, it is a bad design because the output tubes' current will fluctuate some, causing the filaments of the two preamp tubes to become either too hot or not hot enough, which will negatively affect the tone. An easy way to recognize the 1474 is that it has a rectifier tube and four 6L6 output tubes. This one will never sound as good as a modified Model 1484, which does not use a rectifier tube and has only two 6L6 output tubes.

In fact, besides recommending the 1484 Twin Twelve amp, I am also recommending the four output tube six ten inch speaker amp called the 1485. Both of these models can easily be spotted because neither uses rectifier tubes.

Both models were built with what looks like surplus parts to me. I say that because the power supply on both versions each use two voltage multipliers stacked on top of each other for the main B+ power supply. This is very unusual and though it will work, I have never seen it done before. There are other amps with voltage doubler circuits such as the Music Man, but never two voltage doublers stacked. My guess is the transformers were cheap and available and the designer was looking for a way to make an amp from what he could get. Both the 1484 and the 1485 use identical schematics with identical part numbers, except the four output tube model uses twin output transformers—one for each pair of output tubes.

When shopping for the amp, be aware that the Model 1485 used four output tubes connected to *two separate output transformers that each require a load at all times*. If someone isn't paying attention, it is very easy to arc the tube sockets and/or damage the transformers. There are two separate output jacks on this amp. Each output is connected to three of the six 10" Jensen speakers. It is common for someone to accidentally plug a speaker into only one output jack and forget that both jacks must always have a proper speaker load. If someone ran the amp with only one speaker jack plugged in, then the other transformer would not be connected to a load and the amp is most likely damaged. An easy way to tell if this has happened is to remove the amp chassis from the cabinet and look at pins 2 and 3 on each output tube socket. If the amp has been run without a load, there will be black, carbonized arcing marks on the output tube sockets between pin 2 and pin 3. The arcing electricity chars the socket material between pins 2 and 3, thus making it black and prone to future arcing. If the amp has marks between pin 2 and pin 3 on any output tube socket, then you can be sure that the amp was run without a load and the possibility exists that the transformer may be arcing internally. You can turn the amp up loud and play it and if a transformer is arcing, you will hear some intermittent and loud popping sounds. It is not likely the two output tube Model 1484 has ever been run without a speaker load as it only uses one output transformer.

When you get your amp, it will most likely need a cap job, which is a simple matter of changing every electrolytic capacitor in the amp. I would recommend stock values. This will cost your around $100-$150. The main filters are 100 uf at 150 volt. Since I keep the 350 volt 100 uf caps in inventory, that is what I use; but the 150 volt stock version will work fine.

The reverb on these amps sucks severely and there is no hope without extensive rewiring. You don't need it anyway. Forget the reverb. If you are recording, you don't want reverb during the performance because you are married to it. It is much better to experiment in the mix. And if you are doing live performances, there should be enough live room sound that you don't need reverb there either. But if you absolutely must have it, mike your amp and put reverb in the house mix.

Both the 1484 and the 1485 have an unadjustable bias supply voltage that won't go high enough to tame down the output tubes. This is a design flaw that must be addressed. With the stock biasing, the tubes are underbiased and idling very hot. The modifications I recommend will cool down the idle on the output tubes enough to make the amp have more headroom and better touch responsiveness. You will be able to get a clean tone from the amp—perhaps for the first time.

These modifications are too complicated for the novice, so unless you are a tech, I recommend bringing a copy of this article and the amp to your technician and have him perform the modification.

FIRST MODIFICATION—INSTALL A VOLTAGE DOUBLER CIRCUIT WITH TRIM POT FOR THE BIAS SUPPLY.

You will need a few parts to perform this mod.

2—100 ohm ½ watt resistors

2—1N4003 silicon rectifiers

2—100 uf at 100 volt capacitors

1—50K cermet element pot

1—1.2K ½ watt resistor (maybe)

Look at Figure 1. This is the stock setup before the voltage doubler mod. Look at Figure 2. This is the circuit after the modification. After the parts are installed as per the schematic in Figure 2, you will need to have the bias set for the amp. Remove the output tubes and set the

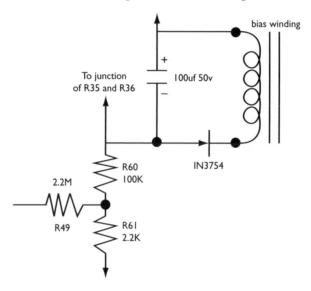

Figure 1.

cermet pot for a nominal setting so that there is approximately -48 volts from pin 5 of any output tube to ground. This will get the bias setting "in the ball park" so that when you get ready to set the bias, the tubes are not trying to run away. Once this has been set, you may install the output tubes and adjust the output tube bias with the amp on; so that the output tubes are drawing plate current of about 50 mA per side on the four output tube Model 1485 or 35 mA plate current per side on the two output tube Model 1484. If the tremolo is weak, you may need to change R61 from the 2.2K stock value to a 1.2K.

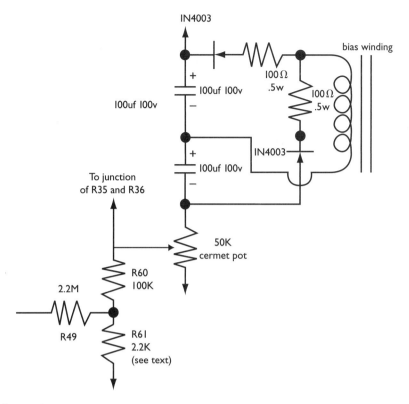

Figure 2.

SECOND MODIFICATION—IMPROVE THE GAIN

Both the 1484 and 1485 suffer from low gain. There are three places in the amp to add a capacitor to voice the amp and improve gain. Look at the schematic diagrams.

To increase the gain of channel 1, a capacitor must be placed across R 22. I would start with a small value of anywhere from .68 uf to 10 uf. If you are using an electrolytic capacitor, remember to always have the minus side going to the "grounded side" of the resistor and the plus side of the cap goes to the "tube side" of the resistor. The bigger cap will give more bottom. The smaller cap will have less boominess and more cut. You want to have good bottom, but not so boomy. This is where experimentation and listening tests with your guitar and playing style makes for the best recommendation.

In a like manner, the gain from channel 2 can be improved considerably by adding a bypass cap across R 23. You may want to

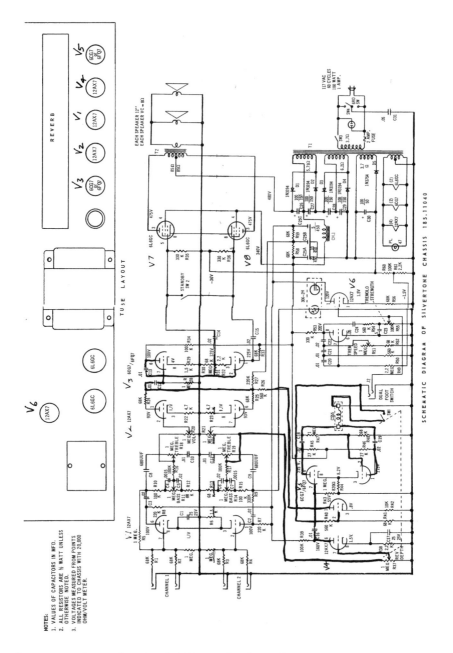

have one channel voiced differently than the other one so you will have two different sounding channels. You could even voice them differently and then bridge them together with a Y cord on the input and then blend the channels with the separate volume controls.

To simultaneously increase gain on both channels, place a capac-

All About Vacuum Tube Guitar Amplifiers

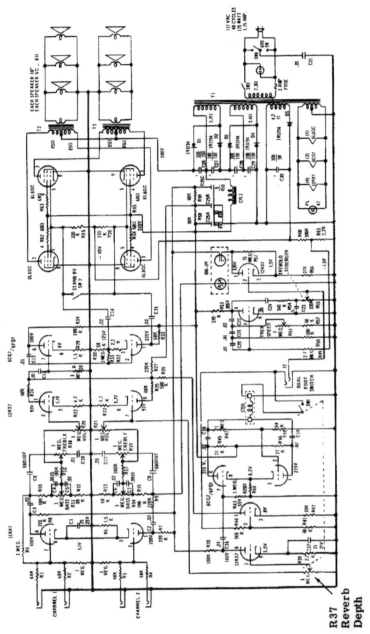

itor across R29. You can try a couple of different values and listen to how it sounds. I would start with a 10 uf at 25 volt. If it is too bassy sounding, change to either a 5 uf or a 1 uf. The 1 uf will have loads of slice, maybe even too much.

AMPLIFIER BUILDING

All About Vacuum Tube Guitar Amplifiers

BUILDING A TUBE GUITAR AMPLIFIER FROM SCRATCH

For most guitar players, the fantasy of building one's own guitar amplifier will come up sooner rather than later. If you have ever blue-printed a hot rod engine or built a motorcycle from scratch, you already know the thrill of creation. There are parts, there is you and then—there is tone. For most men, it is the closest they will ever come to actually giving birth.

Building an amplifier requires more planning than actual building. Taking one's time to double and triple-check everything as you go will assure a successful conclusion. If you are not very handy, you would benefit by enrolling a helper that is. Teaming up with a friend will give you both more confidence. You can split the cost and both will learn from each other's participation. If you are good with projects, you may not need a helper, but either way, I encourage you to take your time and do more planning than actual building. Building a homebrew tube guitar amp isn't a contest to see how fast you can finish. It is an opportunity to take your time and get it right, and learn about tube amps at the same time. Remember the A, B, C's of good amp building: **Always Be Checking.**

PLANNING

Perhaps the most foolproof way of building an amp is to build one of the many kit amps that are on the market today. As a first time builder, the kit amps will take much of the guesswork out and it will come with layout drawings and wiring diagrams.

If you are not building a kit amp, the second choice would be to build one of the early Fender tweed designs. With the early tweed amps, the schematics and layout drawings are very easy to obtain and the replacement chassis are readily available.

When I built my first amp, such was not the case and a custom-made one-off prepunched chassis costed several hundred dollars. Besides a custom chassis, one could build an amp using an old chassis, or you could even have a sheet metal shop make you a five-sided galvanized 18 gauge steel pan for a few bucks and then punch the chassis holes yourself.

If you are not building a kit amp, it is very important to make drawings of the wiring of the amp before you start building. I like to do this using a piece of posterboard. Using the chassis as a drawing template, center it on a posterboard and outline the chassis. You want to make a life size drawing. Once you get the chassis drawn, cut a piece of graph paper to represent the component board and draw all the components on this piece. You may wish to re-draw this several times until you are certain the components are spaced properly and everything fits. Since it is life size, you can place the parts directly on the graph paper to determine correct component spacing.

Once the component board design is drawn correctly, you will want to place the component board drawing onto the posterboard layout drawing and draw in the rest of the wiring connections. Look at your layout. All of the 120 volt A.C. wiring should be on the opposite side of the chassis as the input from the guitar. You don't want hum to appear on the input circuit. If you are not experienced with layout design, building an old Fender design that already has a layout drawing might be a better way to go.

Besides making a drawing, you will need a schematic and a list of parts. It is important to double-check the schematic against the layout drawing and make sure the drawing is 100 percent correct. Also, you do not want the signal path of the circuit to ever cross back over itself; otherwise you may end up with an oscillator/boat anchor and not a guitar amplifier. When you later assemble and wire the amp, this double-checked drawing will be your guide.

BUILDING THE SUB-ASSEMBLIES

Most of the time, it is best to build some of the sections of the amp as subassemblies and later drop the sub-assemblies into the chassis. For example, a component board is easily built outside of the chassis. There will be some jumper wires underneath the board and possibly some leads coming off the board. Building it first as a sub-assembly makes it easy to build and easy to check. On some amps, there may be a smaller component board for the negative bias circuit. This too, would be best built as a sub-assembly. Perhaps the amp uses a cap pan assembly which could also be made as a sub-assembly. On amps with multiple input jacks that have components wired to the jacks themselves, it is a good idea to make these as sub-assemblies. You can even use the amp chassis as a jig to hold them in place while you are building the sub-assembly. On some chassis, you could mount the jacks such that they are turned around and temporarily mounted backwards—going outside of the chassis. This will allow you to do the wiring of the jacks easily without being cramped for space. You can mount them the correct way later after the sub-assemblies are complete. So identify the sub-assemblies and build them first, before the amp chassis wiring is actually started.

WIRING THE FILAMENT CIRCUIT

When I build an amp, the first thing I install in the amp chassis is the tube sockets. The first wiring I do is the filament circuit. The filament (heater) circuit is the 6.3 volt circuit that goes from tube to tube and whose purpose is simply to provide heat for the tubes. This circuit is only used to power the heaters inside of each tube. On most schematics and most layout drawings, the filament circuit is not shown! It is assumed that you know it exists!

There are many different ways to wire a filament circuit. You could ground one side of the filament from each tube socket and daisy chain a wire from socket to socket for the other filament lead—like they did in the thirties and forties (most hum); or you could run a twisted pair from socket to socket and ground one side as was done on many smaller amps (not quite as much hum); or you could run a twisted pair from socket to socket and use either a grounded center-tap or a pair of 100 ohm resistors to make an artificial center-tap (quietest way). Regardless of which way you do it, it is better to wire the filament circuit before anything else is installed in the amp.

When I built my first amp, I didn't know how to wire a filament circuit, so I looked at a Fender Super Reverb and used it as a model.

The pair of filament wires will have much less hum if they are twisted pairs going from socket to socket. You can easily see this if you look at an existing tube amp.

INSTALLING THE SUB-ASSEMBLIES

Once the sockets are installed and the filament circuit is wired, it is time to mount the sub-assemblies. If you are using a Fender-style eyelet component board, there needs to be a plain insulator board underneath to insulate the connections from the chassis. I have also seen amps built with small stand-offs to insulate the component board from the chassis.

WIRING THE SOCKETS

At this point, you have a chassis with sockets installed, filament circuit wired, and sub-assemblies mounted. Now it is time to start wiring the sockets. Start with V1 and wire it, referring to your drawing. Then go to V2, and V3, etc. and continue until all the sockets are wired. Make sure and keep all wires as short as possible. The amp will perform better with less wire.

WIRING THE JACKS AND POTENTIOMETERS

Next, I would start at the jacks and wire them in and then wire the potentiometers. This is where you want to be very careful with grounding. If you are grounding to the back of any pot, make sure and buff or sand the surface to be soldered with either sandpaper or a Dremel tool. If you are using a brass grounding buss (recommended), then you may want to make all the ground connections to the buss before the pots and jacks are installed. You will need a larger than normal soldering iron to get a good solder joint on a brass buss. I use a 180 watt iron, like plumbers use for soldering copper pipe, for making brass grounding buss ground connections.

WIRING THE TRANSFORMERS, FUSE AND SWITCHES

You are almost done now. Install the transformers, fuse and switches and we are ready to test it. To make a neat installation with the power transformer, I like to twist pairs of secondary wires. For example, the two red wires that normally would go to the rectifier tube or the solid-state rectifier could be twisted neatly and cut just the correct length for a beautiful appearance. Likewise, the two green filament wires coming from the transformer should be twisted neatly and the two yellow rectifier filament wires also will look nice as a twisted pair cut to exact length. Depending on the actual design, there may be more than one transformer wire that needs grounding. If this is the case, twisting the wires will look good and make for a professional looking build. Make sure and use the big soldering iron for those chassis grounds.

LET'S DO SOME TESTING

Never ever plug a newly built amp into 120 VAC. If there is short, you could witness a small mushroom cloud when the transformer burns up quicker than you can realize what is happening. You must always start an amp with either a current limiter or a variac. I detail how to make a current limiter in my second book, *Tube Amp Talk*. It is basically a light bulb wired in series with an extension cord. When we use a current limiter, if there is a short in the amp, then the light bulb will light up and no damage will be done. This allows you to be forgiven for any building sins without the consequence of a smoky room and the stench of a burnt power transformer. If the amp is a fixed bias design, power up without power tubes and check to make sure there is adequate negative voltage on the grid of the output tube sockets. Install the power tubes, set bias and debug. Now plug in and rock out!

All About Vacuum Tube Guitar Amplifiers

ASSEMBLING A TUBE AMP IN 20 STEPS

When the tube tone bug bites, she bites hard. And when you are really digging that tube tone the most, sooner or later, you are bound to get the idea of building your own amp. Roll your own? Just think of the guitarists with little or no experience that have rolled their own and ended up with great sounding amps! It's true, he who cuts his own firewood really is twice warmed, and homemade tastes better than store bought!

Assuming you already have a design idea and have drawn a layout on paper, there is a sequence that works best for actually assembling an amplifier. Thinking linearly, it's as easy as one, two, three. For example, you would want to mount the transformers towards the end of the assembly sequence because they are heavy and won't actually be wired until the last anyway. The transformers will get in the way if you mount them first. You want to always wire the filament heaters before any of the other wiring is done. This way, the filament wires can be dressed close to the chassis, away from all the others and near the chassis (ground potential).

There are certainly many ways to skin a cat, but here is the sequence I use when I build an amp:

STEP 1 Begin by cutting, drilling, nibbling or punching any and all mounting holes into the blank chassis. You want to get all the metal work on the blank chassis done **BEFORE** anything else is done because you don't want to risk accidentally getting metal filings embedded into any of the circuitry, sockets or components. By punching all holes first, you can clean and deburr the punched blank chassis before installing any of the components.

STEP 2 Install the tube sockets in the pre-punched chassis. Many old vintage amps used sheet metal screws to attach the sockets. I have used sheet metal screws also, but I have also had excellent results with pop-rivets or (miniature) bolts and nuts. Sheet metal screws sometimes strip whereas the pop rivets or small bolts do not come off. The miniature bolt and nut works the best if you paint the threads of the bolt with fingernail polish after you have tightened the nut. This prevents the nut from coming loose.

STEP 3 Install Pilot light assembly (assuming you are using a 6.3 volt pilot).

STEP 4 Install your twisted pair filament heater wires. These are the 6.3 volt wires that go from pilot light to tube to tube, etc. They need to be twisted and mounted as close to the chassis as possible. Let's say for simplicity that I was using a black wire and a white wire to hook up my filament heaters. I would start at the pilot light and connect a black wire to one pilot lead and a white wire to the other pilot lead. Then I would twist the pair and go towards the output tube socket closest to the power transformer. The black wire and the white wire would each go to a different pin on the tube socket. For the sake of this example, let's use pins #2 and pin #7 as the filament pins on the output tube sockets. Those are the correct pins for most octal socket output tubes (6L6, 6V6, 5881, EL34 6550, KT66 or 7581A.) For example, I like to take the pair of wires and twist them up coming from the pilot and then have one wire, perhaps the black wire, connect to pin #2 of the output tube socket. Yes, I terminate the black wire on pin #2, but I also solder another black wire to that same pin #2 so I will have both a black and a white wire to twist as a pair. I continue twisting and work around until I get to pin #7 and then I have only the white wire connect to pin #7. Besides terminating the white wire on pin #7, I also solder another white wire to pin #7. This wire is used to twist with the existing black wire and together they make up my twisted pair. Now I begin twisting, always in the same direction and I work my way around to pin #2 of the next output tube. At pin #2, I connect only the black wire. In a like manner, I twist the black and white pair some more until I work my way around to pin #7, in which case the white wire gets soldered to pin #7. When you finish wiring the filament circuit to the output tubes, you work you way over in a like manner to the preamp tubes and go from one to the next. The heater filament pin outs are different on the preamp tubes than the output tubes, but you get the picture.

NOTE: When wiring a 12AX7 style tube (12AY7, 12AT7, 12AU7), there is something cool going on here. The 12AX7 tube is a 12.6 volt tube. So how can you use it in a 6.3 volt filament circuit? The tube has a 12.6 volt filament heater, but the heater is actually center-tapped (pin #9). One may hook up this filament in a humbucking fashion; this would require only 6.3 volts. The ends of the filament heaters are pins #4 and pin #5 and the center-tap is pin #9. You would simply hook the filament ends together (pins #4 and #5) and count that as one filament wire connection—let's say black. Now take pin #9, the center-tap, and count it as the other filament wire. Any hum induced in the first half of the filament is also induced in reverse phase by the other half. When you mix the out-of-phase hum with the in-phase hum, they phase-cancel each other, resulting in a hum-free 12AX7. This is one of the main reasons the 12AX7 is the most popular preamp tube ever used for guitar amps.

STEP 5 Build the sub-assemblies. Depending on the design, you may have one or more circuit boards, rectifier boards, component boards or capacitor boards used for the amp. It is always best to build these boards as a sub-assembly outside of the amp and then install them later.

STEP 6 Install the remainder of the jacks, switches, fuseholder, and other chassis-mounted components.

STEP 7 Install the preassembled sub assemblies into the chassis.

STEP 8 Mount the power transformer in the chassis, but don't hook up anything yet. You want the power transformer in the chassis so you can do the chassis grounds as described in the next step.

STEP 9 Do all your chassis grounds. For this, I use a large soldering iron that we affectionately call "Big Bertha." It is a 180 watt iron—like the ones used by plumbers to solder copper pipes. This will get the chassis hot enough to make a perfectly soldered chassis grounds. I like to heat the chassis up first where the solder will go but preheat it without the actual wires. If you try to heat the wires with the chassis, by the time the chassis is hot enough, the insulation on the wires will be burnt. So I heat the chassis with the iron, and after it is hot, I put the wire/wires to be soldered on the spot and reheat. This makes a beautiful ground connection.

STEP 10 Starting at the input jack, wire up everything on that panel of the chassis. This would generally include input jacks, volume control, tone controls, etc.

STEP 11 Starting at the preamp tube associated with the input circuit, wire every tube socket sequentially. When you finish the first preamp tube, move to the second and third, etc until you have worked your way completely down the line of tubes.

STEP 12 Stop and double-check everything you've wired against a schematic. You may have left off a jumper wire, or perhaps something is supposed to go to two points and you only are using one. In any case, it needs to be double-checked before you continue.

STEP 13 Install and hook up the output transformer and choke.

STEP 14 Install filter caps if you haven't already done so. Some amps have a separate filter cap pan with filter caps in it.

STEP 15 Hook up your power transformer leads to the appropriate places.

STEP 16 Install the A.C. cord. The green ground wire gets connected to the chassis. It is always best to attach that using a lug on the green wire and a small bolt and nut. Again, you would want to use fingernail polish to keep the nut secure.

STEP 17 Go back and check everything out once more. Do a really good and slow visual inspection. Look for missed solder joints, bad solder joints, metal filings, stray solder, or anything else that looks amiss.

STEP 18 Tube it up.

STEP 19 Plug the amp into a current limiter and turn it on. If you don't have a current limiter, make one! You could easily blow a $200 transformer if something is wired wrong! A current limiter will assure you that nothing will burn up. (If you don't know how to make a current limiter, there are instructions on how to make one on page 328 of my second book, *Tube Amp Talk for the Guitarist and Tech.*) You are checking for shorts. If there are shorts, the current limiter will light brightly and you will need to troubleshoot any shorts before doing any further adjustment. If there are no shorts, check the grid of the output tube (pin #5 on a 6L6, 5881, 7581A, 6V6, EL34, KT66, KT77, KT88, 6550). You would use a voltmeter and check for negative voltage between pin #5 and ground.

You want to make sure you are getting negative bias voltage. In the case of a cathode biased amp, you need not worry about bias negative bias voltage on the grid of the output tubes.

STEP 20 Adjust bias and play (hopefully). If it doesn't play, go back to Step 17 and check everything once more. A really slow visual inspection will usually find what's missing. If the amp is not passing signal, you will need to check voltages on the tubes. Make sure you have plate voltage on all the tubes. If all the tubes have voltage and there is still no signal, you may need to trace the signal path from input to output and see where it stops. Tracing the signal path can be done with a signal generator and a signal tracer. This is also detailed in *Tube Amp Talk for the Guitarist and Tech* on page 331.

All About Vacuum Tube Guitar Amplifiers

COMMON WIRING MISTAKES AND THEIR CORRECTIONS

After having serviced thousands of amplifiers, I am amazed in the mistakes of lead dress and wiring that is common in what we all consider to be popular amps: tweed amps made in the 50s, black Tolex amps made in the 60s, English amps, American amps. Aliens and either very old or very young women, often times working for sub-standard or minimum wages, wired many of the amplifiers. These people wired amplifiers per drawing specification. Assemblers, for the most part, did not have a clue about how an amplifier works, lead dress, or how it should have been wired. However, a drawing spec does not always give all the information needed.

HEATER FILAMENT MISTAKE

If you check the heater filament winding on a push/pull amplifier, the output tubes should be wired "in phase." The pins for the heater of common power tubes (6V6, 6L6, EL34) are pins 2 and 7. That is to say that pin 2 from one output tube should connect to pin 2 of the other output tube and pin 7 from one output tube should connect to pin 7 of the other output tube. If this is the case, the tubes are said to be wired "in phase." Why is phase important? Because push/pull amplifiers have the output tubes feeding opposite ends of a transformer primary. When the tubes are wired with the filaments in phase, any hum that's induced in one tube will also be induced in the other tube. Since they're feeding opposite ends of the transformer (which are actually "out of phase" in relation to each other), the hum will cancel; thus creating a quiet amplifier with a minimal amount of hum. If the heaters are wired "out of phase," the hum will be induced in both tubes "out of phase," with respect to each other. Since the tubes are feeding opposite ends of a transformer primary, then the transformer will reverse phase of one of them and cause the hum to double rather than cancel. This will cause a low-frequency hum, 60 Hertz, that will occur at all volume levels; and, of course, is particularly noticeable when the volume is turned down or off.

Here is an easy way to check the phase of your filaments. Connect an A.C. voltmeter to a heater lead of one of the power tubes (let us use pin 2 of a 6L6 in this example.) Now connect the other meter lead to the same pin of another power tube. It should read zero volts. If it reads zero volts, then your heaters are "in phase," and no other action is needed.

However, if your meter reads 6.3 volts in the above example, your heater leads are "out of phase" and should be reversed on one of the power tubes. It does not matter which tube you reverse the heater connections, but one socket must be re-wired with the heater supply leads swapped. In other words, you will want to move the wires going to pin 2 and place them on pin 7 and move the wires from pin 7 to pin 2. When you repeat the meter test, you should get zero volts.

If your amp has more than two output tubes, continue checking in a like manner, swapping wires when necessary, until all output tubes have the heaters wired "in phase." You should be able to connect your meter to pin 2 of one output tube and get of zero A.C. voltage reading when placing the other meter lead on pin 2 of the other output tube/tubes.

COMMON LEAD-DRESS MISTAKES

Perhaps you have noticed on many mid to late 60s Blackface Fender amps a certain raspy, non-musical, high-end sound when the amp is turned up. This problem is particularly noticeable when using a bridge pickup or turning the treble up on the amp. It is caused from a parasitic oscillation occurring in the amp. Usually, this oscillation is not very obvious other than the fact that the amp sounds crappy when turned up. I have noticed this on most Blackface Twins and Super Reverbs. It also occurs on Bandmasters, Pros, Deluxes and Showmans; but is particularly noticeable on amps with reverb. It is caused from improper lead dress. The grid wire feeding the first preamp tube is simply too long and it spans across too much other circuitry.

HOW DO WE CORRECT THIS PROBLEM?

I have a way that will usually correct the problem; however, it is an unconventional method. I replace the wire that leads from the input jack resistors to the grid of the first gain stage tube. There are a couple of resistors on the input jack (68K) that are bridged together. One end of the wire is attached to those two resistors. It goes under the board, comes out of the board, and attaches to pin 2 of the first gain stage tube. For the normal channel inputs, this would be this would pin 2 of the first preamp tube. For the vibrato channel inputs, the wire would terminate on pin 2 of the second preamp tube.

HERE'S HOW TO DO THE MOD:

Unsolder pin 2 of the preamp tube and unsolder the wire leading to the two 68K resistors that are mounted on the corresponding input jack. We are going to replace this wire with a shielded cable. A microphone-type, shielded cable will work just fine. The shielding, however, is hooked-up in a most unconventional fashion. On the end that connects to the input jack resistors, we do not use the shielding. We clip the shielding off and shrink-wrap it, so that it does not connect to anything on that end. The center wire of the shielded cable connects to both the input jack resistors on one end and pin 2 on the other end—just as in a stock setup. On the tube end of this wire, we terminate the shielding, **NOT TO GROUND, BUT TO THE PLATE** of that same triode section. The plate will be pin 1 on a 12AX7. We will do this on both the normal and vibrato channel input circuits.

If the problem persists, and the amp still has that raspy, non-musical, trashy sound, then we will need to carry this to the next level. The next level will be to re-wire the wire that connects the wiper of the volume control pot (center lead) to the grid of the second gain stage. On each channel there is a wire going from the center lead of the volume pot to the second gain stage for that channel. The second gain stage is going to be pin 7 of the first preamp tube for the normal channel and it will be pin 7 of the second preamp tube for the vibrato channel. This modification is done in a similar fashion as the input jack wire was done. We use a shielded cable. The end near the volume pot has the shielding clipped-off and unterminated and shrink-wrapped for safety. The center wire of the shielded cable connects to the volume pot wiper on one end, and connects to pin 7 of the appropriate preamp tube on the other end. The shielding, on the tube end of the wire, **CONNECTS ONLY TO THE PLATE** (pin 6). This should be done on both channels' circuitry.

Once these modifications have been performed, the trash when playing at high volumes and with the treble turned up should disappear and pure tone becomes available, perhaps for the first time since the amp was built.

OTHER COMMON MISTAKES

Other common mistakes are found in the grounding, improper soldering, wrong component values and in the length of wires. For starters, shorten every grid wire in the amp as short as possible. The grid wires are going to be pin 2 and pin 7 on a 12AX7-style preamp tube and pin 5 on the output tubes (6L6, 6V6, EL34).

In addition, on output tubes, many amps feed the grid with a small value resistor, either a 1.5K, 2K, 5.6K, or even a 1K. The placement of this resistor is critical to the sound of the amp. It should be mounted so there is zero lead length between the body of this resistor and the actual pin 5 of the output tube socket. Sometimes, assemblers will leave this lead length a ¼" or ½", not knowing that it makes a difference. This should be reinstalled so there is zero lead length between the body of the resistor and pin 5.

Visually inspect solder joints and components. I had an amp in my shop in which one of the input jacks had never been soldered. This amp was thirty-something years old and the input jack had a wire crimped in its lead, but had never been soldered! This was not obvious because it was hard to see unless you were specifically checking it.

COMMON GROUNDING MISTAKES

Every amplifier has filter caps as part of the power supply. The first section of filter caps will be the filtering for the output tube plates and actually feeds the center-tap of the output transformer. The second section will be the filtering for the screen supply voltage of the power tubes. Since current is entering the power tubes from the cathode, and the cathode is grounded; then both the main filters and the screen filters should ground in the same physical spot where the cathodes of the output tubes are grounded. The cathode will be pin 8 for a 6L6/6V6 or EL34 style tube.

The filter caps that feed the preamp tubes should have their grounds near the cathode resistor grounds of the related tubes they feed. This will assure that current being pulled from those tubes will not be altered/modulated by the current that is being pulled by the output tubes.

When current from the output tubes develops a signal across any chassis resistance, the result could be oscillation. This oscillation could manifest itself in many forms. It could be a low-frequency oscillation. It could sound like a vibrato, or it could be a higher frequency oscillation making the tone bad. On the other hand, you could even have such a high frequency oscillation that the amp would seem to cutout at times because the oscillation would be a higher frequency than what humans could hear.

The B+ ground (either a power transformer center-tap or the ground-side of a bridge rectifier), the main filter cap grounds, the output tube cathode grounds and the output tube screen supply filter ground are best grounded at the same point. With such a ground, current feeding the output stage would not interfere or modulate the current feeding the preamp stage.

All About Vacuum Tube Guitar Amplifiers

WOOD MATTERS IN COMBO AMPS

When I was in the process of moving from my old factory into the new factory, I noticed many small pieces of different types of scrap wood that needed to be thrown away. Since there were new employees around, I wanted to make a point of the different tonal qualities of different woods. We took similar size pieces and had each person hold one up. I went around and knocked on each piece. Everyone could easily hear the differences.

I am not an expert on wood, but I know what I hear. The difference between 80-year-old pine and new pine is astounding. I have heard from others that the trees grew differently centuries ago. In addition, the ozone layer was different as was the atmospheric conditions. Perhaps this is so. Tonally, the 80-year-old pine had a sharp focus that was percussive and defined. With the new pine, the tone was muddy and unfocused by comparison. I would even say it sounded somewhat boxy and nasally by comparison.

This reminded me of the hardwood test I did a couple of years ago while working on the Trainwreck Climax project. At that time, there had never been a Trainwreck combo amp—only Trainwreck heads. Yet, every Trainwreck was made in hardwood. The trick was to find out what hardwood sounded the best as a combo. We had heard other hardwood combo amps from other manufacturers and were worried that there were no good sounding hardwoods.

I took the tonal approach (as always) rather than the scientific. I simply had my cabinetmaker make a cabinet out of every type of hardwood he could get his hands on. Certain types of wood were excluded, because we had already heard certain types of woods from other manufacturers and knew those would not work. For example, we did not have a maple or a hickory cabinet done.

For simplicity and expense, we had the cabinets made into 1x12" cabs. When the cabinets were ready, we simply loaded them with the same type 12" Kendrick speaker, and got our Trainwreck Climax prototype chassis out and began listening. As a control, we had one cabinet made from pine, mainly because that would give us some type of reference to compare.

Although this testing was expensive, I learned a lot about wood that I would not have known by any other way. I learned there are certain woods that should never be used for combo amplifiers. For example, I would never build a speaker cabinet from mahogany or lacewood. Those cabinets were the worst. They were muddy and blurry, with no definition. Boxy and nasally, they defined bad tone. You could have had a great sounding circuit and those cabinets would have ruined the overall tone.

Some woods actually sounded slightly better than pine. For example, the poplar sounded better than pine, as did the willow. I found out later that Buzz Feiten makes his famous Feiten cabinets from poplar plywood (we used solid wood in our testing). The poplar was by far the ugliest wood I have ever seen. Its dull greenish tint with no figuring defined drab. The willow was nice, both tonally and aesthetically. I do not know why no one has used willow for a cabinet. It sounded great (much better than pine) and was very pretty.

After testing the beautiful lacewood, we were thinking the prettier the wood, the worst it probably sounds. There was one wood I was reluctant to try because it was so pretty. It was an exotic wood from Brazil called Canarywood. No wood this beautiful could sound great, I was thinking. Actually, this stuff was prettier than the lacewood. Yellow with reddish/burgundy ribbons running through it made it the eye-popper of the bunch. We had tested all the other cabinets, so the time had come to test this obscure wood from the Brazilian rainforest. When the sound came through that cabinet, the three of us in the room burst out laughing. Could this be for real?

The canarywood was at least 3dB louder than any other wood tested. The focus of this wood was uncommonly clear. Every note in the chords was well defined and clearly discernable. Even with closely voiced chord (such as major 7ths and minor 9ths), every note was clearly focused. We did notice that the canarywood was much denser than the other woods. For example, the same cabinet in canarywood weighed eight pounds more than the identical pine cabinet. My theory is that the wood is so dense that it doesn't allow the note to break down into sub-harmonics. These sub harmonics would have a tendency to muddy up a tone. Perhaps this is why the 1x12" open-back cabinet made from Canarywood had the punch of a 4x12" closed-back. I theorize that the energy that came out of that speaker went to reproducing the fundamental frequencies coming from the speaker instead of breaking down into sub-harmonics and thus diffusing the punch.

Go with me to 1999 when I introduced the 2410 Ten Year Commemorative limited edition amplifier. I had a chance to obtain some 80-year-old pine and thought it would be perfect for building a vintage correct amplifier. There are many people building vintage style amps, but for some reason the wood has been overlooked and taken for granted. A piece of this 80-year-old wood looks almost petrified. It is so hard, that it will ruin a brand new planer blade after building about five cabinets. It is so hard; you can hardly sand it. If you try to sand it, it will not sand evenly. It actually comes out looking like it was sandblasted, rather than sanded.

This wood was compared to new pine. Oddly enough, the 80-year-old pine had clarity and punch very similar to the canarywood. Antique pine was much louder than new pine—just like the canarywood. Although pine is considered a softwood, when it is very old and dried out, it performs in a guitar amp more like a hardwood. Can time transform softwood into a hardwood? It appears so in our sound testing.

The antique pine has a sharpness and clarity that is impossible to find in new pine. I am not talking about a subtle nuance that takes Eric Johnson an hour to discern. My tone-deaf sister-in-law could hear what I am talking about on the first try. With new pine, it seems like the energy from the note is divided to make the fundamental and several muddy sub-harmonics. The edges of the notes are blurry—lacking definition. With the antique pine, it seems like all the energy is focused into making the actual fundamental frequency it is given. This sounds better by itself, but is even more noticeable when playing in a band. The antique pine cuts through a mix much better than the new pine. Oddly enough, the canarywood and antique pine were very similar.

In a way, I wish I had not found this out. Once you know something like this, you cannot just keep on doing the same old things. Now I will have to go to Brazil and find an 80-year-old house built of canarywood. I guess everyone needs a challenge.

TROUBLESHOOTING

All About Vacuum Tube Guitar Amplifiers

TROUBLESHOOTING TUBE AMPS WITH A VARIAC

Sometimes it becomes necessary to use unconventional troubleshooting techniques in order to achieve a particular goal. For example, the other day, I had a Vox AC50 in my service department. Besides the amp needing a basic overhaul, the amp would not come on. There was no power getting to the amplifier even though it was plugged in and turned on. I traced the voltage coming from the wall and I was losing it in the voltage selector switch. On vintage Vox amplifiers, there is a voltage selector to choose between 115, 160, 205, 225, and 245 volts. Since this particular voltage selector switch was broken and not making contact, there was no easy way of knowing which leads were which. I was going to bypass the switch electronically, but leave it installed for cosmetics. So the question is: How do I know which primary lead is which? I could have gone the labor-intensive approach and dismantled the selector switch. There must be a better way.

Remember, when a power transformer is designed, it has a particular number of turns on the primary that equates to so many volts per turn. This "volts per turn" is used to determine how many turns need to be on each secondary in order for the voltages to be correct on the secondary. Generally speaking, all the windings of the transformer will have the same number of volts per turn but it will be the number of actual turns per winding that determine the actual voltage that develops on each winding.

FINDING THE VOLTAGE TAPS OF A POWER TRANSFORMER

So here is the solution. I unplugged the amp and removed the output tubes from the amp so there would be almost no load on the transformer. I attached a Variac to the 6.3 volt filament heater terminals on one of the output tube sockets (#pin 2 and pin #7 as this amp uses EL34 tubes). I carefully adjusted the Variac to 6.3 volts. Then I got an AC voltmeter and measured from the one primary lead of the transformer that terminated at the switch, to each of the primary taps that were terminated on the voltage selector. When I found the lead on the voltage selector that measured 115 volts, I knew it was the one I needed. I then hooked that lead to the power switch and bypassed the selector. I left all the other taps on the voltage selector as they were except the 115 volt winding.

A transformer doesn't know the difference between a primary and a secondary. All it knows is turns ratio and volts per turn. By hooking a 6.3 volt Variac source to the 6.3 volt winding, the transformer operated in reverse and the secondary was used as a primary and the primary as a secondary. Actually, with a Variac rated for enough current, I could have replaced the tubes and actually played the amp with the Variac hooked up to the 6.3 volt filament winding and the regular primaries not connected to anything!

CHECKING AN OUTPUT TRANSFORMER FOR A SHORTED PRIMARY

You can perform this simple test to check a push-pull output transformer to make sure it's primary has no shorts. Let's say you are working on a push-pull type amp that uses two 6L6 or EL34 output tubes. You suspect something is wrong with the output transformer and you want to check it out. Turn the amp off and remove the output tubes. You are going to connect a variac to the output of the amp. This is actually the secondary of the output transformer. In most cases, if the amp has an output jack, you will need to plug a ¼" plug into the output jack to defeat the shorting switch on the output jack. Most amps have a shorting jack for the output jack as a safety feature. You want your variac connected directly to the output of the amp. You will then set your Variac to 1 volt.

Next, we are going to connect the meter across the primary of the output transformer. The primary will terminate on the plates of the output tubes (pin #3 on a 6L6, 6V6 or EL34 style tube.) Take your meter and place one lead on pin #3 of one output tube and place the other lead on pin #3 of the other output tube. Your meter is set for A.C. voltage. With the variac still putting the 1 volt into the speaker jack, you will get a reading on your A.C. voltmeter. Write this reading down. You will need this figure later.

Next, you find the centertap of the output transformer and move one of the meter leads from pin #3 of an output tube socket to the center-tap. The voltage measured between the center-tap and pin #3 should equal exactly half of the voltage you wrote down earlier. If the voltage is not exactly half, then some of the turns on the primary are shorted.

Next, double check yourself by repeating the test but with the other lead going to the centertap and using pin #3 of the other tube socket. Each side should be exactly half of the "plate to plate" voltage for the tranny to test as good.

FINDING THE IMPEDANCE OF AN UNKNOWN TRANSFORMER

Sometimes it becomes necessary to test a transformer for primary impedance. Suppose you have a Showman chassis and you aren't sure if it is a Showman (8 ohm) or a Dual Showman (4 ohm). Or let's say you attend a garage sale and they are selling a box of output transformers for $2, but there is no information about the impedances on them.

By performing a few tests, one can easily figure out the primary impedance of an output transformer. For starters, most push-pull output transformers have 5 or more wires. The common color code is brown and blue for the ends of the primary, red for the centertap of the primary, black for the common lead (ground) of the secondary, and some other color for the hot lead of the secondary. If there are other secondary taps for different speaker impedances, you will have those wires too.

By using an ohmmeter and checking for continuity, one can easily, by trial and error, determine which wires go with each other on a winding. For example, if you find continuity between two wires then those two will either be the ends of a single winding, or taps associated with that winding. So you want to use that ohmmeter and determine the primary and secondary. Again, the primary will almost always be brown and blue with a red centertap. The secondary will usually consist of the rest of the wires. Primary wires are generally much smaller in diameter than the secondary wires because remember, the secondary is very low impedance which means there is mucho current flowing. The primary, being very high impedance, has low current and therefore smaller wire is used.

If there are multiple wires for the secondary, you need to know which two wires have the largest D.C. resistance between them. One of these will be the common and one will be the tap for the highest speaker impedance output.

Once you determine which two wires have the highest D.C. resistance on the secondary, you want to connect your Variac to these two wires and set the variac for 1 volt. If you are performing this test to a transformer in an amp, make sure you defeat the shorting switch on the output jack if there is one.

Now, using your A.C. voltmeter, check to see the voltage on the ends of the primary. Simply ignore the centertap and go plate to plate (usually brown to blue). If you are checking a transformer in an amp, remove the tubes and attach your meter on the plate lead of the output tube sockets (pin #3 on a 6L6, 6V6 or EL34). The ends of your primary are already soldered to these leads so it is convenient to use the socket leads as a connection for your meter.

When you get the A.C. voltage reading from your voltmeter, this number will be the "turns ratio" of your output transformer. In order for this reading to be accurate, the variac must be putting exactly one volt into the secondary of the transformer. You can take your meter and connect it to the secondary to verify that the variac was putting out exactly one volt. If the variac was not putting out one volt, adjust it to exactly one volt and repeat the test.

Now you have the turns ratio. There is a formula for determining primary impedance. You will square the turns ratio and multiple by the intended output impedance. This will give you the primary impedance.

For example, you want to know if your Showman is 8 ohm or 4 ohm. You do the test by putting 1 volt on the output of the amp and find your transformer primary has 15.3 volts A.C. on it. So you know the turns ratio is 15.3 to 1. You square the turns ratio and get 234.09. You assume the output impedance of 8 ohms and multiply the 234.09 by 8 and get 1872 ohms for the primary. Fender operated their two output tube amps at around 1875 ohms so you know you have an 8 ohm output.

You could have multiplied the 234.09 by 4 ohms and gotten 936. This is where some common sense applies. No one every runs four 6L6s at 936 ohms, so you know it is not a 4 ohm transformer.

Fender usually used around 4,200 ohm primary on a pair of 6L6s, around 1875 on a quartet of 6L6s, and about 6,600 ohms on a pair of 6V6s. Marshall was about the same or slightly less. They would use around 4,000 ohm primaries on the EL34 amps with two tubes and perhaps 2,000 ohm on four output tubes.

In the case of the unknown taps on the secondary, check the D.C. resistance to the ends. The wire with the least resistance to one end is the first tap and that associated end is the highest impedance tap. The remaining lead will be the smallest tap.

All About Vacuum Tube Guitar Amplifiers

TUBE AMP FIRST AID FOR THE WORKING GUITARIST

When you are playing in a band, there are times when your amp may need attention and it is not possible or practical to bring it to a tech. You need the amp to perform right now. In the same spirit that a automobile driver should know how to change a flat tire, or jump start the engine, there are certain servicing and troubleshooting tips that every guitarist needs to know. For example, you get on stage, turn your amp on and there's no sound. Now what? Or perhaps the amp starts making noises half way through the first set. Or during a song, the fuse in the amp blows. Do you know the correct responses to get the amp working normally?

We are going to give you the correct responses, but first we need to be prepared for the gig. Get yourself a small briefcase or gym bag and put together your own tube amp first aid kit. Bring it to every performance. Certainly, if you play a guitar, you most surely bring extra strings to every performance. A string could break for no apparent reason and so could your amp. Here are suggestions on what to put in your first aid kit.

FUSES — Keep a box or more of the correct type, size, amp rating and voltage rating (For example: Slo-Blo, Littlefuse, 3 amp, 125 volt fuse). If you need more than one kind, get a box or two of each.

OUTPUT TUBES — I like to keep at least one set that is matched to the set in my amp. This way, if a tube malfunctions, I can replace the set without having to bias. This is really convenient in a gig situation.

PREAMP TUBE — One or two extra preamp tubes just in case one goes noisy or microphonic.

RECTIFIER TUBE — If your amp uses a rectifier tube such as a 5AR4, bring an extra as a backup.

PHILLIPS SCREWDRIVER — Many minor repairs can be made with this device.

EXTRA 9 VOLT BATTERIES—For testing speakers, speaker polarity, or footpedal replacements.

NEEDLE NOSE PLIERS—Many minor repairs can be made with these.

MULTIMETER—This can be used for simple troubleshooting, biasing, or testing.

GROUND-LIFT ADAPTERS—These are used to stop ground loop induced hum.

ADDITIONAL PATCH CORDS—If a patch cord malfunctions, you don't have time to fix it. It is best to have several extras—just in case.

FLASHLIGHT—Stages are dark. The inside back of an amp is also very dark.

ANYTHING ELSE—That you feel you must have that I have not included in this list.

THE FUSE BLOWS—You are playing a gig and your amp stops playing. A fuse has just blown. What do you do? Get your box of fuses out of the first aid kit because you may need a few to troubleshoot the amp. Make sure and use the correct type and amperage rating. Most amps use Slo Blo fuses. If your amp uses Slo Blo fuses, do not use regular fuses as they will blow all the time for no real reason.

It is important to know that when a fuse blows in an amp that has otherwise been working fine, in almost every case, **either a power tube went bad or there is a problem with the rectifier.** (The rectifier may be a tube, but in some amps, it will be a diode.) When a fuse blows, it is NEVER a bad fuse!

Here is the quick way to troubleshoot.

Remove both (or four depending on the design) output tubes

Replace the fuse with the correct value and type

Turn on the amp and after it warms up, put it in the play mode.

IF THE FUSE DOES NOT BLOW AT THIS POINT, one of the output tubes is probably bad. To find out which one, simply replace **only one** of them in the amp. Turn on the amp. If the fuse blows, that tube was bad. On the other hand, if the fuse does not blow, it is probably the other tube. (You could verify this by replacing the other tube and see if the fuse blows.) We removed the output tubes and replaced them one at a time to see which one caused the fuse to blow.

IF THE FUSE BLOWS WITH BOTH (OR FOUR DEPENDING ON THE DESIGN) OUTPUT TUBES REMOVED FROM THE AMP, MOST LIKELY YOUR RECTIFIER TUBE OR RECTIFIER DIODE IS BAD. If you have a rectifier tube, remove it from the amp, install a new fuse and turn on the amp again. If the fuse still blows without a rectifier tube and without output tubes, it is almost a sure bet your power transformer is bad. Of course, if removing the rectifier stopped the fuse-blowing problem, then replace the rectifier tube with a fresh tube and you should be good to go.

IF THERE IS NO RECTIFIER TUBE AND THE FUSE BLEW AFTER REMOVING BOTH OUTPUT TUBES, THEN THERE IS A GREAT CHANCE THAT ONE OR MORE OF THE RECTIFIER DIODES ARE SHORTED. Although it would be difficult to change diodes at a gig, that would be the next step. If you are troubleshooting and you do not have any diodes handy for replacement, you could simply remove the rectifier diodes from the amp and replace the fuse. Turn on the amp. If the fuse blows, there is a 99.99% chance that your power transformer is bad. If the fuse does not blow, replace the diodes with new ones and that should fix the problem. When replacing diodes, I like to use better ones than stock. Most manufacturers use 1N4007 diodes that are rated at 1 amp and 1000 PIV (peak inverse voltage). I like to use either a 1N5399, RL207, RL257, or 1N5408. These are the 1.5 amp, 2 amp, 2.5 amp and 3 amp versions respectively of the 1N4007. Remember, diodes are heat sensitive, so using one rated for more amperage will be more dependable—especially when the amp is heated up hot enough to barbecue some beef ribs.

NO SOUND—You turn on your amp with the cabling correct, and there is no sound! You know there is power, because the pilot lamp is shining brightly. Knowing what to do at this point could possibly save you an output transformer! Here is what to do:
STOP PLAYING!
Serious damage can occur to a tube amp if the amp is played with a disconnected speaker. If you operate the amp without a speaker correctly hooked up, then it is likely that arcing will occur—possibly in the output transformer and possibly across the output tube socket. Never play through an amp that is not producing sound, unless you want to run up a huge repair bill.

CHECK ALL CONNECTIONS, ESPECIALLY THE SPEAKER. Naturally, the "no sound" problem could be with the guitar or cabling, so we want to check this. Also, make sure the speaker is connected properly. This includes looking at the lead on the speaker itself to make sure BOTH wires are connected to the speaker. This also includes checking that the speaker is plugged into the speaker output jack and not the external speaker jack. (Note: Many amps have the main speaker output jack shorted with a switch to protect from accidentally running the amp without a load. That is why on Fender Blackface amps, for example, the speaker jack must be used. This prevents the shorting jack from shorting the output.)

LOOK AT THE TUBES. Are they all in the socket? Sometimes a tube will fall out of the socket during transit to a gig. Do a visual check to make sure all tubes are securely in the socket.

CHECK TO MAKE SURE THE SPEAKER IS RESPONDING. You could touch a 9-volt battery to the speaker plug (or speaker leads as the case may be) to see if the speaker is responding. If there is a ¼" plug that plugs the speaker into the amp, try unplugging this and touch the battery's + and - terminals to both the tip and sleeve of the plug. You should hear a "pop" sound coming from the speaker when the battery is connected. If there is no sound; then the plug, the speaker cable or the speaker is bad.

If there is no "pop" when the battery touches the tip and sleeve, try putting the battery directly on the speaker leads. (There are two leads on the speaker and two on the battery. In this case, polarity makes no difference. We are just checking to see if the speaker is responding.) From this test, you should be able to determine if the speaker or the plug is the offending problem.

A NOISE DEVELOPS — You are playing along then unexpectedly... snap crackle pop. You think it is a tube but which one?

IF THE NOISE OCCURS WHEN YOU ARE NOT RUNNING A GUITAR THROUGH THE AMP, you can troubleshoot (using the omission technique) to find which tube is noisy. Start with the amp "on" and in the "play" mode. Begin removing preamp tubes — one at a time — starting at the preamp tube closest to the input jack. When you remove the first tube, notice if the noise was affected. If not, try removing the next tube. Continue removing and listening until the noise stops.

IF THE NOISE IS STILL THERE AFTER REMOVING ALL THE PREAMP TUBES, the noise is not coming from the preamp stage. It is coming from the output stage.

WHEN A PREAMP TUBE IS REMOVED AND THE NOISE DISAPPEARS, then either that tube is responsible for the noise, or the noise is coming from the circuitry—in the circuit before that particular preamp tube.

IF THE NOISE OCCURS ONLY WHEN YOU ARE RUNNING THE GUITAR THROUGH THE AMP, you can troubleshoot (using the substitution technique) to find which tube is noisy. To do this, take a known good preamp tube and substitute it with another preamp tube in the amp. Hopefully, the preamp tubes will be the same type, in which case you could take the one pulled and use it to substitute in the next socket until the offending tube is found.

BUYING A
VINTAGE AMP

All About Vacuum Tube Guitar Amplifiers

10 THINGS TO CONSIDER WHEN BUYING A VINTAGE AMP

Occasionally the opportunity exists to acquire a vintage amp for much less than market value. We have all heard of the farmer story. Farmer Brown kept his '55 Telecaster and low power Fender Twin in the closet for 49 years and he simply wasn't going to sell unless he gets at least $500 for the pair. Yes, he wants to see if you really want it or not! Or we have all heard the "pawnshop story" or the "mother of the deceased Viet Nam vet story' where a nice amp went for small dollars. When such an opportunity exists, it may not matter if the amp is exactly what you are looking for, you can always trade up for something else, or sell it to a friend that did not get such an opportunity. Yes, when the deal is sweet enough, you want to buy it regardless of whether you need it or not.

But if you are looking for a specific amplifier, and it is not that "give away" pricing we all would enjoy having, then there are ten points to consider.

1. **WHAT WILL THE AMP BE USED FOR AND DOES THIS AMP MEET THOSE CONDITIONS?**

Everyone has his or her own circumstance. You may simply be a collector and it may not matter how loud the amp plays, but if you are a player, you will want to consider where you normally would be playing this amp. You may be gigging in a small club, you might be playing at home, and you might be recording. Certainly a 5 watt Gibson GA5 will most likely not be loud enough if you are doing small club gigging. Sure, you could mike any amp to get whatever volume, but if you were playing at club volume, an amp that is loud enough by itself to play a club would be more desirable than an amp that *must* be miked. On the other hand, you would not want to purchase an 80 watt Fender Twin to play in an apartment.

If you need reverb and the amp in question does not have reverb, then you must decide if you want to invest in a stand-alone reverb or look for an amp that includes reverb.

2. WHAT KIND OF SOUND AM I GOING FOR? AND DOES THIS AMP SOUND LIKE THAT?

Different vintage amps offer different tones, depending on the design, speaker configuration and circuitry. For example, if you are going for a quick attack and fast response, you are not going to find it in a 5E3 Deluxe. The Deluxe will compress off the front the note and on fast passages, the amp will not be able to keep up with the player. If you are going for a clean sound, you will need an amp with enough headroom for it to "hold together" at the volumes you will be playing it. On the other hand, if you want more breakup, you might choose a lower powered amp that has to be dimed or almost dimed to get to the volume you require. This will maximize break up.

3. WHEN WAS THE LAST TIME THE AMP WAS OVERHAULED?

Every 6 to 10 years, a tube guitar amp must be overhauled. This involves replacing all the electrolytic capacitors and any other components that are drifted out of tolerance. If the amp is in disrepair, one must consider what it will cost to overhaul the amp to determine the real cost of buying the amp. If the amp is in need of an overhaul, perhaps the seller could be negotiated down to compensate for the expense required to put the amp in tiptop condition. A typical overhaul costs anywhere from $100-300 plus the cost of any tube replacement.

4. ARE THE TRANSFORMERS ORIGINAL AND WHAT IS THEIR CONDITION? SPECIFICALLY, ARE THE TRANSFORMERS RUSTY?

Non-original transformers will devalue a vintage amp, so you will want to look at the transformers and determine their originality and condition. It is very important that the transformers are not rusty. Rust causes interconductivity of adjacent laminates that results in eddy currents that rob a transformer's efficiency. This can also lead to the transformer overheating and self-destructing. Any amplifier with non-original or rusted transformers should have its price adjusted accordingly.

5. WHAT IS THE CONDITION OF THE SPEAKER?

Is the speaker original? Is the speaker blown? Has it been reconed? All of these things affect the value. It could cost $60-$200 to recone a blown speaker. JBL speakers cost about $200. A Bassman or Super Reverb with four blown speakers could cost quite a bit to replace or re-cone. It is fine if you don't have an original speaker if the speaker in it sounds good and the selling price reflects the lack of an original speaker. You don't want to pay top dollar for a vintage amplifier and then find out the speaker is a blown, reconed, or non-original. Of course, if you buy the amp to play and it has an original speaker; you may want to store the original and replace it with a good sounding new speaker. This will prevent you from blowing the original and devaluing your vintage piece by several hundred dollars. It is best to store the original speaker so you can wail without fear of blowing the speaker. If and when you ever decide to sell the amp, you can offer it with the original speaker—thus retaining the amp's value.

6. WHAT DOES THE CABINET LOOK LIKE, COSMETICALLY?

Certainly a pristine condition cabinet is worth more than a road dog, but there are other problems with cabinets that should be considered. For example, it is common on tweed amps for the wood to crack between the top of the sides and the mounting hole on the sides of the top. Sometimes it will be cracked with the tweed still holding it together. This can be repaired, but it is tricky and generally not a job for amateurs. The tweed would have to be carefully heat-gunned off the top; the wood would need to be glued and clamped, and then some hardwood dowels would need to be installed horizontally for re-enforcement.

The baffleboard should also be inspected to make sure there are no cracks. A cracked baffleboard will rattle at loud volumes and to replace one will usually require replacing the grill cloth too.

7. WILL THE AMP NEED TO BE RE-TUBED? AND WHAT TYPE OF TUBES DOES IT USE?

Most people buy a tube amp because it has great tone. If you have a tube amp with weak or used up tubes, you must replace them to get the tone you desire—otherwise there is no point. New tubes can cost mucho dineros to replace—especially if you use new old stock. If the amp is equipped with great sounding American old stock, consider that a huge bonus because it would certainly cost a bundle to buy new ones and replace them yourself. I have seen dealers remove the good tubes from an amp and replace them with "pulls" just before selling it. This will affect the deal in their favor. You end up paying much more for the amp than you thought when you replace the "pulls" with something better.

8. IS THE AMP MISSING THE LOGO, TUBE CHART, SERIAL NUMBER, ETC.?

With certain vintage logos selling for $50 to $150, is it any wonder that some unscrupulous dealers will remove the logo from a vintage amp before selling the amp. They can then turn around later and get money for the logo.

I would be very suspicious of any vintage amp that was either missing the tube label or had an altered serial number.

More than once, I have seen counterfeit cabinets and even counterfeit amps. If you buy a 1958 Bandmaster and it is missing the tube label, perhaps the cabinet is a repro and the price should be negotiated accordingly.

9. WHO IS THE PERSON SELLING THE AMP?

Perhaps we should have made this the #1 thing to look at. When you buy anything from anyone—especially through a non-personal medium such as ebay, mail order, or phone—there must be a level of trust. In some cases, you are buying sight unseen. Reputable dealers have a good healthy selfish reason to be honest with you; they want to sell you something again next month! Dishonest dealers or individuals may not be as motivated. On ebay, there is feedback, but one must be careful to scrutinize the feedback. I once saw a scamster on ebay with several hundred transactions and a 100 percent feedback rating. Something just didn't add up (a U-47 Telefunken Neumann Microphone in mint condition for a $1,500 buy it now. It was a private auction and you had to email the seller first. When you emailed him, he claim to have been ripped off in previous ebay transactions and wanted a Western Union Wire transfer immediately. And then he would let you "win" the auction.) I looked really close at the feedback and eventually discovered the guy had bought several hundred very small purchases in the last month or so from various people on ebay. Each item was $1 or so, so it cost him two or three hundred to get himself a 100 percent feedback rating showing several hundred transactions. Upon further checking, he had never actually "sold" anything on ebay.

10. **WILL THE AMP HAVE THE RESALE VALUE TO JUSTIFY ITS PRICE AND THE MAINTENANCE COSTS?**

When purchasing a killer piece, you may think, "This is a piece I would never sell"! Never say never. If you ever do sell it, wouldn't you like to know your vintage amp retained its value or even appreciated in value? This is where you want to be judicious. For example, you would not pay top dollar for a Premier Stand-alone Reverb unit, spend $300 overhauling it and expect to sell it for what you have in it. That unit was never that popular and though cool, it would never be worth big bucks. On the other hand, getting into a used Blonde 6G15 Fender Reverb for $700 and spending another $300 for an overhaul will be a rock solid investment and will not be worth less later.

I had a guy send me a flood damaged 5F6A Bassman that he bought for $500 and had it completely restored for another $1,000. His investment will surely be protected for many years to come and it will probably appreciate in value.

This "appreciation in value" is a good term to master, as it is a logical argument to share with your "significant other" concerning your amp acquisitions. I have used this on my wife for years and she falls for it every time. I am still amazed at how much money we "earned" on our investments by me buying more amps!

All About Vacuum Tube Guitar Amplifiers

EVALUATING A VINTAGE TUBE GUITAR AMP

Whether we were sizing up a potential acquisition or simply admiring a friend's amp, most of us have had the occasion to look at unfamiliar vintage tube guitar amplifiers from time to time. The question comes up: What am I looking for? Certainly when evaluating an electric guitar, one would inspect the neck, notice what type of pickups, style of neck joint, manufacturer, etc.; but with amplifiers, the inspection procedure may not be as obvious.

When initially inspecting an amp, start out by identifying the amp. What is the brand? What is the model? Who manufactured it? Perhaps it has a logo or label. What is the model number? Is this an amp you recognize? Does it resemble any other amps that you are familiar with?

There were many amps that were manufactured by one company and sold by another under a private label. For instance, Valco Manufacturing made amps for Gretsch, National, Supro, and others. They were made by Valco, but had the logos of other companies on them. Marshall made Park amps. Danelectro made many Silvertone, Sears and Roebuck guitars and amps. Montgomery Wards and Western Auto also had their store brands that were made by other manufacturers.

When I look at an amp that is unfamiliar, I like to look at the tubes. There will be preamp tubes, power tubes, and usually a rectifier tube. The tube compliment will give you a good idea about wattage. For instance, an amp that has one preamp tube, one 6V6 output tube and one 5Y3 rectifier is going to be around 5 watts. It will never be a loud amp. A pair of 6V6s will usually give you from 15 to 25 watts. A pair of 6L6s will be anywhere from 25 to 40 watts. On the other hand, an amp with four 6L6s is probably going to be 80 or more watts and will be very loud. Also of interest, does it have a tube rectifier? Generally, smaller wattage amps use 5Y3 or 5V4 rectifiers while more powerful amps use GZ34/5AR4 or 5U4 rectifiers. Higher wattage amps usually have a solid-state rectifier.

While you are looking at the tubes, what kind of tubes are they? Are they American-made old stock or current foreign-made tubes? What type of tubes does the amp use? Eventually, every tube amp will need new tubes and when that day comes; will you be able to obtain the tubes easily? For instance, an amp that uses 6L6s will be fairly easy to obtain replacement tubes in the future. But let's say the amp uses 7591 output tubes. Where are you going to get 7591s and at what price? Obscure output tubes that are not in current manufacture may be nearly impossible to find and very expensive, if and when you find them. If you get a bargain on the amp, but spend four hundred dollars to re-tube it with obscure NOS American tubes, is it worth it? Tube types that have not been manufactured in years would include the 7027A, 7591A, 6973, 6EU7, 6SC7, 6SL7, 6SN7, 6146B, 6BK11, 6K11, 6C10, 6CG7, 12DW7, 6U10, 6D10, 6SJ7, EF86/6267, ECF82, 12AY7, and 12DW7. Common, current day manufactured tubes would include the 12AX7/ECC83, 5881, EL34, 5AR4/GZ34, 5U4, 6550, 6L6GC, 12AT7, and 6V6. These tube types will probably be very easily available for many years.

Next, let's check out the speaker. Does it look original? Can you identify the brand? If the speaker is in bad shape, it will cost another forty-five up to as much as two hundred dollars have it reconed with correct parts. Certainly a four ten-inch speaker amp with bad speakers could cost several hundred to get the speakers sounding right. Likewise, an amp with a couple of JBL 12" speakers could cost as much as four hundred dollars to have them both reconed with correct parts. All of this must be considered, because if you get a wonderful deal on the amp and have to spend several hundred dollars on speakers, you must consider any additional expenses when evaluating the amp's worth.

Avoid amps with field-coil type speakers, unless the speaker sounds really nice. Field coil speakers cost mega-bucks to have repaired. A coil rewind will probably be two hundred or more and then a re-cone is additional. Not only that, correct parts for the re-cone are nearly impossible to find, so when you do find them, the price of the re-cone will be fairly expensive. There are no current manufactured field-coil speaker replacements. You could have it converted to a permanent magnet speaker, and this will involve a slight re-design of the power supply. Any way you look at it, a field-coil speaker that sounds bad is going to cost mucho dineros to get happening.

Now it is time to look at the transformers. Are the transformers original? Are they rusty? Rusty transformers are not desirable. Rusty transformers can have eddy currents, which zap power and produce excessive heat. Having a transformer rewound can also cost a couple of hundred dollars or more. If you can get to them, try smelling them. If the transformer smells burnt, it is either bad or going bad. New transformers are difficult to find, and depending on the exact transformer, very expensive to rewind. For instance, an Ampeg B15N power transformer is no longer available. These transformers are potted in epoxy. Therefore, if you get it rewound, the rewinder will have to soak the transformer in a special solvent for a week to dissolve the epoxy before the transformer can be rewound. To make a long story short, you could spend three hundred dollars to have it done right. In such a case, a substitute transformer could be used, but the amp's value will go down as well.

Lastly, let's listen to the tone. Is the amp noisy? Loud crackling could alert you to possible internal arcing. Arcing can occur anywhere, but arcing in the transformer will cost the most to repair. Tube socket arcing is easy to repair, but sounds exactly like transformer arcing. Play a "B flat" on your "D" string and listen for unnatural harmonics. Play the chromatic higher and the one lower while listening to the harmonic. If it sounded "out-of-tune" or unnatural, then the amp probably needs a cap job. This could cost anywhere from eighty to two hundred dollars—depending on the actual amp and how many capacitors it has in it.

If the amp has a lot of distortion at a low volume, it probably has one or more bad coupling caps. These are not very expensive to replace, but be informed that the distortion characteristics will change when the bad parts are replaced.

Sometimes the bad components are responsible for the sound. In other words, if you like the distortion characteristics of an amp, and that tone is achieved because of leaking coupling capacitors, worn out tubes, and low voltages from leaking filter caps; then don't expect to have that same tone when the components are replaced with non-defective parts. On the other hand, if the defective parts are left in the amp, be aware of the risk associated with leaving those defective parts in the amp. For instance, you may have several filter caps that are leaking so badly that all of the supply voltages are extra low. The extra current that is leaking from these defective caps puts excessive stress on the power transformer. This extra current could overheat the transformer, thus risking burning it up! This condition makes the tubes have less gain, but more distortion and a darker overall tone. Replacing those defective components would increase gain, dynamics, clarity, high-end, and reduce compression.

And finally, don't be afraid to spend a few bucks to restore a vintage amplifier. Virtually all vintage tube amps are point-to-point wired with paper bobbin transformers. Modern day tube amplifiers with point-to-point wiring and paper bobbin transformers can easily cost a couple thousand dollars and even more. So, if you can get a vintage amp that you like and spend another few hundred to get it back to tip top condition, you will have not only made a bargain, but you will have cast your vote to restore some of the cool stuff that is a part of our culture.

Restoring a worn out piece could easily put another 20 years of life into what could have ended up as garbage. The pride of ownership, and the pride of preserving some of our generation's hip guitar amps and having that special amp perform as it once did, could easily be reward enough to get that amp put back in tip top condition.

All About Vacuum Tube Guitar Amplifiers

SELECTING THE RIGHT TUBE GUITAR AMPLIFIERS

Just as there are many different sounds available from different styles of electric guitars, there are also many different tones available from different styles of tube amplifiers. With an electric guitar, the amplifier has more to do with the overall tone than anything else. I can get a better sound with a $5,000 amp and a $500 guitar than with a $5,000 guitar and a $500 amp. The amp is not a mere "accessory" to use with the guitar. Let's get some perspective here. The amp is what makes an electric guitar "electric." Take away the amp and there is no electric guitar. You need an arsenal of cool sounding amps to compliment your cool guitars.

When selecting the right tube amplifiers, we need to look at our needs and playing styles and choose the appropriate ones. You need many different amplifiers just as you need many different guitars. Just as a single-coil pickup sounds different from a humbucking pickup and hollow bodies are different from solid bodies—cathode-biased amps sound different from fixed bias; Class A sounds different than Class AB; high wattage sounds different than low wattage; ten-inch speakers sound different than twelve inch speakers; open-back speakers sound different from closed-back; etc. If you live in an apartment or a neighborhood, you would need a very small practice amp to avoid confrontations with family members and neighbors, yet also need an amp loud enough to gig or jam with friends. If you are like most of us, you will find that you have many amplifier needs to be satisfied.

How Much Gain? Gain has to do with preamp amplification rather than output amplification. Depending on your playing style and type of music, you will have to decide on how much gain you need. An amp with multiple preamp gain stages can get loads of gain, whereas a simple vintage tube amp will probably have only one or two stages of gain.

If you are playing rock or heavy metal, you might opt for higher gain tube amps such as a Laney, two-input Marshall, early Boogie, or any tube amp with multiple stages of gain on the preamp. Higher gain amps distort by having one preamp tube overdrive another preamp tube. This type of distortion is more homogenized sounding with less dynamics and works great for heavy metal and some types of rock music.

On the other hand, for playing blues or rock-a-billy, the lower gain commonly associated with a vintage amp or vintage amp reissue might give the results you seek. Most vintage amps have one or two stages of preamp gain. Some examples of this would be: the Vox AC30 (one stage), the Vox AC30 Top Boost (two stages), the 5C3 Deluxe (one stage), the 5E3 Deluxe (two stages), the 5C1 Champ (one stage), the 5F1 Champ (two stages), and the 5F6A Bassman (two stages).

Lower gain amps distort by turning the volume up and letting the output tubes overdrive. Because the preamps of these type amps do not have all that much gain, slight changes in picking attack can alter the distortion characteristics dramatically. For example, one can go from clean to distortion by changing how hard the string is picked. This is very useful for most styles of lead guitar.

WHAT TYPE OF BIAS? There are two types of biasing arrangements used on tube amplifiers. Even though all tubes must be biased somehow, when we talk about biasing, we are almost always talking about output tube biasing. Preamp tubes are almost always cathode biased, Class A, single-ended. In fact, I have never seen a fixed bias preamp tube or one operated in push-pull, or one operated in Class AB.

Of the two output tube bias arrangements, cathode biased (a.k.a. self-biased) and fixed bias, there is a difference of tone and envelope. The cathode biased amp will have less punch, less volume, but more sustain and more of a singing quality in general. The cathode biased amp, since it is more compressed, works well for low volume playing, where fullness at a lower volume is required. Use a cathode biased amp for low volume jazz or just low volume practicing. Cathode biased amps seem to go great with slide playing.

When playing with a pounding drummer or just a loud band, the fixed bias works better. Fixed bias has the attack and punch needed to cut through a mix. Where it lacks in sustain, it more than makes up in edge and dynamics. Not only that, but fixed biased amps are generally much louder than their cathode biased counterparts.

In general, lower volume amps are usually cathode biased. For example, the tweed Deluxe and Champ are cathode biased. Almost all EL84 of 6BQ5 amps are cathode biased.

Some examples of fixed bias amps would include the Bandmaster, Showman, Super Reverb and Twin Reverb; all Kendrick amps over 20 watts; all Marshalls 50 watts and over; Hiwatt, Orange, Laney and most Boogies.

CLASS A OR AB? Class A isn't better. Class AB isn't better. They are different from each other. Actually most amps that are called Class AB actually operate in Class A at lower volumes. The only difference between Class A and Class AB is that the Class A design has the output tubes always "on". The output tubes never rest. In a Class AB design at low volumes, the output tubes operate in Class A (meaning the tubes are always "on"). When the preamp gain pushes the output tubes such that they turn "off" (go into cutoff—in other words the preamp gain voltage is high enough to turn the output tubes "off" during a portion of the input cycle), then the amp is said to be operating in Class AB. Why would anyone want to operate in Class AB?

For one thing, Class AB designs are more efficient and much louder than their Class A counterparts. Class AB amps generally have more gain and are punchier than the Class A. For a Class A amp design to remain Class A, the preamp gain must be small enough not to drive the output tubes into cutoff. If the preamp drives the output tubes into cutoff for any portion of the input cycle, then the amp is considered Class AB. Almost all high gain amps are Class AB. All single-ended amps (one output tube) are Class A. Some push-pull low gain amps are Class A. Because the tubes are "on" all the time, Class A designs are usually rounder sounding. There is no "kink" in the waveform.

Class A amps generally are less dynamic and perhaps mushy by comparison to their Class AB counterparts.

Here's an interesting experiment. Take a Class AB amp such as a Blackface Princeton or a Kendrick 2410 and remove one output tube. Turn the volume down to a low level but not "off," and play the guitar through the amp. You are listening to Class A. Notice how round the tone is. Advance the volume slightly and play some more and listen to the tone. Continue advancing the volume in small increments while playing and listening. Eventually, you will get to a point where the amp sounds horrible. When you reach the volume level where the tone sounds horrible, you have just reached the point where the output tube was driven into Class AB. The gain was so strong that it pushed the output tube into cutoff for a small portion of the input cycle. Since one tube was already removed, you could easily hear when you reached cutoff. This sounded bad, because part of the wave was being chopped off.

At this point, you could replace the output tube you removed and listen to the amp again. This is what Class AB sounds like. It doesn't sound horrible anymore, because with both tubes installed, one of them is "on" when the other reaches cutoff and vice versa. Make note of where the volume control was set when you reached the "threshold of Class AB." Go back and forth with the volume control and listen to it both as Class A and Class AB.

How much power? Higher power tube amps generally have bigger power supplies and therefore more headroom. For example a Fender Twin with its volume set on "3" may be much louder than a Champ turned all the way up. The Twin would still be loud and clean at that loudness, yet the Champ would be broken up. The Twin has more headroom. It just doesn't break up as easily.

When choosing an amp for a particular application, consider how loud you will be playing, how you want your amp to break up, and how much power the amp should have in order to achieve this. John Fogerty once told me that he was partial to 40 and 50 watt amps and did not like their 80 to 100 watt counterparts. He said it takes too much to breakup a 100 watt amp and he is right. There will be times, when playing at home, that even 25 watts is too much. This is another reason why you should have many amps that do different things.

Combo or Head/cabinet? There are arguments in favor and against both the combo format and the piggyback style configuration.

Originally, the piggyback style configuration came from trying to run a speaker cabinet as a closed-back design. Tube amps need air-cooling, so to get a closed back speaker design, it made sense to separate the chassis from the speaker cabinet. Closed back speakers are great for long-throw applications because the sound comes out in a more narrow (acute) angle. Closed back designs generally have more bass.

Open back cabinets, on the other hand, generally have very wide dispersion angles, but without the long-throw-ness. Open back designs are so much better sounding on stage. The sound is coming from the back of the cabinet and the front with the dispersion angles bigger. It will sound great on stage.

The head/cabinet idea is nice to use different speaker cabinets with different heads. Also the head/cabinet style, though more stuff to carry, generally is lighter in weight per piece. This makes it nice for those forty-something weekend warriors that don't have roadies schlepping their amps.

All About Vacuum Tube Guitar Amplifiers

HOW TO BUY
A VINTAGE AMP

So, you're really itchin' for that bitchin' vintage tube amp tone, and the time has come to fess up to your "user/addict" status and score yet another "piece."

The wife/girlfriend/main squeeze is keenly aware of how inexpensive guitars and amps are compared to German sports cars... and younger women. But how do you know what you're getting, and how do you keep from getting ripped off? And what amp is best for your needs? It all comes down to how much value something can add to your life. If you actually "play" or gig with your amps, your perspective will, no doubt, be different than a collector's, whose eye is on cosmetics more than sound quality.

One thing is certain—the value of an amp is enhanced if it sounds great. If one distills the essence of a tube amp down to a single word, it would be "sound," because its sound is its essence. So, there are certain aspects of an amp you will want to become aware of when considering a purchase.

BUYER BEWARE

Perhaps the most important part of any sale is the seller and their motives. Most dealers and manufacturers are very honest. They built their business by giving people fair value, so it's difficult to imagine some of the attitudes found with internet sellers. The internet offers a cloak for sellers to get rid of junk by carefully wording their ads so they're not responsible for your dissatisfaction. When shopping eBay or some other internet sites, you have to learn the translation of internet jargon. "This piece powers up, but I don't have a guitar to test it" can translate to "I had this thing to three shops, and although the last shop was able to make it come on without blowing a fuse, it still will not pass signal. I was told that it could be repaired if I wanted to buy a $200 output transformer, a cap job for $200, and a set of tubes for another $150." Or, "Looks original to me, though I'm no expert," could translate to, "Anyone could spot this as a fake from 200 feet across the convention center, but I'm playing dumb to try to dump it for as much or more than the market will bear."

So what's one to do? For one thing, you don't have to buy anything. Just say, "No" to dealing with a "no responsibility, no return" seller. Or better yet, deal with someone you know. If a deal seems too good to be true, it usually is. There is no free lunch, and there is no tooth fairy.

WHAT KIND OF AMP DO YOU WANT?

There are so many types of tube amps: Class A, push/pull, cathode-bias, fixed -bias, Class AB, single-ended, low-wattage, high-gain, etc. The first step to purchasing a vintage tube amp is to decide what types you wish to own. Here's a breakdown.

LOW-WATTAGE SINGLE-ENDED

If you intend to play your amp in an apartment or home, a low-wattage amp may make more sense. You'll make fewer enemies, as the sound pressure level isn't overwhelming.

Generally speaking, low-wattage amps have one output tube. All single-output-tube amps are Class A single-ended, and in general, they're cathode-biased rather than fixed-biased. This type of amp generally has three tubes—rectifier, power, and preamp (plus a fourth tube if the amp has vibrato or reverb). From such an amp, the sound is usually very compressed, with a spongy response. If turned up all the way, sometimes these amps are so compressed that the front of the note is "clipped off," which is a bummer when you play fast. There are many examples of these low-wattage tube amps; the Fender Champ, Fender Harvard, or the tweed Princeton. But there are plenty of single-ended/low-wattage vintage amps by other makers, as well, including the Gibson GA-1 RVT or GA-5, the Vox AC4, and Ward's Airline. Even the Danelectro/Silvertone amp-in-case tube amps were configured as four-watt Class A single-ended! It's a unique sound. Turn it up for the spongier and more compressed settings—perfect for playing a raunchy slide guitar lick. Or you turn it down some—so the threshold of breakup allows for more headroom. But don't think you'll get a pristine clean tone. You might be able to clean it up a little by turning down the volume on the guitar, but it won't be clean.

Remember, too, that amp manufacturers test their amps using the guitars they manufacture. This explains why Gibson amps never have as much gain as Fender amps. Gibson guitars have always had hotter pickups, so its amps were designed with less gain.

HIGH-WATTAGE PUSH/PULL

If you're gigging, —and especially with a drummer—you'll want enough "oomph" to compete with a band. This requires a more powerful amplifier. To produce more output power, an amp needs more output tubes to drive the output stage. The most efficient way to use more output tubes is by configuring a pair or two pair as push/pull.

A single-ended amp cannot be biased into Class AB, but the push/pull amp can be (and usually is) biased into Class AB—giving it even more power and greater efficiency. An amp such as this could have anywhere from 25 to 120 watts, depending on how many and what type of output tubes are used.

Bigger amps are much louder, so you won't be endearing your family, dog, or neighbors by cranking it up and letting her rip. But if you need to be heard over a drummer, a higher-wattage amp is the way to go.

MEDIUM-WATTAGE PUSH-PULL

Between the small and large lie the medium-sized amps. Generally equipped with a pair of 6V6 or a pair of EL84 tubes, examples include the Fender Deluxe, Vox AC15, and Gibson's BA-15RV, GA-20, GA-19, GA-18, GA-14, etc. These amps are desirable because they can be loud enough to gig (perhaps mic'ed through the PA) without emitting enough volume to make your wife go shopping or your dog lunge for your jugular if you play it at home. A medium-sized amp is also nice at a jam session.

YOU NEED THEM ALL

It's true. You really need at least one of each type of amp, so you'll be covered no matter the situation. And of course, you'll need backups for each! But when purchasing a tube amp, there are certain things to be aware of.

THE AMP'S CONDITION

A tube amp is made from components that will eventually wear out and malfunction. A tube's tone deteriorates with time. Tubes are like guitar strings—they sound good when new. But eventually, their tone will deteriorate. And if played long enough, they'll break! When a seller has spent several hundred dollars on an overhaul, they'll certainly note it in the amp's description. So it follows that if a seller doesn't mention a recent overhaul (or re-tube), it hasn't been done. So figure the amp will need an overhaul and new tubes, which could vary from $200 to $600 depending on the type of amp, how many electrolytic capacitors it uses, and the number and type of tubes it uses. A bigger amp, such as an Ampeg SVT, with its 14 tubes and countless capacitors, could easily cost $600. As the number of tubes and wattage goes up, the cost of the overhaul generally goes up.

When purchasing an amp for gigging, it's smart to buy a silverface Fender and have it converted to blackface specs. I've always thought converted silverface amps sound better than the actual blackface. When the circuit is duplicated, the big difference is the transformers, and CBS transformers are better-sounding. The silverface-to-blackface conversion costs about $75. Silverface amps can be purchased for several hundred dollars less than their blackface counterparts.

Of course the logical conclusion points to the reality that one would need all types of vacuum tube guitar amps because each has qualities to be admired—the low wattage amps working great for practice or in quieter playing situations, the higher wattage amps working best for jamming with a drummer, while the medium wattage style amps adapt nicely for doing studio work, gigs or practice.

Because vintage tube amps are old, they are usually deteriorated and we looked at the importance of fresh, new, electrolytic capacitors in the amp. Just as one would surely want fresh, clean oil and a new battery in their vintage '59 Ford T-bird, so would they want fresh caps in their vintage amp. New caps will store energy better and deliver this energy better to the tubes when they need it. The new caps are essential to having the amp sound lively. And we also mentioned that if a seller doesn't brag about a recent overhaul/cap job on his amp that it probably needs one and has not been done in a long time.

It is amazing how little service work is ever done on an amp. The amp can sound like dog doo and no one ever brings it to the shop unless it is blowing fuses, catching on fire or completely dead. So a tune-up to do a cap job/overhaul and new tubes could run $200 to $600 — depending on what kind and how many tubes/caps/etc. the amp needed to bring it up to snuff.

LET'S PUT IT ALL IN PERSPECTIVE

Since the first vacuum tube amp was introduced, good quality vacuum tube guitar amps have always been somewhat expensive when compared to what other consumer goods items sold for in their day. For example, the Gibson EH 185 sold for $185 in the 1930's! It had a small laptop guitar included with it, but $185 was quite a bit of change in that day. A new car could be bought for around $1,000. (Analogy 6 amps = 1 car.) My grandmother bought a new two bedroom home in 1929 for $2,000. So people paid a premium for a guitar amp in the early days. (Analogy 12 amps = 1 car.)

Fast forward to the mid sixties. You could buy a new full size Ford automobile for $2,500 or less. A brand new 45 watt tube amp with reverb (i.e. Blackface Super Reverb) went for around $425. (Analogy 6 amps = one car.) Today, a new Ford may cost around $25,000. So if a new 45 watt all tube 4x10 combo amp with reverb costed around $4,250 it would be a similar value now as it was in yesteryear. Today, a new handwired tube guitar amp could cost perhaps two to four thousand dollars. A vintage amp could be re-furbished for a fraction of this amount. You will have to consider other costs besides the cost of overhauling and re-tubing. It may need a speaker or a transformer or something else that would justify lowering the sales price to you!

If you are not willing to spend money to bring an amp up to snuff, don't buy it. You would be better off to hold out and pay more money for an amp that has already been redone by someone else. There is no point in owning a vintage amp in disrepair. It won't sound good; if you play it anyway, you are likely to damage it and the whole point in owning a tube amp is to enjoy superior tone.

SPEAKERS

If the amp has good sounding speakers it is worth more so you can bet it will be priced accordingly. You can expect to pay more. When buying a vintage amp with an original speaker, since the vintage amp is old, the speaker is also old. There are many problems that could occur with an old speaker. The speaker could lose its efficiency from a number of causes including the magnet losing its strength. If the speaker is weak, the amp will not sound its best. Voice coil rub could ruin the sound of a speaker even though it will still make sound.

Whether the speaker sounds good or not, is it original? If it is original and it sounds good, expect to pay a little more for the amp because that's will have collectible value. If it doesn't sound good, it will cost you to replace the speaker with something else or have the old speakers reconed. You can use additional cost as a bargaining point when negotiating a price. You should get the amp for less if the speaker sounds bad and you will have to spend money to replace or repair the existing speaker.

If the old speakers can be reconed by someone that knows what they are doing, then getting an original reconed is a great way to go. When a speaker is reconed, it is like a new speaker because all of the parts are replaced except the basket assembly. You get a new spider, new voice coil, new cone, new dust cap, new tinsel wire, new gasket. If the speaker is not original anyway or if you just don't like the way it sounds, you may prefer just to go ahead and purchase a new speaker to replace the old one. It really depends on how much it costs and how hard it is to get your old one reconed compared to the cost of a new speaker. Each situation is unique.

TRANSFORMERS

If an amp does have original transformers and they do sound good, expect to pay a little more. The collectibility factor will enhance the value when everything is sounding good with the original transformers. However, if you are a player, you may or may not wish to own original transformers. Certain amps are known to have not-so-great transformers. For example, a Fender Pro Reverb or a Bandmaster uses very small and thin sounding transformers. Suppose you are a player and someone is selling a Pro Reverb amp with a blackface Bassman output transformer. That amp would actually be more desirable to you, yet a collector may be "put out" that the transformer is original. So you have to look at your particular situation and decide what floats the boat in your harbor.

When inspecting a transformer, the first thing I look for is rust. Are the laminates rusted? I am not talking about surface rust on an end-bell or mounting bracket because that isn't a problem. I am talking about on the edges of the laminates that the transformer coil is wound around. If those laminates are rusted, it is likely that the transformer is conductive between adjacent laminates and the eddy currents will zap away the power on the secondary of the transformer. Don't buy such an amp unless the selling price is so low that you could afford to part with yet a few Franklins replacing or rewinding the transformers. If you get it replaced, it is going to cost a couple of hundred or less. And if you get it rewound, it may cost more because all rewinds are one-offs. For whoever is rewinding it, the set-up time to do a one-off has to be included in the one price so one-offs will always cost more than an off-the-shelf transformer that has most likely been wound in quantity.

It is a good idea to inspect the amp chassis and get a·good look at the mounting and the soldering of the transformers. If you look closely at the chassis and you see where someone has disturbed the solder joints that connect the transformer leads to the circuit, or if the mounting bracket doesn't appear original, or if you look at the part number on the transformer and compare it to a schematic or layout diagram and see it does not match, expect to pay much less for the amp. The amp's value is considerably less without the correct transformers and the value of the amp is not as much as one that does have the correct and not rusted original transformers.

REVERBS

Some vintage amps have reverb and others don't. If you are looking at an amp with reverb, it is important to listen to the sound quality of the reverb. You are not just listening to see if it "reverbs," you are listening to see if it is a good sounding reverb. It should not be boingey or raspy. It should be smooth with no rattles. The reverb control should have an even and smooth response as it is turned up.

If the reverb doesn't work well, it could cost you another $50 to $150 to get it working right. When considering the sales price, one must consider the reverb quality and how much value it adds or subtracts from the price.   A new reverb pan sells for around $45 or so. A new reverb transformer could cost around $40-$50 and a new old stock driver tube could be another $25 to $50 depending on the brand used. Again, be willing to pay a little more if everything is working good and you will not have to spend money making the reverb sound good.

EPILOG

Almost any vintage amp can be restored to its full glory for much less than the cost of a new similar amp. If a vintage tube guitar amp sounds good, has a good sounding original speaker and original transformers with a wonderful sounding reverb, it is worth considerably more than one that needs everything replaced or repaired. Use the repairs as a bargaining tool for negotiating a favorable selling price, and deal with people you trust.

All About Vacuum Tube Guitar Amplifiers

A DIRTY DOZEN OF THE COOLEST GIBSON AMPS

There were many models of Gibson tube guitar amplifiers made over several decades. My Gibson Amplifier Master Service Book which was printed in the 60s, lists 153 amplifiers and schematics. Besides the 153 amps listed, there are variations of those. Occasionally I will have a Gibson amplifier come into my restoration shop that the schematic is not listed in my Gibson Amplifier Master Service Book, so I would suspect there are over 200 models of Gibson tube guitar amplifiers out there. This can be confusing as some of these are not that great and some of them are super cool.

Almost any vintage amp you find is going to need an overhaul, and you need to have some knowledge as to what the amp is going to sound like when it is overhauled and serviced properly.

In general, the Gibson amps never seemed to be quite as popular as Fenders. As a rule, the Gibsons had less gain, less power and in general were cleaner sounding. Of course, there are exceptions to every rule. Here are some unsung heros made by Gibson. If I left out your favorite one, please forgive me. There are others that are super cool, it is just that these 12 models are my personal faves.

GA-5 — This was Gibson's version of the Fender 5F1 Champ. The schematic would be identical, except some of the resistor values were slightly different. The cathode resistors for the 12AX7 were 2.2K whereas the Fender used 1.5K. Also the Gibson used 220K plate load resistors. Like the Champ, it had the 8 inch speaker except it was 8 ohm (rather than the 4 ohm output impedance of the Fender Champ). I always thought the 8 ohm was a better way to go, mainly because the turns ratio of the output transformer was less which would allow for less coupling loss. This is a great recording amp and works very well for slide or harmonica.

GA-6-Lancer—This was Gibson's version of the Fender 5D3 Deluxe. It is almost identical. It is missing a bypass cap on the output tube, but one could add a 50 uf/50 volt bypass capacitor in parallel to the cathode resistor of the output tube for best performance. This one used the P12R Jensen speaker. There was also a .001 uf parasitic suppression capacitor connecting the grids of the power tubes. This would phase cancel high frequency parasitic oscillations. I would remove that cap, mainly because I am not one for remedial circuits, and spruce up the layout to get rid of any parasitic oscillations. Usually shortening all grid circuits and cleaning up all grounds is enough to stabilize these amplifiers.

GA-14-Titan—This one was exactly the same as the GA-6-Lancer except it used a 10" speaker (P10R Jensen) and had the bypass cap on the cathode of the output tube. It used a 20 uf cathode bypass capacitor. Both the GA-6 Lancer and the GA-14 Titan worked great for blues guitar and harmonica.

BA-15RV Bell 15RV—You have heard this amp as it was used by Ted Nugent on "Cat Scratch Fever." The circuit closely resembles a 5E4-A tweed Super—but with a cathode biased output stage and reverb added!! Although this amp had only a single speaker, there was a 4x10 version of the amp whose model number was GA-45RV (completely different from the GA-45RVT). If you can get your hands on on of these babies, get an overhaul and name it.

GA-16-T—This amp duplicated the circuitry of the 5E9-A Tremolux except it only had one channel and used a 10" speaker. Or you could say it was like a 5E11 or 5F11 Vibrolux, but with a cathode biased output stage. It came stock with the Jensen P10R speaker. This would make a great recording amp. It used slightly more negative feedback than the Fenders, but if the 47K feedback resistor was changed to a 56K, it would have identical gain characteristics of the Fender tweed Vibrolux.

GA-70 Country and Western—This 25 watt amp (35 watts full clip) was Gibson's lower powered version of the 5E6 Fender tweed Bassman. The schematics were almost identical, but the voltages were less, and the Gibson used a the lower gain 12AU7 preamp tube for the phase inverter. This amp used an Alnico magnet 15" speaker. The Gibson literature states "Unusually clear, bell-like treble with amazing reserve volume and sustaining qualities. Instantaneous response with a tone that bites through, yet is pleasing to the ear and free of distortion." I couldn't have said it better myself.

GA-77—This amp is virtually the same circuit and speaker as the GA-70. Both the GA-70 and the GA-77 were way ahead of their time. These amps were designed in 1953!

GA-78 Bell 30 Stereo—This amp was nothing like the Bell-15. It was actually two amps in one cabinet. Each channel was a separate amp with its own preamp circuit, output stage and speaker. The first channel, and only the first channel, has tremolo. The circuit design included a stereo input jack that would actually connect both channels via a ring, tip, sleeve—$^1\!/_4$" stereo plug. One could use a stereo guitar with this amp.

If you didn't have a stereo guitar, you could plug into either channel and turn the stereo/mono switch to the mono position. This would allow sound to come out of both speakers. Also, this amplifier had phono input jacks for each channel, so one could hook up their stereo record player if they so desired! It is better than Hi Fi!

GA-78RV Stereo Maestro 30—This amplifier is basically the same as the GA-78 Bell 30 Stereo, except it has reverb! Imagine how nice those stereo records must have sounded with that reverb function. I'm sure this would impress all the girls.

GA-79 Multi-purpose—This one is the same as the GA-78-Bell Stereo. It has all of the same circuitry and all the same features. It even has the RCA phono inputs to attach a stereo record player. The GA-78, GA-78RV, and GA-79 all used 6BQ5 output tubes. For absolute best tone, replace these output tubes with 7189A output tubes. These are the military spec version of the 6BQ5/EL84 style tube.

EA-10 Deluxe—This was sold as an Epiphone but of course, it was a Gibson amplifier. This had a similar preamp circuit to the 5E6 Bassman or the 5E7 Bandmaster, but it included a middle control (very hip). The output stage had a paraphase style phase inverter and cathode bias output stage like the 5D5 Pro. Also like the Pro, it had a 15" Jensen speaker (P15P). It used the GZ34 rectifier with a 12AU7 phase inverter tube and 6L6GB output tubes. Both preamp tubes were 12AX7.

EA-30 Triumph—Consider this as a dead nut copy of the Fender tweed Deluxe (5C3) with only two exceptions. It is missing the cathode bypass capacitor on the output tubes and it has a 250 pf cap connecting the grids of the output tubes. This 250 pf capacitor is a parasitic suppression circuit that phase-cancel parasitic oscillations in the higher frequency range. I would get rid of that suppression cap and simply spruce up the layout to stabilize the amp. Also, I would add a cathode bypass capacitor which would bring out the full richness of this fine amplifier.

GETTING THE MOST FROM YOUR AMP

All About Vacuum Tube Guitar Amplifiers

TOP TEN AMP TONE TIPS

Almost any guitar and tube amplifier will sound pretty good, but to get the real killer tone; one must pay attention to small detail. When numerous small details are summed up, the difference could be the difference between so-so tone and killer tone. You will play better when you sound better and that's what we are talking about: "sounding better." Besides a string vibrating on a piece of wood, there are many things that affect what we actually hear. From the strings, to the cable, to the amp, through the circuit settings and the tubes, into the speaker and inside the acoustics of the room—yes, it all makes a difference. Even the room where the sound occurs matters—tonally. Here are some tips that I have found could be the difference between good tone and great tone.

1. PLACE THE AMP WHERE IT SOUNDS ITS BEST. Amplifiers cannot make sound in a vacuum. They make sound in a room full of air. Proper amp placement is critical for best tone. On an open back cabinet, the sound is coming from both the back and the front. For instance if you are in a club, such that the stage is fairly close to the floor, you may find elevating the amp, so that the sound travels above people's heads, will give you better coverage. Check out the rest of the room. Is there a hard wall facing the bandstand that you do not want your tone to bounce against? Too "live" a room can muddy-up the sound. Some players have their amp facing the back wall (in back of the bandstand) so that the sound can bounce off of the back wall. This is especially effective if you have an amp that is highly directional or otherwise needs some diffusion in order to blend with the entire band's mix. When Stevie Ray Vaughan played at the Armadillo World Headquarters in Austin, the soundman had him lay his amps on the floor so that the speakers were facing the ceiling. He had a microphone over the amp. It was the only way that they could blend the sound into a mix. His amp was so directional that he was actually overpowering an 8,000-watt P.A. system!

One of the best small amp setups I ever heard was from Austin blues legend, Jim Talbot. He would put his Deluxe Reverb on a chair tilted back and facing his back. Actually the amp was aimed at his vocal mike. When he was singing, the amp would be facing his back and the sound would diffuse before the audience heard it. When he wanted to take a solo, he would step away from the mike – leaving a clear path from the amp to the mike. The vocal mike would bump up his lead over the PA and with the amp pointing towards the ceiling, the sound would hit the ceiling and bounce over the audience's head. It was a great sound indeed. He told me that he learned of this trick from an old black blues player in east Austin.

2. MAKE SURE THE SPEAKERS ARE MOVING IN UNISON. When using an extra speaker or using a combo amp with multiple speakers, it is important that all of the speakers are moving together the same direction at the same time. In other words, we want all the speaker cones to move forward and then backward, in unison. If a speaker is out of phase, its cone will be moving forward when the other speaker's cones are moving backward. In this case, you will lose bottom end and you will lose tone. As a matter of fact, it would sound better if the offending speaker was disconnected from the system. Phase is important when you are using more than one speaker. This would include a 2x12 amp, a 4x10 amp, or using a multi-amp set-up.

Correct phase can be checked with a 9-volt battery. Unplug your speaker and touch the terminals of the battery to the speaker plug. Connect the plus of the battery to the tip of the plug and the minus of the battery to the sleeve of the plug. As you connect, look at the speaker cone. When the battery is connected, the cone should move forward on every speaker in the cabinet. If the cone moves backwards, then the wires feeding signal to that particular speaker should be reversed. Swap the plus with the minus. Check it again. The individual speaker cone should now move forward. Continue checking each speaker in the cabinet until you are sure they are all moving the same direction at the same time.

3. SET THE OUTPUT TUBE BIAS FOR BEST BREAKUP. Depending on how the output stage is biased, the headroom/breakup characteristics will be affected. Generally, cooling-off the plate current will give more headroom. That is, if you're going for a very clean sound you would want to bias your tubes on the cold side so that they would have less idle plate current. On the other hand, if you would like your

amp to break up more quickly, you would want to bias with more idle plate current. I remember setting up a couple of Kendrick amps for Jimmie Vaughan. He likes a lot of headroom. He does not want his amp to break up until it's on about 7 or 8 and he plays it on 7 or 8. In order to achieve this, the output section had to be biased fairly cold. We were running about 20-25mA plate current per output tube. This gave him the headroom he needed. We could have taken the same amp and biased the tubes to draw 35 mA plate current per tube causing the amp to breakup at around 4 or 5 on the volume setting.

4. USE GOOD SOUNDING OUTPUT TUBES. Players use tube amps because they sound better. Wouldn't it then follow that if you are going to use a tube amp to get better tone that you should use good sounding tubes? With rare exception, stay away from any tubes made in China. Different types of output tubes, and, yes, different brands of output tubes have different tone. For instance, a Tung Sol 5881 has a very creamy, milky top end. It is not very sharp on the high end, like some other tubes may be. A Phillips 5881 is voiced more like a JBL speaker. It has a very crisp, chimey top end—not creamy at all—with a crystal clear bottom and a transparent mid. The Tung Sol, on the other hand, has much more mids. Here you have two tubes, both are 5881s, with entirely different tones. Selecting the proper brand of tube is critical to obtaining the tone you want.

5. CHOOSE THE RIGHT SPEAKER FOR YOUR SOUND. The thing that actually vibrates the air has a sound of its own. Different speakers have holes at certain frequencies and humps at other frequencies. Some of these frequencies may or may not be desirable. For instance, if you want a British-sounding breakup, you would not choose an Altec-Lansing or JB Lansing speaker. On the other hand, if you were going for a very chimey sound with a lot of clarity in the top end, then you wouldn't want a British-sounding speaker. Some speakers grind on the bottom end and other speakers are clear on the bottom end. Mid-range content is also a consideration. For instance, the VOX Bulldog speaker and the Kendrick Brownframe speaker have a lower mid-range hump that creates a kind of a barking quality, very similar to a cocked back Wah pedal kind of sound. This is desirable for most blues and rock tones because it adds "meat" to the fundamental note. This sound may not be desirable for someone going for a glassy, "James Brown" funk sound.

6. LISTEN WITH YOUR EARS AND NOT YOUR EYES. How many times have I seen a player set the knobs to a particular number because he had played another amplifier and liked the setting on the other amplifier and yet he adjusted the new amp to the same numbers? You will not get the same results. The reason is simple: the taper of pots can vary, the actual value of the pots will vary some, even if the manufacturer is the same. So, one must listen to the sound and adjust according to what sounds good, not according to the number one thinks should sound good. You've got to trust your ears. Your sound should inspire you; that's the test.

7. PAY ATTENTION TO YOUR CABLING. All cables have capacitance. In fact, a cable is a capacitor by virtue of the definition of a capacitor being two conductors, separated by a non-conductor. Capacitance softens high-end. The total amount of capacitance of a cable equals the capacitance per foot times the length of the cable in feet. That is, a cable with 10 pF per foot capacitance would have 100 pF of capacitance if it were 10 feet long. If it were 20 feet long, it would have 200 pF and a 50 foot cable would have 500 pF. As the capacitance goes up, the high end goes away.

When recording, it is best to use the shortest cable you can. So, if you're trying to get that chimey, crystal-clear top end and you're using high capacitance, 30 foot cable, then forget it, it just won't happen. If you're using a pedal board, you may find using the shortest patch cords available will keep you from losing high end.

8. SET UP YOUR GUITAR CORRECTLY. There are ways of setting-up a guitar that will alter your tone. For instance, the type of strings used. If you were going for a mellow tone, you would not want to use stainless steel strings because they are too bright sounding. On the other hand, if you're going for an overdriven sound, then you may elect to use a hotter pickup, something with a little more output. Your fret size, the height of the fret, and your playing style makes a difference. I pick strings really hard and do a lot of bending, therefore I use a higher action than most players and I use a taller fret. Some people have a lighter touch, perhaps they don't bend as much, or use a lighter set of strings—in which case a shorter fret and closer action would be more appropriate. The pickup adjustment makes a difference. Get your pickups too far away from the strings and you will not have adequate signal to drive the amp. However, if you get them too close, the magnetic field could interfere with proper vibration with the string. The type of pickups matters because pickups, like strings, have a voice.

9. EXPERIMENT. Don't be afraid to try something different. All great players, known for their great tone, constantly experiment with different tonal ideas. Try different speakers with different amps. Buy different tubes. Try different pickups. Try different biasing. Try placing your amps in different spots. You are likely to find that experimentation is the only way to polish the tone to your exact liking.

10. BE WILLING TO SPEND SOME TIME. Good tone takes time. All great players spend hours and hours and, yes, days and days working on their tone. For instance, players who are already acknowledged for having great tone, such as: Eric Johnson, Jimmie Vaughan or Billy Gibbons, continue to spend time paying attention, listening, experimenting, and paying the price of good tone.

All About Vacuum Tube Guitar Amplifiers

FINE TUNING YOUR AMP SETUP

Living in central Texas near Austin, the music capital of the world, affords me the opportunity to hear many live bands and see many great guitar players. It is rare to come across a guitarist with his amp setup sounding its best. In most cases, I will see a player that has mastery of his instrument, yet his overall sound is lacking. I can only imagine what a huge difference it would make if a player's amp setup did him justice.

It is nice to play well, but there is more going on than depressing a string on a fret. An electric guitar must have an amplifier for it to sound good. Even if the amplifier in question is a quality tube amplifier, one must set it up right to "get the tone." You could have five amps, all of the same brand, and yet have five completely different sounding amps—depending on the quality of the tubes, the quality of the speakers, the biasing of the output tubes, the condition of the filter caps, bypass caps and coupling caps.

For the guitarist, fine-tuning the amp may seem to be the job of the tech. But if you don't have a tech, or want to DIY, here's some basic guidelines about how to fine-tune your amp setup.

The four parts to any tube amplifier are the power supply, the preamp, the power amp, and the speaker. We will take a look at each part and fine tune each part. When this is done, the entire system will be functioning on a new level.

POWER SUPPLY—The most obvious part of the amplifier is the power supply. It consists of the power cord, power transformer, rectifier, filter caps and dropping resistors. It could be said that a guitar amplifier is simply a modulated power supply that drives a speaker. That being the case, the power supply is the core of your sound.

The CHOICE OF RECTIFIER—whether tube or solid-state—will affect the blossoming quality of the amp. If you need more "meow," you could use a softer rectifier. If you need more "front edge" of a note, a more efficient rectifier such as a 5AR4 or a solid-state rectifier may be the ticket. If you are a fast player, you will not like a soft rectifier because by the time the rectifier recovers from the initial attach of the note, you are already playing another different note. It is as if the amplifier can't keep up with you! Faster players need less "sag" or "meow" and more punch.

Consequently, a bluesier player might prefer a softer rectifier to add some character to his sound. This is where it gets tricky. What sounds good to you alone in a room make not work with a pounding drummer. If your drummer plays loud, you will need more "front edge" of the note so that your sound will punch through and not become buried by the drums.

If your POWER TRANSFORMER is rusted, there is a possibility that it is being bogged down with eddy currents. If this is the case, the transformer should be replaced. I don't care if it still makes sound, it will not sound its best if eddy currents are dragging it down.

The filter caps need to be fresh and of good quality. I change mine every six years in every amp I own whether they need it or not. It is like changing oil in your car every 4,000 miles just because. Old filter caps may function yet not perform. We are not talking about "just getting by" here. We are talking about getting superior tone. Good filter caps keep the note you are playing in tune! They pull everything together so you don't get that ugly, out-of-tune subharmonic that is caused by weak filter caps. They also supply power to the other tubes and when you are bearing down, you need caps that will continually deliver the power your tubes demand and not "wash out," tonally. The filter caps need to be of good quality because even new ones can sound bad if they are poor quality. Stay away from the Illinois brand (which is made in Taiwan) and the LCR brand. These imported filter caps sound bad even when they are new. If you have a new Marshall with LCR caps or a new Fender with Illinois caps, change them at once. You cannot afford to screw up your sound by having ugly and out-of-tune subharmonics underneath your tone. Use Sprague or Tech Cap or any other quality American-made capacitor. PREAMP—This is the section of your amp that amplifies the weak signal from the guitar pickups so that the signal is strong enough to drive your power amp. It consists of preamp tubes (usually 12AX7) and a few components to make the preamp tubes work. These components are basically resistors, coupling capacitors, and bypass capacitors. Unless you are getting excessive hiss or noise, the resistors are probably just fine.

BYPASS CAPS, which are found in the cathode circuit of each preamp tube, can go bad and need replacing. Typical values are 25 uf at 25 volts or 10 uf at 16 volts. These should be replaced with good quality, American-made capacitors. I always replace these when replacing filter caps because they are electrolytic type capacitors and don't last forever. COUPLING CAPS rarely need replacing. When they die, they still function, but pass a minute amount of D.C. electricity. This tiny voltage could be as little as .25 volts, yet it is enough to throw the biasing scheme off on the downline preamp tubes and make the amp sound choked and ratty. These caps can be checked easily enough by unsoldering one end (not the end connected to the plate of the preamp tube, but the other end), and checking between the lifted lead and ground with a voltmeter set for D.C. volts. You should get zero volts. If you get more than .25 volts, the coupling cap must be changed. Notice that on some amps (mid 60s Fender, for example), the tone caps are sometimes arranged as coupling caps. The same holds true for these caps as any other coupling cap. If it leaks D.C. voltage, change it.

The preamp is where the amp gets its gain. It is also where the tone controls are usually located. The exact gain characteristics can be fine tuned by substituting other lower gain PREAMP TUBES. For example, if you had an amp that used a 12AX7 preamp tube and you wanted to get a cleaner sound, you may want to substitute a 12AT7 or a 12AY7 for the 12AX7 in order to reduce the gain and consequently clean everything up some.

Preamp tubes generally last longer than power tubes, but replacing worn ones with unworn ones can improve your tone. Also, the actual quality of the preamp tube is important to achieve good quality tone. For example, a 12AX7WA Sovtek, though very quiet, still lacks fullness. This tube would work fine as a vibrato oscillator, but you would not want to use it in a socket that was directly in the signal path. You want to use good sounding brands of tubes that are relatively unworn, for best tone.

POWER AMP—The power amp section of the amplifier would include the phase inverter, the output tubes, and the output transformer. The PHASE INVERTER, though part of the output stage, still uses a preamp tube. The preamp tube that works the hardest and wears out the quickest is the phase inverter tube. These are usually run at higher voltages than any preamp tube and they have a larger input and output than the other preamp tubes. I like to change the phase inverter tube at the same time as the output tubes.

To get your OUTPUT TUBES sounding their best, you need new, good sounding output tubes. Output tubes are like tires on a car. They can last a long time if you don't drive them hard, and if you drive them hard, you will be amazed at how quickly they wear out. If you are playing the amp really hard, you should change tubes at least once a year. It goes without saying that you should reset the bias setting of the output tubes every time the output tubes are changed. The output tubes simply will not sound their best unless the bias is readjusted. The actual bias setting can be varied slightly to make the amp perform more like you desire. If you are going for a squeaky clean tone, you may want to adjust the bias so that the output tubes only draw about .025 A (same as 25 milliamps) of current per tube. On the other hand, if you are going for a quicker breakup, you will want to idle the output tube hotter (as much as 45 milliamps per side.). When you idle them hotter, you simply start them out running so fast that it just doesn't take much to kick them into overdrive.

The OUTPUT TRANSFORMER is critical to your sound. If you have a transformer that is multi-tap with various impedances, the highest impedance setting will sound the best (provided that you have the appropriate speaker load impedance.) There are two reasons for this. First, higher impedance settings have a lower turns ratio that has less loss. And secondly, when a tapped transformer is wound, the highest impedance setting is the whole transformer. All other taps are a percentage of the transformer being used. A 16 ohm transformer with a 4 ohm tap is only using half of the transformer when the 4 ohm tap is selected. A whole transformer performs better than a half transformer.

SPEAKER—Sound is made from vibrating air. Sometimes overlooked, we must remember the SPEAKER AND SPEAKER ENCLOSURE is what is actually vibrating air. The speakers are your amps final filter. If your speaker is inefficient, then the amp will not have the sports car feel.

The speaker is mounted in an enclosure that affects the sound. There are different types of cabinets. A CLOSED BACK cabinet is highly directional and the sound is coming out of the front in a rather acute angle. People near the side of the amp won't get any sound sent their way. Perhaps aiming an amp like this at a back wall would help disperse the sound. The sound would hit the wall at an acute angle, yet bounce back at complimentary angles, thus increasing dispersion.

The other type of cabinet design, the OPEN BACK, has a large dispersion angle in the front and sound comes out of the back too. This type design will get the best stage sound. Miking the amp and putting it in the house mix will help give you a bigger sound. You will simply move more air by using the PA speakers too.

Choose SPEAKERS that have enough brightness to define the note, yet not enough to cause brittle breakup when overdriven. You want the speakers to have enough clear bottom-end to be satisfying, yet you don't want boominess.

MULTI-AMP SETUP—Many players are using more than one amp in their setup. This is fine as long as all the speakers are moving the same direction at the same time. I once heard a player with mastery of his instrument but he sounded thin and weak because he was using a multi-amp setup that was done incorrectly. He could have improved his sound by turning one of his amps off!

It starts with making sure all the speakers in one amp are moving together. This can be checked with a simple 9-volt battery. Hook the + of the battery to the + of the speaker cabinet and minus of the battery to minus of the cabinet. All the speakers should move forward. Any speaker that does not move forward should have its individual leads reversed. These are the wires going to the individual speaker. You would reverse them at the speaker lead end.

Once you get the speakers in a particular amp all moving together, you would then need to have the amps synchronized. It doesn't matter that the speakers in each amp are hooked up correctly; you need to make sure the amps are working together. On some amps, when a positive signal is applied to the input, a positive signal comes out of the speaker. On other amps, when a positive signal is applied to the input, a negative signal comes out of the speaker. You want to make sure the amps are working together.

Here's the easy way. Place both amps (or speaker cabinets) facing you and play a little. Now turn one amp facing backwards and play a little. One configuration will sound great and the other way will sound horrible. If it sounded great by having both face you, you needn't do anything. If they sounded great with one (speaker cabinet) facing backwards, you would need to reverse the leads going from the amp to the speaker cabinet on one of the amps. That way, both amps will sound great when facing you.

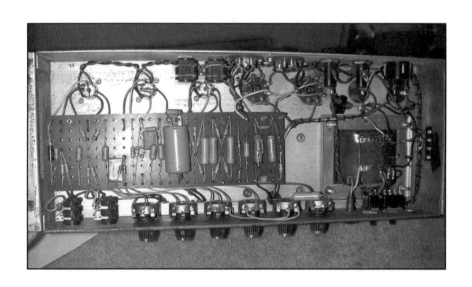

All About Vacuum Tube Guitar Amplifiers

SIX TONE KILLERS YOU SHOULD AVOID

All of us agree that when our guitar tone is inspiring us, we play better. It is so easy to play well when your guitar tone is the source of your own inspiration. But in order to do this, it is necessary for your amplifier to perform well. When your amp tone isn't up to snuff, neither will be your performance.

Getting great tone can be elusive to the guitarist that doesn't know what can kill his tone.

Few players are aware of the amp problems that threaten to ruin their tone. It is easy to avoid the tone killers when you use the "Tone Killer" list as a checklist of what to avoid.

WORN OUT TUBES

Just because a tube lights up, does not necessarily mean it sounds good. Vacuum tubes are like guitar strings—they sound good when they are new and they wear out "tonally" over time. The problem is they wear out so slowly that one cannot tell from day to day that they are deteriorating tonally. But yet, when a fresh set of tubes are installed in the amp, a huge tonal difference can be instantly noticed.

I remember once, when I was getting ready to do some recording with my band. My drummer was changing heads on his drums and asked me when I last changed tubes in my amp. Although I had not noticed any tonal deterioration, it had been about a year since I had retubed, so I went ahead and retubed, anyway. What a huge difference it made. My amp had much more liveliness, yet before retubing, I had not noticed anything was lacking! If it has been a year or more since you retubed your amp, perhaps you too will be surprised at the difference it makes.

INCORRECT OUTPUT TUBE BIASING

An output tube will perform its best when it is biased properly. The problem here is that as the tube wears in, its performance characteristics are constantly changing. You could have a new set of tubes biased properly and a month later, the idle current will have drifted. If the tube drifts towards too little idle current, the amp will lack power, lack dynamics, and sound cold. If the tube errs towards too much idle current, then the amp will lack headroom and will break too early.

Biasing an amp is not rocket science. There are several bias meters available on the market today. Some of them work by plugging them into an output tube socket and then installing the tube back into the meter. These are convenient to bias the amp before a gig. Or you can go the "old school approach" and learn how to bias with an ordinary digital multimeter. I have written extensively on this in my first book, *A Desktop Reference of Hip Vintage Guitar Amps*. My advice is to learn how to bias your amp and bias it often.

DETERIORATED OR FOREIGN FILTER CAPACITORS

Your amp requires D.C. power to operate. The wall voltage supplies A.C. power. So every vacuum tube amplifier has a rectifier circuit to change A.C. to D.C. When this is done, the D.C. is pulsating D.C. Most amps have a full-wave rectifier, so the frequency of the pulses is 120 Hertz, which is twice the frequency of the A.C.

To change from pulsating D.C. to pure D.C., filter capacitors are used. These smooth out the D.C. so that the 120 Hertz does not interfere with the notes you are playing. It requires a great deal of capacitance to do this job, so electrolytic capacitors are used. Electrolytic capacitors do not last forever. They are nearly identical to a battery. Like a battery, they store electricity. They also have a plus and minus polarity and depend on an electrochemical action to do their job. And like a battery, they do not last forever. The capacitor manufacturers suggest replacing them every six to ten years. If your filter caps are old, there are several ways this will affect tone. First, there will be an ugly, out-of-tune sub-tone that is not a harmonic of the note you are playing. This is most noticeable when playing a B or B flat and most noticeable when playing loud. And secondly, bad filter caps can cause low voltage in the power supply, so the tubes are not getting the voltage they need. This manifests itself with low power, lack of touch sensitivity and over-compression. If it has been 10 years or more since your amp's filter capacitors have been replaced, get some fresh American-made filter caps. I use Sprague ATOM caps because they sound better than other caps. Stay away from Taiwanese caps such as the IC brand as these sound blown—even when they are new.

If you buy a new Fender reissue anything, it is advisable to get a cap job immediately after you leave the store. These amps use foreign caps that do not work properly, resulting in that ugly, out-of-tune sound we spoke of earlier. If you own a Blackface or Silverface amp, and you intend on playing it loudly, it is advisable to increase the value of the two main filters from 70 uf to 220 uf. The 70 uf stock value was selected at a time when players were not playing very loud. When you play loud, more current is used and for more current, you need more filtering.

CHEAP GUITAR CABLE

So you got yourself a new three thousand dollar boutique amp and a five thousand dollar guitar. Are you going to connect the two using a ten dollar Chinese guitar cable? Guitar pickups put out a signal that is high impedance, so capacitance is a huge problem. A capacitor is two conductors separated by a non-conductor, so a guitar cable IS a capacitor. Your guitar cable has two conductors that are separated by a non-conductor. Put a capacitance meter to the tip and sleeve of the guitar plug and you will get a measurement of how much capacitance for your particular brand of guitar cable. A cheap cable could have as much as ten times the capacitance of a high-end cable.

Your guitar signal is A.C. and although a capacitor will block D.C. electricity, it will conduct A.C. That means the more capacitance, the more your signal is shorted out before it gets to the amp. There are certain guitar cables that are designed to have almost no capacitance. These cables get the signal to your amp with more integrity. The result is more clarity, focus, better bottom-end, janglier top-end, and in general a fuller tone.

The difference is instantly noticeable, even to the untrained ear. All cables will have some capacitance. The actual amount for any given cable will be measured as capacitance per foot. That is to say that a 20 foot cable of one brand will always have twice the capacitance of a 10 foot cable of the same brand; so you want to keep your cabling lengths as short as possible.

BAD SOUNDING SPEAKER

The speaker of an amplifier is what is converting electrical energy into sound. If your speaker is not great, neither will be your tone. There are many ways that a speaker could malfunction. If a voice coil is rubbing, the speaker will play slightly out of tune when played at a low volume. Sometimes vintage speakers will lose their liveliness. If the cone excursion is weak, so will be the sound.

The solution is to either replace the speaker with a good sounding one, or get your original speaker reconed with correct components.

OUT OF PHASE SPEAKERS

If your amp uses more than one speaker, then speaker phase becomes an issue. What you want is for *all the speakers to be moving the same direction at the same time*. You don't one speaker moving backwards while the others are moving forward. It will phase cancel the sound coming from the other speakers. Nothing can sound worse than an amp whose speakers are out of phase. All bottom-end will vanish, the volume will go, and the overall sound will be thin.

Speaker phase can be easily checked with an ordinary 9-volt battery. Unplug the speaker cable from the amp and connect a 9-volt battery to the speakers. If your amp uses a ¼" phone plug, the tip of the plug goes to the plus terminal of the battery and the sleeve of the plug goes to minus. If there is no plug, then the plus of the battery goes to the lead that is connected to the plus of the speaker. When you connect the battery, you will notice the cone of each speaker will move. The speaker cone should move forward. If the cone of any one speaker does not move forward when the battery is attached, the wires that connect to the offending speaker should be reversed. You want all the speaker cones moving forward when the plus of the battery connects to the plus of the speaker.

Speaker phase is even more of an issue when multiple amps are used. You can tell if the amps are in phase with each other with this simple test. Place both amps side by side and facing you. Play and listen. Now turn one amp around where it is facing away from you. Now play and listen. If the amps sound great facing you, then everything is fine. If they sound great when one is facing away from you, then one of the amps needs to have its speaker wires coming from the amp reversed. If your amp has a ¼" plug feeding the speakers, you would simply swap the two leads inside the plug.

REFERENCE

All About Vacuum Tube Guitar Amplifiers

EVERYTHING YOU ALWAYS WANTED TO KNOW ABOUT FILAMENT HEATERS

Every vacuum tube amplifier, regardless of how big or small, must have a filament heater circuit. The filament heater circuit connects a low voltage from the power supply of the amp to every tube in the amp. Specifically, the low voltage power supply is connected to the part of the tube that lights up, which is called the filament heater. The filament heater; found inside every preamp tube, power tube, or rectifier tube; is a coil of resistive wire—much the same as an incandescent light bulb. When the filament lights up, it provides heat to the cathode of the tube. A tube cannot amplify unless the cathode is hot.

In common language, the terms: filament heater, heater filament, heater or just filaments are all interchangeable terms. The filament is the part of the tube that lights up; the filament circuit is the wiring connecting the power supply to the filaments.

Information about heater circuits is largely ignored in schematic diagrams and amplifier layouts. You almost never see a heater circuit on the layout diagram of an amp. It is assumed that you know about filament heater circuits! I recently hosted an "Amp Camp Seminar" at my factory in Texas, and one of the seminarians brought two kit amplifiers he built that did not work. It was because there was no filament heater circuit instructions included with the kit. This is a subject that is assumed you already know.

Filament circuits for guitar amps are almost always 6.3 volts A.C. There are exceptions to this rule, but about 99 out of 100 amps will use 6.3 volt filaments. In rare instances and especially more often with audiophile tube amplifiers, the filament circuit is rectified to D.C. The reasoning behind a D.C. filament is that if you use A.C. voltage, then you are injecting some hum into the tube and that hum could leak over to the output of the tube, thus adding to the noise floor of the amp. With pure D.C., there is no hum and therefore nothing to leak.

In the old days, amps were not very powerful nor did they have much gain; so the noise floor went unnoticed. Take a 1930s or 1940s amp and you rarely see more than 20 watts output. Not only that, but people didn't "turn it up" like they do today. Little attention was paid to filament hum because it never occurred as a problem—given the circumstances of the day.

Most early amplifiers used a "daisy-chain" style heater circuit. With the "daisy-chain" circuit, a single wire was daisy-chained from one filament lead of each tube socket to the next. This wire was connected to one end of the 6.3 volt heater winding on the power transformer. The other end of each tube's filament was grounded to the chassis—as was the other end of the 6.3 volt supply winding on the transformer. This arrangement was used because it was easy and simple to wire, it didn't hum much and it got the job done.

As amplifiers were later designed with more power and more gain, people started noticing the hum as a problem. It was about this time (mid 50s) that two important designs were implemented to guitar amp design. These two designs, destined to become the standard for guitar amplifiers everywhere, were the "twisted pair filament" circuit and the "humbucking filament."

THE TWISTED PAIR FILAMENT

Instead of wiring the filament circuit as a single-wire daisy chain, a pair of wires is used. The pair terminates on one tube's respective filaments, and then the pair is twisted until it gets to the next tube's filament. Then another twisted pair of wires goes from that tube to the next. So essentially, the filaments are still wired in parallel—just like in the case of the daisy chain, except with the twisted pair, neither wire is grounded and the pair of wires is twisted as they go from tube to tube. The twisting adds inductance, which minimizes hum. In order to have the absolute least amount of hum, the twisted pair filament circuit requires a ground reference that is exactly half way between each side of

All About Vacuum Tube Guitar Amplifiers

the transformer supply winding. There are three ways this can be done.

First, one can use a center-tapped 6.3 volt winding on the transformer. This center-tap, on the 6.3 volt winding, connects to the chassis ground. Since the center-tap is exactly in the center of the circuit, the circuit is balanced and hum cancellation is at its best.

Another way to produce a ground reference that cancels the most hum is to make an artificial center-tap using two 100 ohm resistors. In this design, the 6.3 volt winding on the transformer is not center-tapped. On each side of the winding, a 100 ohm ½ watt resistor is connected to ground. This balances the twisted pair circuit with respect to ground. There are several advantages to this style of circuit. One, if a power tube shorts plate to heater, the ½ watt resistors will blow—thus saving the 6.3 volt winding on the transformer from melt-down. The resistors act as a fuse. Second, two ½ watt resistors are less expensive than a center-tapped transformer winding.

Rather than use a center-tapped filament supply or two 100 ohm resistors, a small value potentiometer can be used in place of the resistors. In this design, a 100 or 250 ohm pot is used. Each end of the pot connects to each side of the filament winding on the transformer. The wiper of the pot is grounded. To get the least amount of hum, the pot is adjusted by ear for the quietest setting.

Note: When a matched pair of output tubes is used, the hum induced in each is amplified equally and the hum is cancelled perfectly with the 100 ohm resistors or the center-tap design. However, if the output tubes are not perfectly matched, the hum is unequal and the center-tapped transformer or the 100 ohm fixed resistors, though helpful, will not cancel the absolute most hum. That is why the hum balancing pot is sometimes used. With the pot, if the output tubes are not perfectly matched, the ground reference can be shifted just enough to maximize the hum cancellation. You will see the hum balance pot design on many Ampeg amps and some Fenders.

HUMBUCKING FILAMENTS

In most guitar amps an A.C. filament voltage supply is used in a humbucking fashion. The 12AX7 tube and all its cousins (12AU7, 12AT7, 12AY7, 12AZ7, 5751, etc) use a special type of filament circuit that is designed to cancel hum within the tube itself. Instead of using a 6.3 volt filament, a 12.6 volt center-tapped filament is used. Instead of connecting the supply to each end, the two ends of the filament are connected together (pins #4 and #5) and these are connected to one end of the 6.3 volt winding. The center-tap of the filament (pin #9) is connected to the other side of the 6.3 volt supply. Since the filament is essentially folded in half, the 12.6 volt filament (150 mA) becomes a 6.3 volt filament (300 mA) humbucking. Any hum introduced into a tube, an opposite hum is also introduced, thus bucking (canceling) the hum within the tube itself. Those clever tube designers!

I know what you are thinking. If the humbucking preamp tube is so great, why don't they use a humbucking filament in the power tubes? In almost all guitar amps, 15 watts or more, the output stage is push-pull. In the push-pull design, any hum induced in one output tube is also induced in the other output tube. Since these tubes are connected to a push-pull output transformer, the hum is phase cancelled; that is, provided you have the output tube filaments wired in correct polarity. Let's say you are looking at an amp with two 6L6 output tubes. The filament heaters on the 6L6 are pins #2 and #7. When you are wiring the filament circuit as a twisted pair, you have to make sure the wire that terminates on pin #2 of one output tube gets terminated on pin #2 of the other output tube. Likewise, pin #7 of one tube should go to pin #7 of the other output tube. You may want to check this on all the tube amps you own. If a heater filament is wired from the factory with incorrect polarity, the amp will still work—it just won't be it's quietest. I have seen thousands of amps come through my shop and about 60% of them have the filament circuits wired with incorrect polarity. This is a case where Murphy's Law defies the law of probability.

When using 12AX7 tubes wired in humbucking fashion (6.3 volts), polarity is not critical since each tube is canceling hum within itself. It is only on the output tubes that polarity is an issue. In the case of a single-ended output stage, you will not be phase canceling hum as in the push-pull design; but then again, if you only have one output tube, the amp is probably not very loud anyway.

A WORD ON DC FILAMENT HEATERS

I recommend staying away from D.C. heaters. Why don't I like D.C. filaments? They are not as reliable as an A.C. supply and the actual voltage is difficult to regulate. When a filament is cold and power is initially applied, the filaments draw five to ten times the normal current for about a second or two. Let's say you are playing the amp for one set then take a break. When you turn off the amp, the filaments cool off almost immediately, but the D.C. rectifier stays hot (thus reducing its amperage capability). You turn your amp back on, the filaments are cold and draw major current for a second and the D.C. rectifier blows. To get reliability, you would need a 25 amp rectifier for a 3 amp filament circuit. And not only that, since the filament heater circuit draws the most current of any part of the amp, a tiny resistance (even a fraction of an ohm) can make a huge difference in the actual D.C. voltage.

SERIES VS PARALLEL

Almost all heater circuits are wired in parallel. With filament heaters wired in parallel, if one tube's filament burns out, the other tubes will still light up and one can easily see which one isn't lighting. If the tubes are wired in series, if one tube dies, they all go out and each tube must be removed and tested independently. Also, with parallel wiring, one needs only use tubes with the same filament voltage. Current is a non-issue with parallel wiring as each tube can draw as much current as it needs to operate. However, with series wiring, only tubes that draw the same amount of current can be used within an amp. With a series design, the filament voltages of each tube are added up to equal the applied voltage. You may have noticed, mainly in the power transformerless amps, that a 50 volt power tube, a 35 volt rectifier tube and two 12.6 volt preamp tubes are used with the filaments wired in series. If you add those voltages, you get 110 volts. Sometimes a ballast resistor is added to take up another 7 volts or so. This circuit was hooked directly to the wall A.C. as a way to build a low powered amp without a power transformer. You may have seen this before as it was used on some of those "amp in the case" Silvertones and many 1960s 45 RPM record players. The one advantage to the power transformerless design was that the amp would work on A.C. wall voltage or D.C. battery voltage!

All About Vacuum Tube Guitar Amplifiers

EVERYTHING YOU ALWAYS WANTED TO KNOW ABOUT CAPACITORS

If you are servicing amplifiers or building project amplifiers, you may have wondered which different types of capacitors are available and what they sound like. You may also have wondered if different brands are better sounding or more reliable than others. Why do they have different kinds of capacitors in the same amp? Tubular foil, ceramic, electrolytic, silver mica, polypropylene and polyester capacitors are some of the types used in tube guitar amps.

WHAT IS A CAPACITOR?

Two conductors separated by a non-conductor exhibit the property called capacitance. This combination can store electrical energy. Since the two conductors are separated by a non-conductor, capacitors cannot pass D.C. electrical current, although they will allow A.C. signal to pass. The non-conductor in between the conductors is called a dielectric. All sorts of non-conductors can be used and each different dielectric will sound different.

Capacitors, also called condensers, oppose any change in voltage. Because of the many electrical properties found in capacitors, they can be used for a variety of different functions in guitar amplifiers.

Capacitors are microphonic. That is right, you can talk into one and hear it coming out of the speaker in a guitar amp. Some capacitors are more microphonic than others. In fact, some of the best microphones in the world are condenser microphones, which are simply a capacitor that is intentionally made to be extremely microphonic.

FILTER CAPACITORS

Every tube guitar amplifier has filter capacitors in the power supply. These are always large value electrolytic capacitors. All tube circuits need a high voltage D.C. power source. The wall electricity is A.C. power, so the amplifier will have a rectifier circuit to change wall A.C. to high voltage D.C. This is done with a step-up transformer to get the wall voltage up, and a rectifier—either tube or solid-state—to change the electricity from A.C. (electrons go both ways) to D.C. (electrons all go only one way). When the A.C. gets rectified to D.C., it actually becomes pulsating D.C., and it's not very smooth. Filter capacitors are used in the power supply to smooth out pulsating D.C. current and make it pure D.C. This requires a lot of capacitance. Remember, capacitance opposes any change in voltage.

Electrolytic capacitors are used where large values of capacitance are required. Electrolytic capacitors deliver the most capacitance in a given physical size than any other type of capacitor. They are also the least expensive price-per-microfarad than any other type of capacitor. These capacitors are made with an electrolyte paste placed between two plates. The plates are rolled up and placed in a cylindrical aluminum container, with one of the plates connected to the aluminum container, which acts as the negative terminal. Sometimes these are made with several capacitors in one container and these are called multisection type electrolytic capacitors or simply: can caps. The multisection types are used to conserve space. These resemble a metal tube.

American-made electrolytics are the best. Most foreign-made electrolytic capacitors are inferior to American-made electrolytic capacitors. Although they may read good on a meter, they allow too much ripple current, resulting in out-of-tune sub notes. Stay away from anything made in Taiwan. Cornell Dublier, Illinois Capacitor, and many other former American-made capacitors are now made in Taiwan and sound nothing like the earlier American versions.

Even the LCR British caps should be avoided, unless you desire an ugly, out-of-tune sub note that is not in tune with the note you are playing. Certain electrolytic capacitors made in Germany perform with excellence and are as good as American made.

Avoid smaller size electrolytic capacitors. If you have two elec-

All About Vacuum Tube Guitar Amplifiers

trolytic capacitors of the same value and one is twice the physical size of the other, the larger one will always sound better.

BYPASS CAPACITORS

Almost all preamp tubes are cathode biased. In order to prevent degenerative feedback and to get the best gain characteristics, the cathode resistor is bypassed with an electrolytic capacitor. These capacitors are typically 10 uf to 50 uf, at very low voltage—usually 25 volts or less. The D.C. quiescent current, coming into the cathode of the tube, travels from ground through the cathode resistor into the cathode of the tube. Remember, D.C. can't pass through a capacitor. A.C. sees a capacitor as a short circuit, or very low resistance path. So, the A.C. component of the signal will travel through the bypass capacitor (and not the cathode resistor) because it is the path of least resistance for A.C. This is why it is called a bypass capacitor, because the A.C. signal current bypasses the cathode resistor and goes through the capacitor instead.

Likewise, the American-made electrolytic capacitors are best for use as a bypass capacitor.

Sometimes, a bypass capacitor can become very microphonic. On a bypass capacitor with axial leads, if the leads are tight, the capacitor can vibrate and resonate on certain notes and this resonance can come through the speaker. I've seen this occur where the amp would sound like a rattle in the speaker and upon further troubleshooting, a resonant bypass cap was the source of the noise!

COUPLING CAPACITORS

When a preamp tube amplifies, there is both a D.C. component and A.C. component on the plate of the preamp tube. Generally, a coupling capacitor (also called a blocking capacitor) is used to block the D.C. voltage and to pass the A.C. signal voltage.

The Fender tweed amps used an Astron brand cap. Later amps used other brands. I like Mallory 150 series caps for preamp coupling capacitors. These are made by taking two strips of foil, which are coated with polyester (used as the dielectric); and then everything is rolled up like a cigarette. They are bright yellow in color. There is also a Mallory 152 series, which is the 600 volt version, that I like equally as well. These are physically larger and work great.

For the phase inverter section of the amplifier, I like to use the Mallory Orange Drop coupling caps. These are called Orange Drop because they are dipped in Orange Epoxy and look like an orange blob, however they are made similar to the Mallory 150s.

In the old days, there were coupling caps made from two sheets of foil separated by a sheet of paper as the dielectric. The foil/paper/foil would then be rolled up like a cigarette and dipped in wax to keep it together. These are true paper capacitors.

Silver mica caps work great and sound wonderful as coupling caps. Typical coupling capacitor values may range from 500 pf to .05 uf.

TONE CAPACITORS

Not always, but in most amps with tone circuits, the tone capacitors are sometimes configured where they are both the tone stack and the coupling cap. In this case, I like the Mallory 150 series polyester tubular foil caps the best. For the treble cap, which is generally a much smaller value than is available in a Mallory 150, I like either a silver mica or a ceramic disc cap. In the silver mica, small mica sheets are used as the dielectric whereas in the ceramic disc, ceramic is used. The Mica sounds smoother while the ceramic is clear and trebly.

GROUND CAPACITORS

On amplifiers with a ground switch, there is a 600 volt .05 uf capacitor going from the 120 volt A.C. power line to ground. These caps are not in the signal path and do not affect tone. Any type cap will work in this circuit. There is a reason why a 600 volt or better value is used. The capacitors are rated in D.C. working voltage. When a cap is connected to an A.C. power source, its D.C. rating should be approximately 5 times the A.C. voltage. So don't replace that 600 volt cap with a lesser voltage. It will blow up.

When I first started shipping Kendrick amplifiers to Europe in 1989, after thoroughly testing an amp, I converted it to 230 volts without removing the ground capacitor. When a conversion to Euro voltage is done, the ground cap (.05 uf at 600 volts) should be removed. Europe uses a balanced A.C. line, so you don't need the ground cap; but since the voltage there is 230 volts, if you were going to use a ground cap, you would need a 1200 volt rated capacitor. After the customer played the amp for about 10 minutes, the ground cap burnt up with a nice little cloud of smoke. It was hard to convince the customer that the amp was fine and there was no problem to fix— especially after the firework show.

All About Vacuum Tube Guitar Amplifiers

ALL ABOUT RECTIFIER TUBES

The three common types of tubes used in a tube guitar amp are output tubes, preamp tubes and rectifier tubes. Preamp tubes and output tubes require DC. voltage in order to operate, however the wall electricity is A.C. A rectifier tube is part of the power supply and its purpose is to convert A.C. wall voltage to D.C. voltage. In A.C. wall voltage, the electricity is moving two directions. A.C. voltage is 60 cycle, which means the electricity changes directions 60 times in one second. D.C. voltage on the other hand, only goes in one direction.

I like to think of a rectifier tube as a check valve. It allows electricity to only pass in one direction. So when A.C. voltage is connected to a rectifier tube, only the electricity that is going in one direction can pass. When electrical current tries to go in the other direction, the rectifier will not let it pass.

THE HALF-WAVE RECTIFIER

Originally, all rectifiers were made from tubes. An alternating voltage was placed on the plate of the tube and the cathode of the tube was hooked up to the D.C. circuit. When the alternating current on the plate changed to positive, the electrons would flow from cathode to plate, but when the alternating current on the plate turned negative (and remember this was happening 60 times per second), no current would flow. Like all thermionic vacuum tubes, a rectifier tube has a heater inside that warms up the cathode so that the electrons are encouraged to flow. Electrical current will not flow if the cathode isn't hot.

In the scenario above, only one half of each A.C. cycle is used because the half that wants to flow in the other direction simply isn't used and doesn't flow. Hence such a device is called a half-wave rectifier.

When the tube changes the A.C. voltage into D.C. voltage, the D.C.

is not smooth. Instead, it is pulsing at 60 times per second and if used like this, will cause the amp to have considerable hum. Therefore, a filter capacitor is used to smooth out the voltage. The cap charges up when the current is flowing and then discharges between the pulses when electricity is not flowing, thus keeping the voltage smooth. I like to think of a filter capacitor as being something like a battery. It stores electrical energy and the pulsating D.C. is absorbed into it, keeping the output voltage steady.

THE FULL-WAVE RECTIFIER

The half-wave rectifier is almost never used in guitar amplifiers because it wastes half of the cycle. Instead, a special rectifier tube is used that has two plates (rather than one.) Each end of a centertapped transformer winding is connected to one of the two plates. When one plate is positive, the other is negative and vice-versa. So when the first half-cycle occurs, one plate is positive and the electricity flows from the cathode to that particular plate. During the second half of the cycle, that first plate changes to negative and no current flows, however, the other plate changes to positive and current flows. So on each half-cycle, current flows from the cathode to one of the two plates. The electricity going into the cathode is always moving in the same direction. This type of rectifier is called a full-wave rectifier. Because it uses both half-cycles, it is very efficient and the pulsating D.C. that it produces pulses at 120 times per second. Even though the wall A.C. is changing direction 60 times per second, since each half-cycle is producing a pulse of electricity, the frequency is doubled to 120 pulses per second.

Of course, one will still need a filter capacitor to smooth these pulses out, but 120 pulses are much easier to smooth out than 60 pulses. And since both halves of each A.C. cycle are used, nothing is wasted.

Full-wave rectifiers are almost always the rectifier of choice for tube guitar amplifiers. Some common types are the 5Y3, 5V4, 5AR4, 5Z3, 5U4.

WHAT IS THE DIFFERENCE TONALLY?

There are two different types of cathode designs that are incorporated in the design of a rectifier tube; namely: directly heated cathodes and indirectly heated cathodes. The directly heated types use the filament heater itself as the cathode. So the heater heats up and electrons are emitted from it. You can see why it is called "directly heated" because the cathode is the heater. An example of a directly heated type would be a 5U4 or a 5Y3.

Indirectly heated rectifier types have a separate heater and cathode. The cathode gets hot because it is near the heater. Some examples of this type rectifier would be the 5V4 or 5AR4. These tube types are sometimes used in tube amps that lack a standby switch. Sometimes referred to as "controlled warm-up time" rectifiers, it actually takes more time for them to warm up than the other tubes in the amp. No D.C. power is produced until the rectifier is warmed up; but by that time, all the other tubes are already warmed up. So it is like having a built-in automatic standby switch.

Because of differences in design, the cathode of an indirectly heated cathode tube can be placed very close to the plates. For example, in a 5AR4, the cathode is only two hundredths of an inch from the plates. This results in very low resistance. On the other hand, a directly heated type such as a 5U4 has the cathode much further away from the plate resulting in much more resistance. The differences in resistance affect the envelope of the notes.

When too many electrons try to pass through a rectifier tube, the voltage output drops and all of the tubes in the amp suddenly get less voltage. This results in a phenomena sometimes referred to as "sag." When the voltage sags, all of the other tubes in the amp have less output temporarily. As the note starts to die and less current flows, the sag becomes less, voltage increases and the other tubes in the amp begin to have more output. Since the amp's output decreases slightly when a loud note is hit and the output increases as the note begins to die, sustain is improved.

OTHER TYPES OF RECTIFIERS

Besides the rectifier tubes mentioned, there are also mercury vapor rectifier tubes such as the 83 type, as used in the original 5F6 Bassman. This tube contains mercury vapor which ionizes when more current tries to flow through it. This has the effect of lowering the resistance as more current flows which will counteract sag. The tube looks cool because the ionized mercury vapor has a purple hue that gets brighter or dimmer with the notes. These tubes were never very popular and the only amp I remember using them was the 5F6 Bassman. It is interesting to note that the 5F6A Bassman changed to the 5AR4.

Besides tube rectifiers, sometimes solid-state rectifiers are used. The solid-state rectifier, also referred to as a "diode," is an electrical check-valve that only allows current to flow in one direction (from cathode to anode). Solid-state rectifiers have very low resistance and can handle much more current than a tube rectifier, which explains why they are almost always the rectifier of choice with the more powerful tube amplifiers. All blackface Twins and 100 watt Marshalls use solid-state rectifiers. Some examples of popular solid-state rectifiers are the 1N4007 or the 1N5399.

WHICH IS BETTER: TUBE OR SOLID-STATE?
Sag is basically the combined effect of current and resistance in the power supply. There are many factors in an amp other than the rectifier that can impact sag and with high current amps, there is usually enough sag without a tube rectifier. Everything you've ever heard by Jimi Hendrix, Eric Johnson, Robben Ford, or Coco Montoya was through a solid-state rectifier. These guitar players use high-powered amps such as blackface Twins and 100 watt Marshalls. These amps have enough sag because there is already some resistance in the power supply and with a lot of current flowing; the blossoming of the sag is there. On the other hand, lower powered amps don't draw as much current and usually sound better with the extra resistance that a tube rectifier provides.

Sometimes you will see two rectifier tubes in an amp. This is done for two reasons. First, two rectifiers will have half the resistance of one rectifier, so the sag is decreased. And secondly, two rectifiers can handle twice the current of one rectifier. For example, a single 5AR4 is rated for 250 mA (same as ? of an ampere). Two of them would handle 500mA or ? ampere of current.

CAN YOU PUT A TUBE RECTIFIER IN A 100 WATT MARSHALL?
Almost every day, someone asks if I can put a rectifier tube in a solid-state rectified amplifier such as a 100 watt Marshall. Less than

five minutes ago, while I was writing this article, a guy called and asked. I asked, "Why do you want a tube rectifier in a Marshall?" To which the guy replied, "To get sustain and sag." Most people are under the false impression that a tube rectifier is always better than a solid-state rectifier. Neither is better. It depends on the particular amp design. If you put a tube rectifier in a 100 watt Marshall, the rectifier would have so much resistance that the amp would overcompress and basically compress-off (is that a word?) the front of the note. If the player was a fast picker, it would be possible for him to outrun the amp! Not only that, one rectifier would have a hard time handling the current of a 100 watt amp and would be unreliable. You could use two rectifier tubes to handle the current and reduce the resistance, but at that point, you might as well have solid-state rectification.

TRICKS TO IMPROVE RECTIFIER TUBE RELIABILITY

Here is a little trick to improve the longevity and reliability of a standard five volt tube rectifier such as a 5Y3, 5U4, 5AR4, etc. You will need two 1N5399 diodes. Look at the rectifier socket. You will see two red wires that go to pin #4 and to pin #6 of the rectifier socket. Remove the red wires. Install the banded end of a 1N5399 diode to pin #4 and the banded end of the other 1N5399 to pin #6. Solder these to the socket. Now attach one red wire to each unbanded end of each diode. Basically, you have put a solid-state rectifier in series with the plates of the tube rectifier. This will not affect tone at all, but it will improve reliability by keeping the negative voltage off the plates, thus cutting the peak inverse voltage that the tube "sees" in half. In the event the rectifier tube shorts cathode to plate, the amp will continue to work and not go down on stage.

All About Vacuum Tube Guitar Amplifiers

TESTS OF CURRENT EL34 TUBES— HOW DO THEY SOUND?

Recently I had the good fortune to test all of the current production EL34 style output tubes. I tested these through a new Kendrick New Joy Zee amplifier with both clean and overdriven setting. There are many different brands currently being manufactured and this is what I found:

THE SOVTEK EL34WXT—This is easily one of the best sounding EL34 style tubes around. It had good output, as much or more than most any other EL34. The clean sound was chimey and rich with plenty of note definition. Think of the clean sound as a full bodied clean with good clear bottom and punchy clear lower mids.

When pushed to distortion, the Sovtek tube actually sounded very Mullardesque, but with not quite as much power as an actual Mullard. The overdriven sound was smooth and creamy—like it is supposed to be. The tube has plenty of bottom and loads of tone. Highly recommended.

THE TESLA EL34—Tesla makes three different versions of this tube. This one is the commercial version. They also make a higher quality military version called an E34L and the E34LS.

It's interesting to note that the British designate military grade by moving the last alpha character to the end to make E34L rather than EL34. The Americans (and apparently the Russians as well), on the otherhand, use a W designation (as in 6L6WGB) or change to a four digit name.

The commercial grade EL34 was stronger than either Russian tube tested. The clean sounds lacked high-end and note definition. It just wasn't as chimey as the others.

When pushed to distortion, the lead sound creamed out—maybe even a little too creamy—almost to the point of losing some note definition when playing chords. The front end of the note compressed nicely and would have been good for recording.

THE TESLA E34L—This is the military version of the tube above. It would probably also be wonderful for high-end audio.

I would describe the clean sound as having more top end with extra clarity—like a 6L6. Think Fendery black face clean with lots of high-end sizzle. These tubes had the highest fidelity of any tubes tested. The E34L Tesla also had more edge than other tubes except for its big brother, the E34LS.

THE GROOVE TUBE TESLA E34LS—This was my favorite of everything tested. Aspen Pitman, of Groove Tubes, has worked with the Tesla factory to improve the regular EL34s. This tube produces more power, because of its specially modified grid design. The grid is gold plated wire and the pitch of the winding is changed to produce more output at the plate. The new heat sink wings attached to the seams of the plates actually improve high power operation. Although Groove tubes literature suggested not to use current draw method biasing, I did it anyway and was careful to add another 10 to 20 percent more current draw. This biasing method worked perfectly and is an on-board built-in feature of the Kendrick New Joy Zee amplifier that I was using to test the EL34s.

The E34LS was easily the most powerful tube tested. It's clean sound reminded me of that curly blackface clean tone—not unlike what you would expect to hear coming out of a pair of matched black-plate RCA 6L6s.

The overdrive sound of the E34LS was creamy, but with good note definition. I cannot imagine anyone not liking this tube—it simply has too much going for it. The dynamic range was wider than other tubes tested.

THE SVETLANA EL34—This EL34 seem to possess a rather thin clean sound. Maybe it wasn't thin, but just lacked clear bottom and lower midrange.

When pushed to distortion, the tone became creamy—much like the commercial grade Tesla EL34.

CONCLUSION

I felt like the Groove Tubes Tesla E34LS and the Sovtek EL34WXT were special and seemed to be a cut above everything else. I compared both of these tubes to some NOS Seimens EL34s and the E34LS was as good or better. The Sovtek was as good tonally, but once again is not as powerful as either the E34LS or the Seimens.

All About Vacuum Tube Guitar Amplifiers

PACKAGING YOUR AMP FOR SHIPPING PROPERLY

With more and more amps being bought and sold on eBay and the multitudes of amps being sent off for service, it becomes more important for the person packaging the amp to be aware of proper shipping techniques. You can insure your packages all you want to, but the claim will be denied if they deem the packaging to be at fault. To compound matters, some of the cool, more affordable amps, such as Danelectro, Sears, Truetone, Airline, etc, are made from particle board and staples, so they are not very rugged to begin with. Sometimes it is best to remove tubes and sometimes it isn't. Sometimes it is best to send the whole amp sometimes just the chassis. Read on.

THE AMP CHASSIS OR THE WHOLE AMP?

If you are sending a combo amp for service and there are no cabinet or speaker problems, you will benefit by sending the chassis alone without the cabinet. This will save a bundle on shipping charges both ways and lighter packages are less likely of damage because the lighter package the less inertia, if and when the shipper bangs it around.

When shipping a chassis alone, it is important to remove the vacuum tubes and wrap the chassis in plenty of bubblewrap. You want to use the correct size bubblewrap. This can be purchased locally or through ULine Shipping Supply (1-800-295-5510). The smaller diameter bubblewrap is not useful for packaging amplifiers. After you wrap the chassis in plenty of bubblewrap, you will want to tape the bubblewrap so that it is taut around the amp. Use sealing tape. This comes in either a brown or clear style. Do not attempt to package an amplifier using scotch tape or masking tape. Although it would probably work, stay away from Duct tape. Sealing tape is the way to go. And don't be skimpy with the tape. Sealing tape comes in different thicknesses and widths. If you are using the thin stuff, you may wish to double-tape everything just for good measure.

The tubes should be packaged separately. If you have tube boxes for the tubes use them, but if they are loose inside the box, it is best to stuff something in the box so they do not want to bounce around. A small piece of bubblewrap or a couple of plastic peanuts will work well for this; or even a paper towel. You just don't want them sliding around in the box. If you don't have tube boxes, you can always wrap each tube individually with bubblewrap.

COMBO AMPS ARE DIFFERENT

With a combo amp that needs cabinet attention or speaker attention; you will need to send the entire amp. Also, with amp heads, it is a good idea to send the entire head because the cabinet offers very good protection and doesn't weigh that much. Most combo amps have the chassis on top so I would recommend wrapping the entire amp with several layers of bubblewrap and then placing the amp in the box upside down. This will put the heaviest part of the amp (the chassis) on the bottom. Of course, there is nothing assuring you that the amp chassis will stay on bottom as the amp could be jostled around during shipment, but you are hedging your bets by having it on bottom; and there will be less stress on the mounting screws that hold the chassis in.

I like to coil up the A.C. cord and wrap bubblewrap around it so that it will not flip around in the box during shipment. I secure the bubblewrap with sealing tape. On most combo amps, if the tubes are held firmly in the sockets and the A.C. cord is wrapped well, it will be OK to ship with the tubes still in the sockets. Leaving the tubes in an amp head will also work fine if the A.C. cord is secure. If on the other hand, the tubes are loose in the sockets. You may wish to remove them from the chassis and wrap them separately as described above.

If you leave the tubes in a combo amp, it is best to add bubblewrap to the inside open back to prevent the tubes from falling out. This will also help stabilize the A.C. cord. You don't want that cord flopping around inside during the amp's journey.

WHAT KIND OF BOX?

A good cardboard box makes all the difference in the world when shipping an amp. Double-walled cardboard with clean and crisp corners is the way to go. Do not use cheap, single wall cardboard or cardboard that is mushy. If you use a bad box, the shipper can use it as an excuse to deny a claim in the event of damage. Not only that, but a quality box can actually prevent damage problems. If there is a music store nearby, you may wish to go by there and ask for a discarded shipping carton. Or you might just go dumpster diving. If the store is a Fender dealer, every amp they receive comes in a box and they usually just throw them away. The box will be sturdy and double-walled, just what you need.

PRECAUTIONS TO TAKE

Never remove a tube from an amp unless you are sure about the tube configuration. I have seen certain oddball vintage amps that use obscure tubes with faded or missing markings on the tubes. If you don't have a configuration label and/or the labels are missing, you will encounter a nightmare to try to determine from the circuit which type of tubes go where. If there is no tube configuration label on the amp, you can always take a Sharpie or other labeling device and write directly on the chassis which tube type goes where. The same thing holds try of a tube. If it is missing its markings, you can easily write on the tube with a Sharpie to identify what type of tube. This could save a lot of time later on and if a person wanted to, they could always remove the markings later.

You never want to use plastic peanuts for packaging an amp. Remember how the breakfast cereal settles in the box even though it was full before shipment? The same thing can happen if you use plastic peanuts to package an amp. When they settle down, there is no protection and if the amp gets damaged, the shipper is not responsible for damage.

There may be an exception to the rule somewhere, but I would advise against taking your amp to a Pack N Ship place. One could develop the false security that they know what they are doing there. I have seen so many amps come from these type of places where they used plastic peanuts that settled, thus causing damage. How many times have we seen packages arrive damaged in our service shop and the customer was stunned that it could possibly arrive damaged, after all they had it "professionally" packaged. That kid working for minimum wage is not going to give your vintage amp the attention it deserves. It's not that he doesn't want to, he just isn't capable in most instances.

SHIPPING INDIVIDUAL SPEAKERS

When shipping a loudspeaker, one must be cautious of the fragile paper cone. I have found the best way. You want a box that is slightly bigger than the speaker. You cut a square of the plywood so that is just fits into the box and mount the speaker on the plywood. Masonite or pegboard will also work equally well. Mounting the speaker on the plywood, prevents movement inside the box. You may need to add a couple of thicknesses of cardboard on the back of the speaker to assure zero movement inside the box.

INSURANCE

When shipping an amp, I recommend taking out insurance for the full amount. It is fairly cheap and if the shipper damages the amp and it is packaged properly, you will get paid for what it is worth.

You can call a shipper such as UPS or FEDEX for a one-time pickup, but they will ask you the dimensions of the package and the weight, so make sure and get that before you call.

WHAT TO DO IF YOU RECEIVE AN DAMAGED SHIPMENT

In the event the shipping company manages to damage a product sent to you, there are a few things you should do.

1. Make note with the delivery person if the box looks damaged. You can write an exception on the delivery ticket if you like or make an exception and have the driver sign it.
2. Take photos—particularly if the box is penetrated or damage is seen from the outside.
3. Keep all packaging material. Do not discard any packaging material until after the claim is made and the shipper has inspected it.
4. Make a claim with the shipper and secure repair estimates.

All About Vacuum Tube Guitar Amplifiers

GAINING TUBE AMP KNOWLEDGE

With the surging interest in tube guitar amplifiers, many guitar players want more knowledge about their own tube amps. In the same spirit a race car driver wants to do his own valve adjustment and oil change, many guitar players insist on learning how to do their own overhauls and perform bias adjustments on their own vintage tube amps. Although vacuum tube technology has long been considered obsolete, there are still many resources available for those that wish to learn more.

MAKE A TUBE AMP FRIEND

Perhaps the quickest way to learn about tube amps is to make it a point to become friends with someone that knows more about tube amps than you. Do you know anyone that is "into" tubes? I once had a neighbor that was a designer for R.C.A. in the 50s and 60s. He even had something to do with the design of the first R.C.A. color television sets. He had many stories to tell and much info to share.

When I first started working with tube amps, I met someone that had been a Hammond organ repairman for several decades. He had tons of guitar amp schematics and other material from the 50s and early 60s, leftover from when he had been an amp tech at a local music store. Since he had quit servicing guitar amps long ago, he sold me about 100 lbs of Fender, Gibson, Ampeg, and Kustom schematics. In fact, my master Gibson schematic book that has about 400 Gibson amp schematics came from him. Seek and ye shall find.

VISIT THE UNIVERSITY

Any college or university will have an engineering library. If it has been around 50 years or more, you can rest assured on finding dozens and perhaps hundreds of books regarding tube technology. In my area, there is the University of Texas. Although I was not a student there, an "off-campus" library card could be bought for less than $50. I spent almost every Saturday there for a year. There were rows and rows and aisles and aisles of tube amplifier books. Most of these books were written in the 40s and 50s. Copies of the original Western Electric writings (bound in loose leaf binders) were even there! Besides books, there was a complete collection of *Audio Engineering* magazines from the 50s. I loved to read those old magazines and check out the ads and articles. I was reading one from 1953 and saw an announcement that a New Jersey Tube manufacturer, Tung Sol, had designed a new tube—the 5881! Also there were dozens of very interesting tube amp projects in those magazines. I made some of those projects and did experiments at home. Specifically, I remember making a negative feedback driven, active, tone control from a project article taken from that magazine.

READ INSTRUCTIONAL BOOKS

Besides libraries, there are many instructional books on the market. Besides the four books I have written, *A Desktop Reference of Hip Vintage Guitar Amps, Tube Amp Talk for the Guitarist and Tech, Tube Guitar Amplifier Essentials*, and *All About Vacuum Tube Guitar Amplifers*, there are many others available. My good friend, Aspen Pittman, has written a few books on tube guitar amps and in his latest book, there is over 800 schematics on CD ROM. Richie Fliegler, Tom Mitchell, Michael Doyle and Kevin O'Connor also have written much useful information about tube guitar amps.

You need to become friends with John Kinnemeyer, at JK Lutherie. Almost every guitar amp book currently in print can be purchased from JK Lutherie. Check his www.jklutherie.com website.

WATCH INSTRUCTIONAL VIDEOS

The Chinese say a picture is worth a thousand words. A video can sometimes be worth a thousand pictures! Pop up a batch of popcorn, kick back, watch and learn. It's better than the Discovery Channel. There are many instructional videos on Tube Guitar amps. I have produced two instructional DVDs—each with four hours of video instruction organized by topic, with point and click navigation.

With a video, you can learn by watching rather than reading. For some, this is the preferred method of learning. You get to see and experience, rather than read and conceptionalize. This is another area where JK Lutherie can provide assistance, as they sell almost every tube guitar amp instructional video in current production.

BUY (AND STUDY) A TUBE MANUAL

Yes, the *R.C.A. Receiving tube manual* (Technical Series RC-19) and the *GE Essential Characteristics Tube manual* are a necessity when dealing with tube amps. In these two important books, you will find thousands of American tubes listed. You can get important information about the tubes, including: pin out, suggested circuits, maximum ratings, output, filament current, etc. It is hard to find an original tube manual, but Antique Electronic Supply in Tempe, Arizona (www.tube-sandmore.com); has reprinted these essential reference books. I keep one of each of these manuals near every phone at my shop and I have two at home. I use them every day.

If you check out the example circuits in these tube manuals, you are likely to find nearly exact schematics of certain vintage amplifiers! Besides everything else, there are also sections to show what tubes will substitute for each other. There is even a cross reference for European tubes in the GE manual.

SEARCH THE INTERNET

They don't call it the information superhighway for nothing! How ironic to use the Internet to research tube technology. Hey, you can save it to hard disk and print it on your laser printer! When you are looking for information, try using www.dogpile.com for a search engine. This search engine will present your question to other search engines simultaneously and you will have plenty of what you are looking for.

Use the Internet to find suppliers, technical information, formulas, used equipment, used books, etc. In fact, there are many good books on tubes that can be found on www.ebay.com. My favorite book, for someone wanting to learn more about tubes, is *Basic Electronics*— Navpers 10087. Yes, it is the old beginner's electronics training manual for the United States Navy. This book takes you from knowing little or nothing and explains such electronics as power supplies, transformers, Class A and Class AB amplifiers, push-pull circuits, voltage dividers, voltage bleeders, etc. and much, much more. I have bought several of these books over the years. I found my first one in a used bookstore for fifty cents. I have bought them on ebay for $10-$15 depending on the auction.

Another important book to find is *The ARRL Handbook for the Radio Amateur*, published by the American Radio Relay League. Although it includes much information about ham radios, this book also has important information about amplifiers, voltage multipliers, power supplies, tube circuits, definitions, etc. You will find electronic theory and formulas written in easy to understand language.

One more book you don't want to miss, *The Radiotron Designer's Handbook* by Langford and Smith and originally distributed by R.C.A. There are several versions of this. The version I have is the Fourth Edition. It is 1500 pages just chock full of everything you can imagine (and probably more than you can imagine) concerning tube circuits. Expect to pay $100 or more if and when you can find one of these.

VISIT INTERNET BULLETIN BOARD

If you are online, check out some of the tube amp bulletin boards. A favorite is the Weber VST amp bulletin board, run by Ted Weber, who is not a relative, but a very knowledgeable designer and good friend of mine. On the Weber VST board (www.webervst.com), you will find links to other guitar amplifier bulletin boards.

The bulletin boards are meant to be interactive. You can post questions or comments and others can contribute answers. Perhaps you can even contribute information to someone that is not as knowledgeable as you!

ATTEND A SEMINAR

I couldn't pass up the opportunity to tell you about the Tube Amp Seminars and amp camps I host. It's a one-day, interactive format, seminar/workshop about tube guitar amps with a small group of participants. All experience levels are welcome. You will find dates and times on the Kendrick website—www.kendrick-amplifiers.com.

SHARE WHAT YOU LEARN WITH OTHERS

I encourage you to share what you know about tube amps with others and I promise you will further your own understanding. Just from putting what you know it into words, you will help clarify your own thinking. Not only that, sharing knowledge with others always opens new questions, which helps keep you growing. When you help another up a ladder, you find yourself closer to the top.

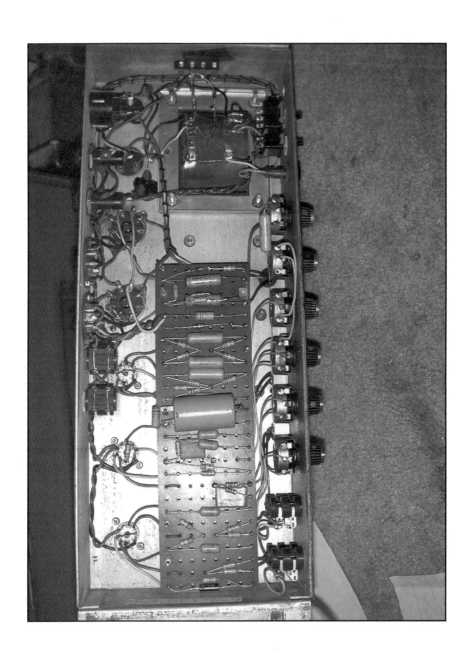

All About Vacuum Tube Guitar Amplifiers

TUBE AMP POP QUIZ

Let's start off with a little True or False quiz! Read each statement carefully and be sure you understand the statement. Answer True or False in the space provided. You even get to grade yourself when done!

1. TRUE OR FALSE ____When a tube amp makes crackling noises, one may remove the phase inverter tube. If the noise stops, the problem is in the output stage.

2. TRUE OR FALSE ____When checking multiple speakers in a push-pull amplifier, a 9-volt battery can be connected to the speaker lead. If one speaker cone pushes forward while the other speaker cone pulls backwards, the speakers are wired correctly.

3. TRUE OR FALSE ____To determine output wattage, simply measure the A.C. voltage swing across a dummy load resistor and divide by the load resistance.

4. TRUE OR FALSE ____When using a tube amp with a multiple impedance transformer such as one that has a 16, 8, 4 ohm output; you should use speakers that allow using the 4 ohm setting—in order to get the greatest power output.

5. TRUE OR FALSE ____ Removing the two outer output tubes of a four output tube amp can cut the output power in half, provided I use a speaker load that is half the normal "four output tube" rated speaker load. Or if the amp has an impedance selector, I simply set the selector to twice the "real" speaker load.

6. TRUE OR FALSE ____ If I double my wattage output power, this will double the volume.

7. TRUE OR FALSE ____ When replacing a 6L6 tube with another 6L6 of the same brand, it is not necessary to re-bias.

8. TRUE OR FALSE ____ When turning "on" a tube guitar amp, it is perfectly OK to turn the standby switch to the "play" mode at the same time as turning the power switch "on."

GRADE YOURSELF

Okay, let's set how you did! Give yourself 20 points for every "False" response.

160 POINTS—You know your vacuum tube guitar amplifiers—end of discussion. There's no point in you reading the remainder of this book.

120–140 POINTS—You are extremely hip when it comes to vacuum tube guitar amplifiers.

80–100 POINTS—You are much hipper than most other guitarists when it comes to vacuum tube guitar amplifiers.

40–60 POINTS—You are much hipper than the average drummer when it comes to vacuum tube guitar amplifiers.

20 POINTS—Few lead vocalists approach your expertise when it comes to vacuum tube guitar amplifiers.

0 POINTS—It looks like we got to you just in time!

QUESTION #1 FALSE. When a tube amp makes crackling noises, isolating the noise can be accomplished by removing the phase inverter tube. This breaks the signal path between the preamp circuit and the output tubes. If the noise stops, the problem is not in the output stage. We know this because the output stage occurs after the phase inverter in the circuit. If we pull the phase inverter tube and the crackling stops, the crackling can only be coming from either the phase inverter circuit itself or any other circuit that occurs before the phase inverter.

QUESTION #2 FALSE. This is somewhat of a trick question. When checking multiple speakers in a push-pull amplifier or any other multiple speaker amp, the speakers should be wired so that they move the same way at the same time. If a 9-volt battery is connected to the speaker load, the speaker cones should either all move forward or all move backwards. If one speaker cone pushes while the other speaker cone pulls, the speakers are wired incorrectly.

One could think of a speaker as an air pump. If the speakers are wired correctly, they are all pushing and pulling in unison. This moves a large wavefront of air in unison and results in beautiful tone. If one speaker is pumping forward while the other pumps backwards, then the air that is moved by one speaker is "sucked up" by the vacuum created by the other speaker and vice versa. This results in horrible tone and virtually no bottom-end.

QUESTION #3 FALSE. This one was almost true. To determine output wattage, one could measure the A.C. voltage swing across a dummy load resistor or speaker load, but this measurement *must be squared* before being divided by the load resistance. Basically we are going to square the voltage and divide this product by the load resistance in order to determine wattage.

Here's the exact textbook procedure: Simply put a 1000 Hz sine wave of 100 millivolts or so on the input of the amp. With a scope hooked up to the speaker or dummy load resistor, turn up the volume until clipping is observed. Now back the volume down until the clipping barely goes away. Using a true RMS A.C. voltmeter, measure the A.C. voltage across the speaker or dummy load resistor. Take this number, square it and divide the product by the speaker's nominal impedance in ohms. (Or in the case of the dummy load resistor, use the actual resistance.)

This will give you the RMS wattage that the amp is putting out, at the onset of clipping. Although the textbooks will tell you to use a dummy load, I am more interested in what the amp puts out with a speaker load, because I will be using a speaker when I play the amp and I will not be using a dummy load.

QUESTION #4 FALSE. When a transformer is wound, it is wound for the highest impedance output. That is to say that a multiple output transformer is wound for the 16 ohms and then the 8 ohm tap is added later by dividing the number of turns (on the 16 ohm winding) by the square root of two (1.4142135). To get the 4 ohm tap, the 16 ohm winding is center-tapped. (This is actually the same as taking the number of turns for the 8 ohm tap and dividing by the square root of two.)

Yes, the 16 ohm winding is the full and entire secondary winding. This being the case, it is the winding with the lowest turns ratio and therefore the least amount of coupling loss.

Had this question been in reference to an output-transformer-less solid-state amplifier, the 4 ohm load would provide the most power. In such a design, unlike a tube amp design, the speaker load is the operating load for the transistors. Less impedance means more current can flow, thus developing more wattage.

QUESTION #5 FALSE. Removing the two outer output tubes of a four output tube amp can cut the output power in half, however to correct the load impedance, use a speaker load that is twice the normal "four output tube" rated speaker load. Or if the amp has an impedance selector, simply set the selector to half the "real" speaker load.

QUESTION #6 FALSE. Most of us think in logical terms and it does seem logical that doubling the output power wattage would double the volume, however; such is not the case. If I double my wattage output power, this will cause only a 3 dB rise in actual volume.

Conversely, when I cut the power in half, I am only losing 3 dB. How much is 3 dB? It is the smallest increment of volume difference discernable by the human ear. It's not very much difference. For example, the volume difference between the number 1 and number 2 input jacks on a Blackface Fender (or even a 4x10 Bassman for that matter) reduces the input by 3 dB.

QUESTION #7 FALSE. When replacing any power tube, regardless of brand, one should always adjust grid bias unless the amplifier is a self-biasing amplifier. There are differences from tube to tube even with the same brand of tube from the same lot number.

A way around biasing every time would be to buy matched quartets instead of matched pairs. Bias the first pair and then when it is time to replace the tubes, you may use the other two tubes of the quartet thus saving the hassle of biasing.

Or you could just learn how to bias your amps.

QUESTION #8 FALSE. I would never recommend turning the standby and power switch "on" at the same time. To do so will put high voltage on the preamp and output tubes before the cathodes in them are warmed up. This "shocks" the tube internally and can cause "cathode stripping." Have you ever seen "dandruff" inside your tubes? It's what used to be the chemical coating of the cathode and for whatever reason is no longer intact.

Some amps have rectifier tubes with indirectly heated cathodes that take longer to warm up than do the output or preamp tubes. These types of amps do not need a standby switch because the other tubes are warmed-up before the rectifier tube actually produces power supply voltage.

Q&A

All About Vacuum Tube Guitar Amplifiers

FENDER

I have a question concerning a reply to a question posed in the October 2003 issue of *Vintage Guitar Magazine* in the Ask Gerald article. The question was concerning the tone of a late fifties Fender tweed Twin amp and your reply said to change the cathode bypass capacitor to a smaller value and also try a .68 uf cap in parallel with the electrolytic. What I would like to know, could this additional cap be added to any preamp tube or just to the first stage before the EQ? I am mulling around in my head a project that involves using a simple design like the 5F2—a Princeton only using a 6SL7 preamp tube and an EL34 power tube (single ended). I would like some input from you.

Yes, you can use this idea on other amps and on other stages. You are simply altering the value of the bypass capacitor (or adding a bypass capacitor) across the cathode resistor. Smaller values let less bottom-end through so you hear it as more mids and highs. Using a bypass cap always increases gain.

Your 6SL7, EL34 amp will probably be a killer but the 6SL7 has a little bit lesser gain than a 12AX7—think of it as similar to a 12AT7. You may want to try the 6SQ7 that has gain more like the 12AX7. Since the tweed Princeton is basically a Champ with a tone control added, there are losses from that tone control that need to be made up. The 5F2A amp design is a very low gain design to begin with and an EL34 needs more gain than a 6V6.

How can we get the gain up? The second stage cathode resistor is unbypassed on the 5F2A, so you could use a 25 uf at 25 volt bypass cap across the 1.5K cathode resistor to dramatically improve gain.

Has anybody played around with adding "dwell" control to the Fender AB763 chassis for better/more reverb control?

It's done every day. There is a 12AT7 reverb driver tube in the amp. This tube is connected to the reverb driver transformer. There is a 1 Meg grid return resistor on the 12AT7 reverb driver tube. The 1 Meg resistor comes from a 500 pf capacitor (and underboard wire which attaches to pins #2 and #7 of the reverb driver tube) and goes to ground. You would remove the 1 Meg resistor and replace it with a pot.

One end of the pot goes to ground and the other end goes to the capacitor—just like it is now. But you would have the wiper (middle lead) of the pot going to the grid (pin #2 and #7) of the tube. The existing wire goes under the board and attaches to the existing 1 Meg resistor and 500 pf cap. You may want to unscrew the board some and cook the solder joint (at the junction of the 1 Meg, the 500 pf and the under-board wire) so you can remove the wire that goes under the board. This wire will attach to the wiper of the dwell pot. It is just that simple.

I own a '66 Princeton Reverb that I am putting in new tubes. This amp has a tube chart stating it is a model AA764 with a date code of "PI" and the rectifier on the chart indicates a GZ34, which is currently installed in the amp. I have come across some information stating this chart may be incorrect and the model should be AA1164. Do you know what tube is specified for the rectifier on this chart? Most schematics suggest the 5U4GB. Every piece of documentation I can find says I should be using the 5U4 in this socket. I know the 5U4 draws twice the amount of current and I want to be sure this is the right tube for the amp (as not to burn up my power transformer). I am also most interested in making the amp sound it's best. It is working now, but I want it to be working at it's best. Most people seem to think the GZ34 is better sounding. Can you clarify?

Upon comparing the schematics I have, the only difference is the rectifier tube. Both use identical transformers and identical circuitry, therefore it is perfectly safe to use whichever rectifier you like.

I will confess that I didn't have a schematic of the AA764 with reverb, the only schematic I have is of the non-reverb model, but they may have made an AA764 with reverb.

Use whichever rectifier tube you like best but the GZ34 gets my vote.

My Fender amps get kind of muddy at high volumes and especially on the low "E" string. Would changing the output transformer help? What do you recommend?

Although the output transformer could be part of your problem, there are many other possible problems that could be the source of your muddiness. You want to check both the output transformer and the power transformer for rust. If the laminated core is rusty in either of these, I would recommend replacing it. When a transformer core gets rusty, it can become conductive from one laminate to the next adjacent laminate. Ordinarily, the laminates are lacquered so that they are not conductive with one another. If they become conductive, the transformer can't tell the difference between a secondary winding and the circuit created by the conductive laminates. So some of the energy that normally would have gone to the real secondary ends up on the conductive laminates and the electricity just goes round and round. (This phenomenon is called "eddy currents.") Since some of the power is being used up to make the eddy currents, there is less power available for the rest of the amp. Also, the transformer will run very hot.

Low plate voltages can cause muddiness. Either a weak rectifier tube or outdated filter caps can cause low plate voltages. If you haven't had a cap job in the last 10 years, it is time. Also, when you get a cap job, make sure to use American made caps. The Taiwanese caps used in new Fenders and by unknowledgeable service shops, will test fine on a low-voltage capacitance tester, however, they will not do the job when operated at the high voltages found in guitar amps. Brand new Fenders suffer from the problem you describe and the problem can be remedied by simply replacing the filter capacitors with American-made filter capacitors.

In the case of solid-state full-wave rectifier amps, such as a Bandmaster or a Twin, if one side of the solid-state rectifier circuit is open or blown, the full-wave rectifier circuit becomes a half-wave rectifier, causing the plate voltage to drop dramatically. In this case, the amp will still make sound, but will lack the high-performance feel.

I have an early Fender Deluxe that uses the 6SC7 preamp tubes. I am having problems finding 6SC7 tubes that are not noisy or microphonic. Should I rewire the sockets to nine pin, so I can use the more common 12AX7 tubes? Is this a worthwhile modification?
There are pros and cons either way. The argument for keeping the 6SC7 is that the amp will be original. If the amp is in an amp museum or if you don't really play guitar and just want to collect vintage amps, by all means, leave it alone. Or if you bought it to sell

later, I would advise leaving it alone.

If you are a player and just want the amp to sound its best, you may want to perform the modification. The argument in favor of the 12AX7 is tube availability. Also, the 12AX7 will give considerably more gain, having an amplification factor of 100, while the 6SC7 has an amplification factor of 70. This may not seem like much difference in gain, but when you figure there are two gain stages, the math ends up as 100 X 100 when using 12AX7 tubes, instead of 70 X 70 using the stock 6SC7 tubes. No circuit modification is necessary, but you will need a nine pin socket and a washer with the inside hole big enough to mount the nine pin socket. This assembly will replace the octal socket currently in the amp. Make sure and keep the old parts in case you ever want to put it back. Here is how to wire it:

6SC7		12AX7
pin #2	goes to	pin #1
pin #3	goes to	pin #2
pin #6	goes to	pins #3 and #8
pin #7	goes to	pins #4 and #5
pin #5	goes to	pin #6
pin #4	goes to	pin #7
pin #8	goes to	pin #9

I recently performed the "Silverface to Blackface" conversion (as outlined in your first book, *A Desktop Reference of Hip Vintage Guitar Amps* on a mid 70s master volume model Super Reverb. The amp has higher output tube plate voltages before and after the conversion than its Blackface counterpart. My question is "How detrimental are these higher plate voltages (505) plate volts versus 460 plate volts portrayed on the Blackface schematics) to the amp's ability to reproduce the Blackface tone and what, if anything, can or should be done to lower these voltages to Blackface specifications? Also, would a reduction in plate voltage change the operating parameters of the now "functional" bias pot which I can presently only adjust to a maximum of 30 mA output tube current?

There is more going on here than what meets the eye. In general, when the idle current of the output tubes decreases, the plate voltage increases. This happens because the transformer has an internal resistance. The actual wire that is on the inside of the transformer has a resistance. When current is run through this resistance, a voltage drop appears across the resistance (This is basic Ohm's law in action—resistance times current equals voltage). As more current is drawn, there is more voltage drop. Conversely, if less current is drawn, there is less voltage drop. The voltage drop is subtracted from the applied voltage and the difference is the actual plate voltage. (In transformer lingo, this difference is referred to as "copper loss.") That is to say that if you idled your tubes hotter, then the plate voltage would drop.

The real problem is that your bias parameters are such that you cannot idle the tubes any hotter than 30 mA. Look on the back of your bias adjustment potentiometer. There is a resistor going from one lead of the pot to ground. This resistor value must be reduced in order to increase the idle current parameter of the bias circuit. Usually a 15K resistor will do the trick. If by chance you already have a 15K, you will need a smaller value to get that idle current up.

When you get the output tube current up to 35 mA per tube, your plate voltage will probably go to around 485 volts or so. This is not much difference from an actual Blackface. I know the Blackface schematic shows 465 volts. In real life, the schematic is usually low and by actual measurement. A typical Blackface measures around 475 volts. If you get 485 volts from your amp, when idled up to 35 mA, that is not enough difference to worry about.

If you find, after getting your idle current up to 35 mA, that the plate voltage is still too high, you can get it lower by using a 5U4 or a 5R4 rectifier tube.

The plate voltage does affect the parameters of the bias circuit. Higher plate voltages want to draw more current (assuming the bias voltage does not change.) Therefore, if you change your rectifier tube after biasing, you will want to re-bias using the new rectifier tube.

I have a 5E6 Bassman that sounds great except occasionally it seems to cut out on me while in use. This is both unpredictable and disappointing because the amp cannot be trusted for a gig. I have brought this to several different amp shops that have checked

and rechecked the amp, but no one can fix it. What should I do?

Your amp can be repaired to play perfectly. It seems to be suffering from a phenomenon known as parasitic oscillation. Parasitic oscillation is a problem that occurs when the amp is feeding back (in this case, at a higher frequency than can be heard) and it draws more and more current so that all the power from the power supply is being used to amplify the inaudible high frequency oscillation. This gives the appearance of the amp shutting down, but in reality, the amp is using all its power to amplify very high frequencies. Why does it oscillate? Old amps usually have questionable grounds, especially on the control panel where corrosion is a factor. Any resistance here and it could become a phantom load! Perhaps humidity has been absorbed in the cloth hookup wire making it more susceptible to leakage. The circuit board, when old, could become somewhat conductive. This could cause a type of coupling (capacitive coupling), where the amp feeds back positively on itself.

Check all grounds and perhaps install a grounding buss. If that doesn't solve the problem, simply rewire the amp, using a new circuit board and new wire. Troubleshooting a parasitic oscillation requires considerable experience and a complete rewire will solve the problem without spending weeks trying to isolate the phantom circuits.

How does the "bright switch" on a Blackface Fender work?

The "bright" switch is simply a switch that connects a small value capacitor (120 pf) from the input of the volume control to the output. A volume control has three terminals. Looking from the back, the left terminal is grounded, the right terminal is the input and the middle terminal is the output. The bright switch connects the capacitor from the middle terminal to the right terminal. Since higher frequencies see the capacitor as an almost dead short, the higher frequency signals bypass the volume control altogether. The volume control attenuates the lows and mids, but the very high frequencies bypass the volume control through the capacitor without much attenuation. We hear this as bright. The control is more effective when the volume control is turned down and conversely, the "bright" switch will have hardly any effect when the volume is turned up. If the volume is all the way up, the "bright" switch will have no effect whatsoever!

When I bridge the two channels on my Deluxe Reverb amp, I get a decrease in volume as I turn the control up to around 4. As I continue to turn the volume control, it will then begin to get louder. This occurs on both channels when I use either volume control. This does not happen when I bridge the channels of my Bandmaster. Thanks for any help you can give.

I am assuming your Bandmaster does not have reverb. The Deluxe, since it has reverb, has one extra gain stage in the reverb channel that the normal channel does not have. Since a stage of gain inverts phase 180 degrees, my guess is when your volume is on 4, you are experiencing maximum phase cancellation. On the Bandmaster, both channels are in phase and therefore you don't have the same situation. To understand phase relationships, think of a sound wave as vibrating "up" and "down" (going "up" first and then "down" second). Now if you had an identical wave, except it started its vibrations as "down" first and then "up" second, and you had both waves on top of each other; then they would have a tendency to cancel each other out. It is like adding minus one and plus one; you get zero.

Can I hook my Super Reverb cabinet to my Blackface Deluxe Reverb Amp?

It is possible, but you would need to rewire the four ten-inch speakers in the Super to reflect an 8 ohm load. A stock Super Reverb is wired in parallel for 2 ohms, and the output of the Deluxe Reverb is 8 ohm. If you hooked this up without rewiring the Super cabinet, the amplifier would think it was operating into a somewhat shorted circuit. Zero ohms is a direct short, so you can see how 2 ohms would look almost like a short for an amp designed to run at 8 ohms.

I would recommend going with a parallel/series wiring. To do this:

1. Wire two speakers in parallel. The plus terminal of each speaker gets wired together and the minus terminal of each gets wired together.
2. Wire the other two speakers in parallel. You should now have two pairs of speakers.
3. Wire one pair of speakers in series with the other pair. To do this, a plus of one pair goes to a minus of the other pair. The minus from the first pair goes to the sleeve of your ¼" speaker plug and a plus from the second pair go to the tip of the plug.

4. Plug the speaker cabinet into the output jack of the Deluxe, set the volume of the amp on 10, and wail!

I own a 6G7-A Bandmaster head. The very early Fender piggyback (blonde Tolex with Maroon grille cloth) Bassman and Bandmaster amps had one 12" speaker in the speaker cabinet, followed very soon by a cabinet with two 12"s. I've read where Fender went to two 12"s because one 12" speaker just couldn't handle the power from the heads, but there is no other technical reason given. I would like to run mine with two 12" speakers (4 ohm). Wouldn't a cab with one 8 ohm speaker need an 8 ohm transformer, and two 8 ohm speakers a 4 ohm tranny? Or, did Fender just make those early piggyback heads all the same (4 ohms)? I have always thought the only 4 ohm speaker ever used by Fender was the 8" found in the Champ models. I've looked at some schematics in your first book, including the Bandmaster models 6G7 and 6G7-A with one (8 ohm) speaker. This used a #45217 output transformer. The next Bandmaster schematic you published is for the AB763 (Blackface) circuit, which shows a 2x12 (4 ohm) #125A6A output transformer. I noticed the 6G6 Bassman has a GZ34 rectifier tube. I noticed the 6G6-A and 6G6B Bassman, plus the Bandmasters beginning with the 6G7 model, have solid-state rectifiers. I don't think that has any impact on the speaker combination, but a 6G6 Bassman may just happen to have a tranny that was intended for a cab with one 12, unless the trannys were all the same. How can you tell if these early piggyback heads are correct for a one 12 or a one 12 cabinet?

If I could find a 125A6A (4 ohm) tranny and swap it for the 45217 (8 ohm) in my 6G7-A head, what other mods would I need to make for it to work properly for 2x12" speakers? Or, could a 125A13A (4 ohm) output tranny from the Bassman heads (6G6-A and after) be used successfully in my 6G7-A Bandmaster head?

I'm beginning to question whether Fender ever actually made any (blonde Tolex with maroon or wheat grille cloth) Bandmasters with 2x12 speaker cabs from '62 to mid '63 as claimed on pg 92 of the *Fender First 50 Years* book. Could the 2x12 cabs have been wired for 8 ohms (vs 4) and used with the 6G7-A heads? Or, was the

125A6A (4 ohm) tranny used only prior to the AB763 circuit?

During the summer of 1964, a band called "The Fabulous Goldenaires" was gigging at the teen club in Groves, Texas. The band had three 6G series Fender piggyback amps—two Bandmasters and a Bassman. All three amps used twin 12" speaker cabinets that were set up in a line with all three amps tilted back at the exact same angle. It looked as cool as it sounded. Sidney Polk was playing a Gibson SG through one of those amps, Dwight Landry played bass guitar through one, and the lead singer used the other one for the PA. I can still hear them doing Otis Redding's "Call me Mr. Pitiful."

The main difference between the 1x12" versions and the 2x12" versions of these piggyback amplifiers was the output transformer. The power transformers were changed slightly when the amp design was running a solid-state rectifier instead of tube rectifier, as in the case of the 6G6, verses the 6G6B Bassman. The output trannies were definitely eight ohm for the 1x12" format and four ohm for the 2x12". You are correct that Fender used 8 ohm speakers for everything except the Champ models.

That would not stop you from using two 16 ohm 12" speakers wired for a nominal impedance of 8 ohms! Keep the original baffleboard and simply make a new baffleboard to accommodate the two 12" 16 ohms speakers. This would leave your amp head with the original 8 ohm output transformer. If you ever want to sell it, you could easily put the baffleboard and speaker back to stock.

How can you tell which output transformer is for 8 ohms or 4 ohms? One way is to check the transformer numbers against the schematics. The schematics showing 1x12" are 8 ohm output transformers and the ones showing two speakers are 4 ohm.

If you are unsure of the numbers, you can always perform a turns ratio test to determine the intended nominal output impedance. With a variac, put 1 volt A.C. on the secondary of an output transformer and then measure the A.C. voltage on the primary. The A.C. voltage measurement on the primary will now be the "turns ratio." If you square the "turns ratio" and then multiply it by the intended output impedance, you will get the primary impedance. In the case of the Fender, they usually ran a pair of 6L6s at around 4200 ohm primary impedance. So working backwards, to find the turns ratio, you could divide the output impedance into the primary impedance. Take your answer and solve for the square root, which will be the turns ratio. It is really easier than it sounds.

For an 8 ohm output transformer—divide the 4200 ohms (primary impedance) by 8 ohms (secondary impedance) and get 525. The square root of that is 22.91. That means if you put 1 volt A.C. on the secondary and you measure 22.91 A.C. volts (or approximately 23) on the primary; you have an 8 ohm output transformer.

For a 4 ohm output transformer—divide 4200 ohms (primary impedance) by 4 ohms (secondary impedance) and get 1050. The square root of 1050 is 32.4. That means if you put 1 volt on the secondary and measure 32.4 volts (or approximately 33 volts) on the primary, then you have a 4 ohm output transformer.

You could replace the output transformer in your amp with a 4 ohm output transformer such as the 125A6A or the 125A13A if you like, but you will need to change the feedback resistor. The 4 ohm output will have less signal voltage and more current to develop the same output power. You need a smaller feedback resistor to keep the performance of the amp the same as with the 8 ohm output transformer. In fact, Fender changed from the 100K feedback resistor on the 1x12" 6G6 Bassman to a 56K for the 2x12" 6G6A.

I have a question concerning a Fender Bandmaster amp that I recently purchased. It is a 1968 model and it has been converted to a Brownface faceplate and it has been installed into a custom-built 2x10" cabinet. There have been no internal modifications made to the amp.

What harm can be done to this amp if I run a jumper from input 1 of the vibrato channel to input 2 of the normal channel? (I plug my guitar into input 1 of the normal channel.)

I recently purchased a new Marshall JTM 45 amp, and in the owner's manual they recommend doing the same thing to pick up extra tonal possibilities. I tried the same thing with the Bandmaster and it really expands the tone characteristics. I do not want to do anything to damage the Fender so I am asking for your help in this matter.

If this technique can't be used on the Fender, can you explain

why it works on the JTM 45 (or at least it is advocated by Marshall) and why it would not/should not be used on the Fender?

The technique you describe to "bridge the channels" can be used on any two channel amp whose preamp channels have the same number of gain stages. It is usually done on two channel amps that have two inputs per channel. You plug your guitar into the high gain input of one channel. Then you take a short patch cord and plug one end into the low gain input of that same channel, and the other end into the high gain input of the other channel. This enables you to run both channels at the same time. By adjusting both channels volume controls, you may blend the tone of one channel with the tone of the other channel and kick everything up a few notches!

This technique works fine on all vintage Marshalls, because both channels have the same number of stages before the signal hits the output stage. This technique works fine in most Fender amps that do not have reverb. For example, it will work fine in a Bandmaster, but not in a Bandmaster Reverb amp. It works great in a Showman, but not in a Twin Reverb. It works in a 5F6A 4x10" Bassman, but not in a Blackface Bassman because the Blackface has an extra stage in the bass channel.

Why doesn't it work in an amp whose channels have an unequal number of stages? When a tube amplifies a signal, the signal changes phase by 180 degrees each time it passes through another stage of gain. If one channel is 180 degrees out of phase with respect to the other channel, then a phenomenon called "phase cancellation" occurs, which actually cancels the signal.

If you have a two-channel amp, but with only one input per channel, you can still "bridge the channels" by using a "Y" cable. It is safe to do this unless you turn it up so much you blow a speaker!

On a Silverface Vibro Champ amp, would switching the power tube and rectifier tube—by mistake of course—cause damage to the amp? I know that I have done it once before, but realized it quickly and turned off the amp and I don't think any damage was done. I did it about a year ago and its worked just fine ever since. What would happen if you ran the amp with the tubes switched for a while?

You wouldn't run the amp long with the tubes switched, because the amp would not make sound at all! To determine the answer to your question, we need to look first at the pin out of the rectifier tube socket and see how that would connect to the 6V6. The rectifier tube uses pin #2 and #8 for the 5 volt filament and pins #4 and #6 for the B+ A.C. voltage. If you plugged a 6V6 into the rectifier socket, the 6V6 doesn't even have a pin #6, so you are safe there. Pin #4 on the 6V6 is the screen, so high voltage on the screen shouldn't present a problem. Pin #2 on the socket has 5 volts A.C. filament voltage which would connect to the 6 volt lead of the 6V6, but no current would flow because pin #7 of the 6V6 would not go anywhere. Pin #8 on the socket would connect to the cathode of the 6V6 so there is no circuit there. There is no problem if you plug a 6V6 tube into the rectifier socket on the Vibro Champ.

Let's look at the this the other way. Let's say you accidentally plugged the 5Y3 rectifier tube into the output tube socket. The 5Y3 uses pin #2 and #8 as the filament. Pin #2 of the 6V6 socket is indeed a 6.3 volt filament, but the other end of the filament is pin #7 on the 6V6 socket and the 5Y3 has no pin #7, so there is no circuit there. Pin #8 of the 6V6 socket goes to the cathode resistor which has no voltage on it. Pin #6 of the rectifier tube is a plate, but on the socket on the 6V6, pin #6 is used as a mounting post. So there is no problem plugging the 5Y3 into the 6V6 socket. So there is no damage done other than the amp will not produce sound.

What do you think about removing a master volume control from a mid 70s Fender. How would you go about it?

Anyone that knows me already knows the answer. I am not in favor of master volume controls in any shape, form, or fashion. Even if the volume control is turned all the way up, it still represents a path to ground which degrades the tone. In the 80s, I experimented with master volume controls and even made a prototype amp with a DPDT switch to take the master volume control "in" and "out" of the circuit. This was done so I could A/B what the amp sounded like with and without a master volume. The "no master" wins every time.

If your guitar playing is substandard and you lack technique, by all means, leave the master volume in there. You can crank the preamp, turn the master down and cover up your lack of technique, mispicking and mistakes. It will cover up the nuances of your playing and if you playing sucks, it will actually improve your overall sound. I describe is as "homogenizing" the tone. But if you are a seasoned player that demands the amp respond to you and your picking nuances, I recommend you get rid of that master before another sun sets.

The difference here is the difference between output tube and preamp tube overdrive. We've covered this before, but to refresh your memory, preamp tubes are all Class A single-ended. This means that the tube amplifies a sound by drawing more or less current. We've all seen a picture of a sound wave where it looks like a "hill" and a "valley." The preamp tube amplifies the "hill" portion of the wave by drawing more current and it amplifies the "valley" by drawing less current. When you overdrive the preamp tube, the "hill" portion saturates—which sounds fine; but when the bottom half of the wave is overdriven by drawing less current, it draws so much less that it cuts off! There are not different degrees of off. Once the tube is not drawing current, it is off and there is no more wave. That is why preamp distortion is so unpleasant to the ear. Play it for a few minutes and you will begin to suffer from listener fatigue. It's because you are only hearing half the note and the other half is cut off!

Now let's look at output tube distortion. The output stage of an amp is almost always operated push-pull. This means there are two output tubes instead of one. One tube amplifies the "hill" portion of the wave by drawing more current. The other tube amplifies the "valley" portion of the wave *by also drawing more current*. As the output tubes are overdriven, both the top and bottom of the wave saturate evenly—resulting in a warm and organic tone. When I talk about an organic tone, I am talking about something that sounds natural and not electronic—like a cello or a violin! Output tube overdrive is soothing to the ear. You never get tired of hearing it.

So how do we remove the master volume in a Silverface Fender? Easy, replace it with a straight wire. The master volume control will have three leads. Looking from the back, the lead on the left is grounded to the chassis. The lead on the right is the input (coming from the preamp section) and the lead in the middle is the output (feeding the output stage). Follow both the input and output leads back to the board and cut both of them. Now connect a straight piece of wire from the two points on the board where these wires were originally connected and you are basically done. You may want to remove the lead wires that connect to the pot. If you are doing this to an amp that has a current reissue counterpart, such as a Twin, you can go to your Fender dealer and buy a Blackface faceplate that doesn't have the master volume hole. Then your amp will look like a Blackface.

I would recommend converting it to a Blackface also while you are at it. This can be done with about $15 worth of parts. The amp will have more of a "sports car" feel after it is converted. Instructions how to do this are in my first book *A Desktop Reference of Hip Vintage Guitar Amps*.

I've got an original Fender Bassman 5F6-A on my bench that I'm restoring for a friend. My question is about the volume control's taper. In your first book, *A Desktop Reference of Hip Vintage Guitar Amps*, you say: "The volume pots used back then were listed on the Fender layout diagram as audio taper; however, they appear to be a much faster taper (get louder faster) than today's audio taper. Perhaps that is why Fender used a linear taper on the Bassman reissue." The Fender parts layout does list these volume pots as audio taper. This amp on my bench has one original pot in it (Volume Bright—I can tell by the pot date code) but it is a 1 Meg linear. I have a 5F6-A of my own (wired stock) which I then checked and my amp has 1 Meg linear pots for both volume controls (with original date codes). Does it make that much difference or should I consider replacing them? Could I mix and match a 1 Meg linear with a 1 Meg audio?

If you are playing the amp loud, then it really doesn't matter if you use Audio or Linear taper pots. The Audio taper is almost always used for volume pots because it comes up in volume slowly whereas the linear comes up all at once. It becomes a problem when a harp player is trying to get it loud without feeding back and he has to split hairs between too loud and not loud enough with the linear taper pot. In this case, the audio pot would be the best way to go. With the audio taper pot, the knob can be fine tuned easier because the pot comes up slowly and evenly. Some manufacturers deliberately use linear taper volume pots so the amp will seem to be more high performance than it really is. When a linear taper pot amplifier is A/B tested with another amp that uses audio taper pots, the linear taper pot amp would seem much louder because the linear taper volume control set to 3 would be about the electronic equivalent of an audio pot set to 7.

I have a '65 Fender Twin reverb, Blackface and I recently read a review on the new '57 twin reissue that stated the amp lacked the "Tone Sucking add-ons of tremolo and reverb." Is this true that these circuits rob the amp of its natural tone and if so, what is the best way to disconnect them in my '65 Blackface twin reverb? Also, how does a variac effect the voltages/currents in an amp when the voltage is turned DOWN to around 94 volts A.C. and what does this do to the output signal in terms of distortion?

Thank God for positive reviewers. You have got to love those that are always looking for the good in something. Rather than saying, "the '57 Twin reissue lacks reverb nor does it have vibrato," he chose to make it into a positive statement—almost turning the missing features into a feature. While it is true that reverb and vibrato can be poorly designed into a circuit such that the circuit would sound better without it, it is equally true that some people prefer reverb and or vibrato. In fact, there are those that would argue that a guitar tone with reverb is better than a non-reverb sound.

With the '65 Twin Reverb, the vibrato is designed using an opto-coupler that loads down the signal somewhat. When the amp was originally voiced, this loss was already compensated for. Should you wish to disconnect it, you need only remove the chassis from the cabinet and, looking at the back of the intensity control, disconnect the wire going to the right lead. This will disable the vibrato and bump up the gain by a few dB.

Having more gain may or may not be a good thing, depending on what kind of sound you are going for. When you disconnect the vibrato, you can pretty much kiss your clean sound goodbye. If you want to disconnect your reverb, you needn't bother, simply use the normal channel and you are there.

Regarding your variac question, consider that 94 volts is about 22% down from the normal 120 volts coming from the wall. Since all tube amps use a power transformer, all the voltages coming out of the secondary side of the power transformer will also be 22% lower than normal. That means your filament voltage that heats the tubes will be less than 5 volts instead of the 6.3 volts required to make the tube work correctly. That means the tubes will not heat up enough to be responsive. I believe a 6.3 volt tube should never be operated at less than 6 volts. To do so makes the tubes very sluggish—almost like they are wanting to fall asleep. Also your plate voltages will be down 22%. This will make the amp lose headroom, lose gain and lose responsiveness. If you are a relatively new player and play rather sloppy, this loss of headroom, gain and dynamics may be just the ticket to mask undeveloped technique. On the other hand, if you have mastery over your instrument, you will most likely prefer having the amp operate at original design parameters.

There is another possibility. A 6.3 volt auxiliary transformer could be wired directly to the 6.3 volt heater circuit in the amp so the tubes will be hot enough to be responsive and dynamic. At the same time, a variac could be hooked to the A.C. plug of the amp. This way, you could reduce the high voltages in the amp without affecting the 6.3 volt heater voltage. You could reduce the operating A.C. voltage to 94 volts, but then you would have to rebias the amp to bring the output tubes back up to proper idle. This setup would be responsive, the overall wattage output would be reduced, you would lose some highs, but the tubes would break up at a lower volume. This may be a cool way to go. I helped Tony Nobles, who writes the Guitar Shop column in this magazine, to set his amp up like this. He experimented for quite a while and experimented with different variac settings. He found himself setting the variac to the 120 volt setting, so he eventually removed the auxiliary filament transformer and variac and went back to stock.

I have a mid 70s Silverface Fender Deluxe (rewired to Blackface specs and totally recapped) that I absolutely adore. The one problem: it's not quite loud enough for some situations. So I've been thinking of modifying an amp, say a Bandmaster or Showman head or a Twin, to take four 6V6s, giving me 40 watts or so and that lovely squishy 6V6 sound. Basically I want just a bigger Deluxe, twice the power and two twelves.

So my question is: what power and output transformers would

work for that? If the extra filament current were a problem I could do without Channel One and pull that preamp tube, and for that matter I could live without reverb and tremolo. For the output, I'd like it to drive two 8 ohm twelves if possible. How could I work out the impedance difference between 6L6s and 6V6s?

So you want to modify a Bandmaster or Showman or a Twin? Great idea.

The filament current will not be a problem because a 6L6 tube draws twice the filament current of a 6V6. If you are taking an amp that uses two 6L6s and changing it to four 6V6s, the filament current for either scenario will be the same. In the case of the Twin, since it has four 6L6s output tubes, you will have more current handling capability than what you need.

As far as output transformers, if you start with an amp that uses two 6L6s, you could run four 6V6s with the same transformer. But remember to use the same speaker load as you did when you were using the 6L6s. In other words, if you took a 4 ohm Bandmaster amp and changed it to four 6V6s, you would still want to use the 4 ohm speaker load. This could easily be done with the two 8 ohm twelve inch speakers you mentioned.

That is not the case when converting a Showman or Twin. Because the Showman and Twin already use four output tubes, to correct impedances in the output stage, you would need to run a speaker load that is twice what the amp would have used with 6L6s. Hooking the 4 ohm Twin up to an 8 ohm speaker could do this. That would cause the reflected impedance on the tubes to be near what the 6V6 wants to see. With the Twin or the Dual Showman, you could run two 16 ohm twelve-inch speakers in parallel to get 8 ohms. If you used the Twin or the Showman, the output transformer has so much iron in it that the amp will have mucho headroom. This would be ideal for that big clean tone. If you used a single Showman, then you would need to run it at 16 ohms. Again, you could use the two twelve-inch 8 ohm speakers and wire them in series to get the 16 ohms needed.

The problem you will have is that of supply voltage. The 6L6 style amps will have considerably more voltage on the main power supply than what you would expect to find in a 6V6 style amp. There are five possible solutions:

1. Change the output stage from fixed bias to cathode bias. This would lift the tubes above ground by however much voltage was developed on the cathode. Since the tube would only see the difference between the plate and cathode, it would place the tube voltage more into operating range. For example, consider a 465 volt plate supply on a Bandmaster that you modified the output stage to take four 6V6 tubes. If you cathode bias the output tubes, a voltage of perhaps 40 volts will develop on the cathode. With plus 40 volts on the cathode and plus 465 volts on the plate, the tube only sees the 425 volt difference. This makes the tube "think" it is operating at 425 volts.

2. Add Zener diode in power supply. A 50 watt chassis-mount Zener diode could be placed in series with the centertap of the power transformer. Depending on the voltage value of the Zener, one could "eat up" a particular number of volts. 50 watt Zener diodes are cheaper than power transformers.

3. Use a dropped Screen voltage supply. Most Fenders are running the screens at nearly the same voltage as the plate. So instead of having the screen voltage supply really high, you would add a voltage divider to drop the screen voltage in half. Of course, you would have to bias for idle plate current draw with this scenario.

4. Use a different rectifier tube. Certain rectifiers tubes have considerable voltage drop. If you have an amp that uses a 5AR4, changing to a 5R4 might drop the B+ voltage by 50 volts! This could be an option, depending on the particular amp you are modifying.

5. Use a different power transformer. This option is the most expensive and most obvious solution.

Starting with the topology of a Blackface Super Reverb, what would be the tube wiring changes to be made when two EL84s would be used instead of two 6L6s. Let's assume, for purposes of this question, that our plate voltages are correct for the two EL84s. What I'm concerned with right now is the pinout and screen/grid resistors and associated wiring of the sockets. I want two-tube, push-pull wiring of EL84s.

It could be done. If you wanted to go cathode bias, you could wire it like the output stage of a Vox AC15 (which is in my first book, *A Desktop Reference of Hip Vintage Guitar Amps* on page 454). It even shows the pinout (except for the heater which is pin #4 and pin #5.) You may want to use a larger value long-tail resistor at the bottom of the cathode circuit on the phase inverter. Stock value is 22K ohm on the Super Reverb, the Vox AC15 used a 47K ohm. The 47K ohm will reduce the gain somewhat as it doesn't take much to drive a pair of EL84s. Also, you can use the existing output transformer if you deliberately mismatch the speaker load to twice what would normally be used. The Super Reverb is usually 2 ohm; so you can run it with a 4 ohm speaker load and the reflected impedance will be 8,400 ohms— which is just about right for a pair of EL84s.

I am having problems with my AA864 Bassman. I've replaced the 500 pf cap on the input to the phase inverter with a .02 uf. I love the extra bottom-end girth, but it motorboats like crazy with the volume and bass set above 7. I've checked all grounds and switched the location of the grounds for the screen supply, etc. Yesterday, I was talking to my friend, Dan Torres, who suggested replacing the pot for the bass control with a 100K pot. He also suggested replacing the two .1 uf caps in the phase inverter with .022 uf. Do you think these are good solutions? If not, do you have any other ideas?

Both suggestions seem valid. Sometimes reversing both primary and secondary of the output transformer will stop it. Sometimes isolating the grounds works.

To isolate grounds, ground the centertap of the B+ winding, the cathodes of the output tubes, the main filter cap ground and the screen filter cap ground all at the same spot and ideally as far away from the input jacks as possible. The filter cap ground for the phase inverter and the ground for the phase inverter cathode circuit should be near each other and in the middle of the amp somewhere.

The filter cap that feeds the preamp section should be grounded near the input jack and near the cathode resistor grounds of the first stages. This grounding scenario should straighten everything out. But you cannot screw it up. When I say ground all the power tube grounds near the B+ centertap ground, that does not mean string a wire from one cathode to the next and then to the chassis ground! It means one cathode has its own ground and goes to the grounding pint. The next cathode has its own wire, etc. Each individual ground terminates to the chassis near the grounding point and none of the grounds are in series with each other.

I have several mid 1960s Fender Blackface amps. To what extent is it necessary to get them out now and then and play them? What kind of "exercise" would be most valuable to keep these amps in proper working order?

When I was 26 years old, in 1979, my best friend was a 65-year-old Jew from New York named Merv Galvin. He liked to tell me his philosophy of life, which he summed up as, "Left to itself, everything turns to sh#*!" There was much wisdom in Merv's words. Let's say you have an automobile and it just sits in your garage and you leave it to itself. It will fall into ruin rather quickly. The suspension will settle; the tires will dry rot; the gasoline will form varnish. In short, it won't last long. The same could be said of an electric guitar amplifier. If you don't play it every now and then, the life of the electrolytic capacitors will be shortened, the resistors will absorb humidity from the air, and the dirt dauber will build his nest on the speaker cones. And we haven't even mentioned spider webs and insect larvae.

At the very least, you need to leave it on for several hours every few months. Play it hard and listen to it some, but leave it on for several hours. Even if you don't play it, if you at least turn it on and leave it in the play mode, it will reward you with a long life between over-hauls. Filter caps will deteriorate if they sit up a long time without D.C. on the electrodes. When you play it, the chemicals in the filter capacitors are stimulated. Remember, the electrolytic capacitors depend on an electro-chemical reaction to do their job. If there is a long absence of D.C. on the electrodes, the caps will deteriorate and not be able to filter out the 120 Hz ripple current that occurs when A.C. wall current is rectified to the D.C. electricity needed to run tube circuits.

Also, humidity can sometimes become absorbed into carbon composition resistors. By running the amp for several hours every few months, the humidity can be "cooked out"; thus insuring the amp doesn't remind you of your favorite breakfast cereal: snap, crackle, pop.

Besides playing your amps to keep them sounding good, you need to also pay attention where you store your amplifiers. Never put an amplifier in a place where it can be subjected to high humidity. Forget about storing them in a damp basement. You don't want to end up with a rusty chassis and rusty transformers! Never put an amplifier where it will be subjected to extremes of hot and cold. The condensation of water vapor that occurs when going from cold to hot could cause irreversible damage. Regarding storage of an amplifier, here is a rule to live by: Never put a tube guitar amplifier where you wouldn't sleep.

I would love to add a standby switch to my 1954 Fender Pro, especially since I run it with a plug-in solid-state rectifier module! To add this standby switch, I do NOT want a new hole in the chassis. I've seen amps with a toggle, 3-way switch for on-off-standby. Is that a feasible modification to consider? If so, how can it be performed? There is an easy way to replace the present on/off switch with a DPDT switch such that it is standby/off/on. First you will need an on/off/on DPDT toggle switch. This is a great way to go because you don't do any damage to your vintage piece. Remember to save the old switch when you do the mod. I would place it in a baggy and staple the baggy inside the cabinet. You will never lose it that way.

Wire one section of the switch so that one set of poles turns the amp "on," regardless of which way you throw the switch. To do this, one lead from your A.C. cord goes to the middle lead of the switch. There is a jumper wire connecting the two end-leads to each other. A lead from the transformer primary also connects to one of the end-leads of the first section of the switch.

On the other set of poles, wire from the middle lead to one end lead as the actual standby switch. You are simply breaking the wire going from pin 8 of the rectifier to the rest of the circuit. In other words, the middle lead could go to the rectifier socket (pin 8) and the end lead could go to the rest of the circuit. This way, the amp would function as "off" in the middle—"standby" when switched one way and "play" mode when switched the other way.

I recommend adding a small value cap going from pin 8 of the rectifier socket to ground. I use a .05 uf at 630 volt.

I want to make a new baffle for my '66 BFSR. What is the best wood to use? What thickness is optimal? Can I make it a floating

baffle? How exactly would this change the sound?

The Blackface Super Reverb already employs the floating baffleboard design. The baffleboard is supported on two edges (left side and right side) while the other two edges (top and bottom), were left unsupported. That meets the criteria for floating. The tweed Fenders were also floating, but the top and bottom were the supported two edges while the left and right side were left to float, unsupported.

Baltic Birch, though not the original baffleboard material, is what I recommend. It sounds clearer and more focused than the particleboard used in the originals. I would use either ⅜" (9 millimeter) or ½" (12 millimeter) depending on the weight of the magnet on the speaker. The ⅜" sounds best, but a heavy speaker will crack it. Actually the Baltic Birch is sold in millimeters.

When making a baffleboard, be sure and cut a 1" wide strip of ¼" plywood and install this all the way around the perimeter of the baffleboard. This is a spacer that keeps the grill cloth from touching the actual baffleboard. If you leave this off, the grill cloth will rattle like a snare drum and you will hear it rattle every time you hit a note. The perimeter strip can be made of any type of ¼" wood as it will not affect the sound.

I have a question concerning an early, mid 70s Twin Reverb (master volume model). What could cause all of the rectifier diodes to go bad? I realize that if one or two go, the rest are going to cascade fail. I don't want to replace the diodes just to have them short out again. I did replace all of the power supply electrolytics, as there was evidence that they were bad. The person who this amp belongs to had not powered it up for a few years, and decided to bring it up to full power slowly (over a few hours) using a variac. Toward the end of this exercise, the amp made a loud buzzing noise, the fuse popped, and then refused to turn on. It kept blowing fuses.

If the amp draws too much current, the diodes could easily blow. That amp uses 1N4007 diodes that are rated for 1 amp of current. A shorted output tube, a bias failure, an incorrectly installed filter cap, or an incorrectly installed bias cap are all legitimate reasons for the amp to draw too much current—thus blowing the diodes. If the amp had a tube or bias failure that caused the tubes to idle full throttle, then the diodes could easily go. If a filter cap or bias cap were installed with incorrect polarity; that too, could cause the diodes to draw more than 1 amp of current.

When an amp blows a fuse, it is a bad idea to keep putting fuses in to see if one will eventually not blow. When an amp blows a fuse, the source of the problem is never a bad fuse. Something inside the amp is drawing too much current. That current is causing something to overheat and by the time the first fuse is blown, whatever is going to overheat is already very hot. As the diodes get overheated, the current and voltage rating of those diodes diminish somewhat. If you keep replacing the fuse without fixing the problem, you are likely to force some type of system failure. A component will blow. And let's rejoice if it is only a few diodes and not a transformer!

You need to replace the diodes, remove the output tubes, turn the amp on and place the amp in the standby mode. Using a voltmeter set for D.C. volts, put the common lead of the meter to the chassis and touch the plus lead of the meter to pin #5 on either output tube socket. You are checking pin #5 of the output tube sockets for negative bias voltage. You should see somewhere around minus 45 volts from pin #5 to ground. Check both sockets. If there is no bias voltage or too little bias voltage, you must troubleshoot that before replacing the tubes.

If you are getting proper bias voltage, the correct troubleshooting procedure is to leave the power tubes out and switch from the standby to the play mode. If the amp blows a fuse, there is likely a shorted filter cap or a filter cap installed backwards with respect to polarity.

If the fuse holds, switch the amp to standby, replace only one of the power tubes and let it warm up. When it is warm, flip the switch to the play mode. If the fuse blows, that particular tube is bad and probably what caused the original malfunction.

If the fuse holds from the insertion of one output tube, then flip the amp back to standby and replace the other output tube. After it is warmed up, place the switch to the play mode. If the fuse blows, the second output tube is bad.

I've converted my tweed Twin to cathode bias output stage according to the method in your book. It worked. What's the optimum current to run on the plates to get the most compression/crunch out of this amp? In just about every test point, I'm

getting higher voltage than what the schematic lists. The phase inverter plate voltage reads particularly high (over 100% higher than listed). The amp seems to perform OK, but I haven't run it for long periods at high levels. Can I safely pull a pair of the 5881s and reduce volume and maintain tone? The neighbors are calling the police!

If you use a 500 or 600 ohm cathode resistor with a 100 uf at 350 volt cap, that should work just fine. You can fine-tune the cathode resistor value with a decade box to have your output tubes draw more or less current.

If your tubes are biased cold, the plate voltage will go up. You refer to the voltages listed in the schematic. The original schematic was written assuming the wall voltage to be 110 volts, which was normal in that day. Nowadays, the wall is no longer 110 volts. It is about 120 volts; which means an original amp's actual voltage will be much higher than the printed 1950s schematic.

Some of the extra voltage is a function of the schematic being based on 110 volts instead of the real wall voltage, but if it is extraordinarily high, that just means there is not enough current flowing through that part of the circuit. A larger than normal cathode resistor on the output tubes or a larger than normal long tail resistor (the 10K in the phase inverter circuit on the schematic), could make the voltage go up.

Also, with regards to the voltages in the phase inverter being high, are you using a 4,700 ohm cathode resistor or a 470 ohm? The 470 is correct. If you used too high a resistor, that too would make the voltage go up on the plates. I have seen people misread a color code and replace that resistor with a 470K or a 4.7K.

If you pull a pair of output tubes to get the volume and headroom down, you must double the cathode resistor value and double the speaker impedance. To keep proper load on the remaining two output tubes, you must use an 8 ohm load instead of the stock 4 ohm load. You must also double the cathode resistance value to keep the cathode to grid bias voltage correct. Since two output tubes draw only half the current of four, then the cathode resistor of the output tubes must be doubled to create the same bias voltage relationship.

What's the purpose of the bypass cap on pin #3 of the preamp tube in the Fender 5E1 Champ and why was it excluded in the

model 5F1? In my Silverface tweed modified champ, I have much higher voltages than those indicated in the tweed schematics. How does this contribute to the amp's performance? And finally, when I set the volume control on ten, and I play hard, I get a strong compression on the attack. It is much too strong for my taste. I'm forced to set the volume at 8 to get rid of this effect. Is there a way to avoid that? Anyway at 8 it sounds great.

The bypass cap was used on most 5F1 Champs even though it wasn't shown in the schematic. The bypass cap improves gain and stabilizes the bias voltage by providing a path for A.C. to enter the tube's cathode circuit without having to travel through the cathode resistor. A preamp tube will function without the bypass cap, however the A.C. current entering the cathode circuit of the preamp tube must do so through the cathode resistor. This extra current will alter the instantaneous bias voltage relationship between grid and cathode. You can try it for yourself so you can see what it does. Unsolder the ground side end (minus) of the bypass cap and listen. Then while you are playing, touch the lead back to ground and notice the increase of about 6 dB in gain. If you remove this bypass cap, you will have less gain and can turn the amp more without overdriving it. It will compress more, even though it has less gain! This is because without the bypass capacitor, fluctuations in cathode current cause the tube's bias to drift towards a colder setting when driven hard. This reduces the gain of the louder notes and we hear this as compression.

It is perfectly normal to have strong compression when the amp is dimed. That setting should only be used if you want that sound. At the risk of sounding smart alec, if you don't like the sound of the amp when you dime the volume control, quit setting the amp that way. If you are looking for headroom, turn it down or remove the bypass cap or both.

I have a tweed Deluxe re-issue amp and wanted your expert opinion of vintage speakers. Are the vintage speakers, both original cone or re-coned, a good choice for a replacement speaker? And are the modern day Alnico versions as good?

An original vintage speaker may work OK but one must bear in mind that an old speaker is old and may have deterioration problems related to age. For example, most vintage speakers used paper bobbin voice coil formers. Imagine a piece of paper that is 40 years old and that has been repeatedly heated and cooled over the years. Do you think it could become brittle? You could get an original vintage speaker that is in mint condition, but there is no getting around the age of the paper voice coil former.

A reconed original may work fine, but unless the exact replacement parts are used, it will not sound the same. It may sound fine; it just won't sound like the original.

As far as modern day Jensen reissue speakers, they are not really Jensens. They are Italian speakers made by an Italian company that paid Jensen to use their logo. In 1989, I tried to get the real Jensen Company in America to reissue the Jensen speakers and they informed me they would never make a musical instrument speaker again. It seems there is much more money in automobile stereo speakers and they were quick to point that out to me. Also, those foreign Jensens use metric parts and original Jensen speaker parts will not retrofit into them. I don't like any of the new Italian speakers with the Jensen logo except the P12N. I think they did a nice job on that particular speaker; however the other speaker models sound too thin and lack the efficiency needed to impress me.

I've been considering adding an extension speaker cabinet to my '64 Fender Super Reverb (via the ext speaker jack on the back of the amp). Before I move forward, I was wondering if you know what ohm rating cabinet I could use with this Super extension jack. The Super itself is wired at 2 ohms, but I was hoping to hook up my 4 ohm, '63 Bassman cabinet (with two 12" speakers in it) to the extension speaker jack on the back of the Super. The amp doesn't specify a load for that jack! I am hoping I don't need a 2 ohm cab for that, I've never even seen such a thing. Do you happen to know?

I would not run anything lower than an 8 ohm cabinet with the Super Reverb's stock wiring. Here is a way you could run the Super Reverb with the 4 ohm Bassman extension cabinet. Rewire the four tens in the Super Reverb to parallel/series. This will make them perform at 8 ohms. Add your Bassman cabinet (4 ohms) and that will bring your nominal impedance down to 2⅔ ohms. The Super Reverb wants to see a 2 ohm load, so the 2⅔ is about as close as you can get and still use the cabinets/speakers you have. Also, you do not want to run the 4 ohm cab without rewiring the speakers, because the stock Super Reverb speaker load is 2 ohms. Add the 4 ohm cabinet across that and you are getting down to $1\frac{1}{3}$ ohms. That is too low. You will be inviting tube failure. Zero ohms represents a dead short, so $1\frac{1}{3}$ ohms is getting pretty low—too low for comfort.

Thank you for explaining the Super Reverb ext. speaker question, but are you saying that an 8 ohm extension cab would be perfectly safe with my current setup (without rewiring anything)? What about a 16 ohm cab? What did Fender have in mind when they put that extension speaker jack on the Super Reverb?

The 8 ohm would work fine. It would drop the nominal impedance down to $1\frac{3}{5}$ ohms which is close enough to the 2 ohm output impedance of the amp. If you ran it that way, the four tens would get $\frac{4}{5}$ of the power and the two twelve cabinet would get the other $\frac{1}{5}$ of the power.

If you rewired the Super Reverb speakers parallel/series to 8 ohms and used a 4 ohm Bassman cabinet, the four tens would get $\frac{1}{3}$ the power and the two 12" cab would get $\frac{2}{3}$ the power.

The 16 ohm cabinet would work as an extension to the 2 ohm stock Super Reverb speaker load, but the 16 ohm cab would only get $\frac{1}{9}$ of the amp's power! You don't want to do that. I really don't know what Fender had in mind when they put the Ext speaker jack on a 2 ohm output amp. Perhaps, since they were using only a few chassis styles to make several different amps, they may have put the ext speaker jack on there just because all the other models had it. I do know that it is not a good idea to run a tube amp at much less than 2 ohms.

Is there a difference between a Silverface champ and the Blackface ones? If so, what? And is it easy to make Blackface again? I want to do the "virtual tweed" champ mods listed in your book, _Tube Amp Talk for the Guitarist and Tech_, and wasn't sure if everything was the same.

You can do those mods easily. The Silverface is either the same or almost the same, except it has better sounding transformers. Most of them had increased filtering on the main filter cap that connects to the rectifier, which I prefer because the bigger cap has the amp hum less and plays in tune better. I'll take a silver Champ over a black one any day. Look on the output tube socket. If there is a small value cap on pin 5, cut it out and you are Blackface. Some of them had it and some didn't.

Can I install a pot to control the negative feedback on a tweed Champ? If so what values are needed? 2. What is the best way to install a switch on the cathode bypass cap on the first stage of a tweed champ—so I can switch it in and out of the circuit? What value resistor is needed to prevent popping noise—and what is the best way to wire the switch—simply lifting the ground of the cap?

You don't want to get too much negative feedback, so I would design it with a pot in series with the existing feedback resistor. When the pot's resistance is zero, you are at dead nut stock. When you open up the pot's resistance, you get less negative feedback. Since the tweed Champ uses a 22K ohm feedback resistor, I select a pot whose value is about 6 to 8 times the feedback resistor. Let's see, that comes to 132K—176K and that is between standard values. So, I would try a 100K or a 250K pot for this. The 100K would probably give you more usable range because after cutting back the negative feedback to a low amount, cutting it lower will not make any audible difference.

I have not tried this modification on a Champ, but on a Fender Super Reverb, a 5K ohm pot works best. The Super Reverb uses an 820 ohm feedback resistor. This is about a 6 to 1 ratio (potentiometer resistance to feedback resistor ratio). However, the Super Reverb is 2 ohm which means the feedback voltage is less than an equivalent 4 ohm circuit. I think I would experiment with the 100K ohm and the 250K ohm pot for the tweed Champ circuit and go with what gives the best range on the potentiometer.

To cut gain on the 12AX7 by disconnecting the cathode bypass capacitor, you only would need to interrupt the ground connection on the bypass cap that connects to the cathode of the 12AX7 (pin #3). Using a switch by itself will probably not pop very much. But if you wanted to make sure it didn't pot at all, you can wire a 50K ohm, ½ watt resistor across the switch. Actually, when you wire it across the switch, it is in series with the bypass cap when the switch is open. This will sound the same as if the bypass cap is disconnected because the 50K resistor will prevent electrons from wanting to take that path. But when the switch is closed, it will short the 50K ohm resistor and put the bypass cap into the circuit. This will increase gain by having the cap in the circuit.

On a 5E2 Princeton, what does the filter choke do? And does it affect the tone? I read that Fender removed it from the circuit on the 5E2-A model and later models. What was the reason?
All chokes are simply a coil of wire that basically resists changes in current. Just like a capacitor resists changes in voltage, the choke resists changes in current. If a choke is configured with a capacitor on either side, as is the case in the 5E2 Princeton, it is called a Pi filter. Fender experimented with this circuit on many models in the late fifties. I don't like it as well as other circuits. One of the problems with that design is that all current going through the output stage must also go through the choke on its journey through the power supply. A choke resists changes in current, so it actually suppresses the output stage dynamics to some degree. If you have a 5E2, it will sound better if you move the output transformer red wire (B+) to the rectifier side of the choke. Fender themselves did this with later designs involving a choke and never went back to the other way. It is still a Pi filter from the vantage point of the screen supply, but looking from the output transformer, it is a simple capacitance input filter. In other words, the current going through the output stage doesn't ever go through the choke on its journey through the power supply.

Also, if you happen to have the 5E2 or any other tweed Princeton, it will sound better without the tone circuit. There is too much loss associated with the tone circuit. There is a .005 cap that goes from the tone pot to ground. Disconnect the ground connection of this cap (while leaving the cap connected to the pot in case you ever want to put it back) and you will have removed the tone stack from the circuit—electrically. But, if you ever want to put it back, it is only one solder joint away from stock.

I've got a Silverface Princeton with a suspected bad power transformer. When I opened up the amp, I found a dead short in the original capacitor can, which burned the 1K ohm 5 watt power resistor in the power supply. I am pretty sure this took

out the power transformer. The plate voltage measures at 457 volts (pin #3 of V5) and 462 volts (pin #3 of V6). The interesting issue is the filament voltage measures zero volts across pins #2 and #7 (V5 and V6) and measures -0.00001 across the pilot light. Just out of curiosity, what damage to the power transformer would physically cause a 0.0V filament voltage?

If the voltage on pins #2 and #7 of the output socket is measuring zero, that doesn't necessarily mean the filament winding on the transformer is bad. It is possible that you measured zero volts because the wiring connecting the two green wires to the circuit may be open or the soldering is bad. The pair of green wires from the power transformer are the actual filament winding. You need to check the filament voltage across the two green wires. One meter lead goes to one wire and the other lead to the other wire. Set your meter for A.C. voltage. If you accidentally set the meter for D.C., you will get zero volts!

If there is zero A.C. voltage between the two green wires, then the power transformer needs replacement. The filament winding is open. But there is more going on here if that is the case. An open filament winding did not happen because of the shorted cap. The filament circuit is a different circuit and does not interact with the shorted cap or the burnt 1K resistor or their associated circuits. I can see how a shorted cap could blow a 1K—5 watt resistor but the zero filament voltage problem is a different problem.

In your second book, *Tube Amp Talk for the Guitarist and Tech*, I read page 402, where a guy restored a 70s Fender. You mentioned, "If you leave the master volume in as stock, and do the Blackface mod, it will load the input impedance of the Blackface style phase inverter." I don't really care for the master volume anyway, and since I did the B.F. mod it sounds like I'd better remove it, correct? So do I just jumper between where the shielded wire goes to and from the master volume pot? Do I need a shielded wire? What about the "pull boost" feature—should I remove that, too?

The two wires going from the board to the master volume pot need to be removed from the amp and a short straight wire should jump between the two points on the board where the wires originally connected. The jumper need not be shielded.

You can dump the boost circuit on the back of the master too. Simply remove the wires that are going to the pull-switch on the back of the master volume pot. Just remove the wires from the amp and that will be sufficient. The "pull boost" simply takes some of the signal that drives the reverb and routes it around the reverb circuit for extra gain.

On a '71 Twin I removed the Master Volume wires, jumpered in a straight wire to bypass it and removed the wires going to the extra gain switch. But the amp has a lot of background noise now (not effected by volume control) and there is major hiss when I turn up the volumes and tone controls. Is this normal? Or do I need to remove the resistors from that extra gain circuitry as well?

Also, when it gets hot I'm getting a sound like brushing your hand over a live microphone. I pull tube V4 and it goes away. I swapped tubes and still there was no difference. Can this be related to removing the master volume or is this something else? "Chopstick" testing led me to the wires going to the tone controls. I resoldered and it seems better, but is still there.

The amp hissed before, but you probably never noticed it because the master wasn't turned all the way up. Since you removed the master volume, it is as if you had turned the master wide open. Now the hiss is noticeable. And besides that, your amp seems to be suffering from parasitic oscillation. This is when a phantom circuit exists in your amp. It can manifest itself many ways. One way is the "brushing up against the microphone" sound. There are several lead dress mistakes made in the Fender design that contribute to this problem, however; most problems can be corrected by shortening the grid wires in the tone circuit, changing the input wire to a shielded type, and changing the volume control wire to a shielded type.

You want to modify both channels to achieve better lead dress. You want to move the tone caps (that is the 250 pf treble cap and the .047 uf and .1 uf cap, typically found in the tone circuit) off of the board and mount them to the pots directly. If you move the slope resistor too, (this is the 100K ohm resistor connecting the treble cap with the mid and bass cap), and mount it between the treble cap and mid+bass cap that are mounted to the tone pots, it is possible to remove two or three feet of grid wire from the amp. Grid wire acts as an antenna to cause parasitic oscillations. Removing two or three feet will make a huge difference in the stability of the amp.

Next, you want to change the input wire to a shielded type. Notice the input jacks and how two 68K resistors apex together and a wire runs from the apex of the resistors to the grid (pin #2) of the 12AX7. You want to change this wire to a shielded wire and ground the shielding only at one end (usually the jack end).

And finally, you want to replace the unshielded grid wire on the volume control wiper to a shielded wire. This is the wire that attaches from the wiper of the volume pot on each channel and goes to pin #7 (the grid) of a 12AX7. Again, you only want to ground the shielding on one end—usually the pot end works best. These modifications of lead dress can mean the difference in an amp that is "acting up" and very noisy to an amp that is church-mouse quiet and totally noiseless.

I have a couple of questions about changing the fixed biased design of a Fender tweed Twin to cathode bias. Your examples in your books are pretty strait forward. I am cross referencing Torres and his chart and formula comes out with a much higher Resistor value, but it is for only two power tubes. I believe the tweed Twin plate voltage is lower than a Blackface. Mine measures about 380 volts. How does this affect the value of the cathode resistor?

Generally speaking, the cathode resistor value for a four output tube amp will be about half the resistance of one used in a two output tube amp. Four output tubes draw twice the current of two tubes, so the resistance of the cathode resistor is cut in half so that the voltage drop across the resistor will be about the same. Remember Ohms's Law: current times resistance = voltage.

If your Twin is 380 volts, it is very low. I've never seen one that low unless the output tubes were idling too high. As the idle current goes up, the resistance of the internal winding of the transformer creates a larger voltage drop across it—which is subtracted from the actual plate voltage. If you remove the tubes, the plate voltage will go up because there is no current going through the resistance of the B+ winding. Conversely, as more current is drawn, the plate voltage goes down.

With a two output tube amp, the 35 mA per tube plate current sounds about right, but when you go to four output tubes, 70 mA may be too much. In actual listening tests, 50 to 55 mA is what most people like the best. It allows some breakup, some headroom and a good feel.

Let's say you are using four output tubes and you want to end up with -48 volts bias. Both output tubes together would draw about 110 mA plate current, so you use Ohm's law to find the resistor value. Current equals voltage divided by resistance. (I = E/R) or .110 Amps = 48 divided by resistance or .11= 48/R

When you solve for R (436 ohms), you round it to the nearest standard value say 400 or 470 ohms. Now you know how much resistance, but you need also know what wattage. To figure the wattage value, you would calculate as follows: voltage square divided by resistance equals actual wattage on the resistor; now double it for safety. 48X48 = 2304. Then 2304 divided by 400 = 5.76 watts. Multiply times 2 for safety and you need a 12 watt resistor. Twelve watt resistors are not very standard, so you will probably go with a 15 watt.

You will need to experiment with the actual resistance value because as your plate voltage goes down, the amount of bias needed to tame the current also goes down and the resistor value needed would become smaller. Nothing beats starting with an educated guess and then experimenting with other values to hone in on exactly how you want it. When it comes to cathode resistors, a smaller resistance value develops less cathode voltage which causes more plate current.

I have a Fender Bandmaster (6G7-A) that I would like to set output tube bias. When I monitor plate current, I am reading 16 mA on one tube and 29.5 mA on the second. The tubes are matched Sovtek 5881WGC. Is there any components I can play around with to get these spot on, taking into consideration this is a cathode bias amp?

Anytime you notice a mismatch in plate current on a set of tubes that are supposed to be matched, you need to know if the mismatch is because the tubes are not matched or because of a malfunction in the amp. First try swapping the output tubes in the sockets to see if the current goes with the tube or stays with the socket. If it goes with the tube, your tubes are simply unmatched. If it stays with the socket, you have got something wrong with the output stage. It could be both!

I modified a CBS Silverface Champ with a tweed conversion kit, and also properly biased the amp. When I turn it on and play my 1984 Stratocaster, the amp sounds great—lower lows and bell like highs with slight break up. Much better than the original sound. But, when I turned the volume up past 4, chords seem to get mushy or washed out. It seems like the higher the volume, the mushier the chord. Is there a modification that allows the volume to be set to 8-10 and get that nice break up, and get a nice chord voicing with out the wash out? Or am I asking too much of this Champ?

There are basically three things you can do to address the problems you mention. Namely, you can beef up the power supply, use a better speaker, and or remove the bypass capacitor on the first gain stage. You mentioned you properly biased the amp. Both the Silverface Champ and the tweed Champ are self-biased (same as cathode biased) and therefore do not require any adjustment! So I am at a loss at what you meant by "properly biased the amp." Remember the Champ is "Class A," so it should idle at about 50 mA of plate current.

The Champ's circuit really needs higher than stock microfarad values in the first two filters on the power supply. The stock values are not helping your "mushy response" situation. I use a 100 uf at 500 volt for the main filter and a 20 uf at 500 volt for the second filter. The main filter is the one that connects to pin #8 of the rectifier tube. Actually on yours, it connects to an eyelet on the board and then the wire goes on over to pin #8 of the rectifier. The second filter is the electrolytic capacitor that supplies the screen voltage. The stock values cannot store enough power to supply the tube when it needs it. Upgrading to the 100 uf cap and the 20 uf not only provides more power when needed, but it will help the tone and keep the amp playing in tune at higher volumes. One purpose of the filter caps is to remove ripple current from the power supply. Ripple current gets worse as you turn up the volume, so the bigger filters hold down the ripple current, thus helping the amp to play in tune. When the ripple current is not filtered properly, it will modulate the note you are playing and make it sound out of tune. Ripple current is 120 Hz— exactly twice the 60 Hz that is coming out of the wall.

The speaker that is stock probably needs replacing. It sounds self serving for me to say it, but the truth is: the Kendrick 8" Blackframe speaker is the best sound you will get out of a Champ. It is made with an original Donal Kapi cone and sounds great, whether playing clean or overdriven. The newer Jensens, which are not really a Jensen product but a product from a foreign company that simply pays Jensen to use their logo, will sound thin and hollow. I recommend staying away from those foreign speakers unless you are going for a really thin sound. Weber VST, who is not related to me in any way even though we have the same last name, makes some excellent products that will sound good in a Champ. When replacing the speaker, remember to use a 4 ohm speaker. An 8 ohm speaker, such as Celestion, will rob a large percentage of the power and therefore should be avoided.

On some tweed Champs, there was a 25 uf at 25 volt bypass capacitor that connects across the cathode resistor on the first stage. Some of the amps had them and some did not. Look at the preamp tube and specifically look at pin #3. There is a wire going from pin #3 to the board. It will attach to a 1.5K ohm resistor. This is the cathode resistor. If your amp has the 25 uf at 25 volt cap across the resistor, remove it. This will cut the gain down a little so that the volume control will have a more usable range as you desire. Try it. If you like it better than stock, leave the cap out. If you like it better with the capacitor across the 1.5K cathode resistor, them put it back in.

And last but not least, don't expect everything from every amp. Try to see where the amp sounds its best and set it there. If it sounds great until you turn it past 8, don't turn it past 8! The idea is to have it be the best it can be for what it is, and a single-tube practice amp is not going to have a pristine clean sound turned up past a certain point. Allow the amp to put it best tonal foot forward, so to speak.

In a pinch, a guy brought in his Fender Bassman re-issue which needed an output transformer. He needed it for the weekend and I only had one of your Kendrick 2410A-8 output transformers in stock. I left the speakers configured for 2 ohms. I know I could have wired 4x8 ohm speakers in series/parallel, but I didn't. Bad? The customer's only comment was that the amp was more responsive and louder. Could it be a result of the 8 ohm transformer? I'm sure we improved on the quality of the Fender production OT anyway,

but I'm still curious as to what your thoughts are on using your 8 ohm Bassman OT with the speakers running at 2 ohms.

The amp may have well been louder, but I am certain it shortened tube life considerably. When you run a 2 ohm speaker load on an 8 ohm output tranny, the reflected impedance on the primary is much less than what it should be. For example, the 8 ohm Bassman style output transformer is a 22.91 to 1 turn's ratio.

When you run it with a 2 ohm load, and do the math to determine the primary impedance, you would square the turn's ratio and multiply by the actual load impedance to determine primary impedance. This is 524.86 X 2 = 1,049.72 ohms. The output tubes are normally running into 4,200 ohms, so at 1,050 ohms, they are 3/4 of the way into a short circuit (a short circuit being zero ohms). This is louder because there is much more current flowing. Don't expect the tubes to last very long.

I'm looking at buying a friend's late 70s Silverface Fender Pro Reverb (high power version), but when I plugged into it, I felt electricity running through my hands from the strings. I shut it off immediately. What could be going on? Also, what are your thoughts on these amplifiers?

Most people just don't realize that the guitar strings on an electric guitar are connected to the amp's chassis. The strings are first connected to the ground connection of the guitar's output jack. This is done by running a wire from the bridge to the output jack ground. While the guitar itself doesn't put out very much voltage, the ground of the output jack is connected to the shielded part of the guitar cable which leads to the amp and connects to the ground connection of the amp's input jack. The amp's input jack ground is connected directly to the amp's chassis; and that chassis may or may not be grounded. So in essence, the strings on your guitar are connected to the amp's chassis! If the amp's chassis is grounded properly, when you touch your strings on the guitar, you will be touching ground and you will not be shocked. However, if the amp's chassis has an electrical charge on it, then so will your strings. If you are standing on ground and you touch these electrically charged strings, your body will complete the electrical circuit and you will have a shocking experience.

I remember playing in bands in the 60s and we would set everything up to play a gig and if we were shocked by the guitar or microphone, we would simply rotate the A.C. plug 180 degrees and that would usually stop the shocks. If I was singing and the microphone had lightning jumping to my lip, I would rotate the A.C. cord on either my amp or the PA and that would stop the problem.

But somewhere around the late sixties or early seventies, I can't remember when, someone came up with the idea of using a three-prong A.C. cord that had a third connection—earth ground. The third wire was actually internally attached to the amp's chassis and the third wire on the A.C. wall receptacle was connected directly to earth ground. When this 3-prong A.C. plug was inserted into a properly wired A.C. receptacle, the chassis itself was directly connected to the earth ground via the third prong. So this setup virtually eliminated the possibility of getting shocked—just so long as the A.C. wall receptacle was wired correctly. The late seventies Pro should already have this 3-prong A.C. plug. And if it is wired correctly, you should not be getting shocked. Something is amiss. In your case, there are four possible explanations why the Pro shocked you. The A.C. receptacle was wired wrong, or there was a 3-to-2-prong adapter on the A.C. cord (which would defeat the third prong), or the A.C. cord on the amp was wired incorrectly, or the A.C. cord was bad.

What do you think of the Fender Blues Junior? Is it a decent little inexpensive tube amp, or an overpriced practice amp?

There are many amps that a person could purchase today. In fact, there are probably more really good tube guitar amps available now than any other time in history. These amps fall into two categories: gourmet and fast-food.

The fast-food amp, designed to have an affordable selling price, will appeal to the masses while still allowing the dealer to earn a substantial profit. The physical look and dealer margin are the primary considerations of this type of design. As long as it makes some kind of sound, looks good, has a substantial dealer margin, and is easy for the dealer to unload, it meets the criteria. With this type of design, in order to keep the price down, corners are cut with regards to components. Cheap transformers, ¼ watt resistors, printed circuit boards with computer ribbon cable and push-on connectors (instead of soldered connections) are common. Taiwanese filter caps, Russian or Chinese tubes and even manufacturing in third world countries with unskilled labor are the norm. The cabinets are always made from cheap woods such as green plywood or particle board. Generally the cheapest possible speaker is used. This entire corner cutting keeps the manufacturing cost and the selling price down while keeping the dealer margin up. If this type of amp malfunctions, the repair technician will likely have to disassemble the entire amp just to replace a resistor, so the servicing cost could actually rival the cost of the amp. Think of it sort of like a disposable razor. It is cheap, you can shave with it, and you can throw it away when it malfunctions. The best buyer for this type of amp is anyone that reads a menu from right to left. Just like MacDonald burgers, the market for this type of amp is huge. Many bedroom pickers or beginners want to have some type of sound for their electric guitar and with limited funds, the selling price of such an amp is appealing.

The gourmet amp could be considered as the opposite of the fast-food amp. Designed up to a standard instead of down to a price, the primary consideration is tone. Price is an afterthought. People that buy the gourmet amp usually read a restaurant menu from left to right. Custom-made transformers, American filter caps, ½ watt carbon composition resistors, point-to-point construction, and top of the line speakers are the norm. Besides the quality of components, cabinets are generally made from superior sounding North American white pine and in some cases even antique pine. The amps are assembled by skilled labor—usually musicians. Because quality ingredients cost many times more than lowest bidder parts, the gourmet amp is rarely sold through a dealer, even though it may be priced at more than the fast-food amp. Most gourmet amps are sold direct, so there is no dealer margin included in the price. This type of amp is usually bought by the seasoned player that cannot be satisfied with fast-food.

To answer your question about my opinion, let me point out that before I was in the amp business, I was in the gourmet meat business for 14 years. I dealt only in the finest quality beef, shrimp, king crab, and lobster. I started building gourmet amps in 1989 because I was not satisfied with any amps sold in music stores. My tastes are simple, the finest will do. If I play a fast-food amp, to me, it sounds broken because I am already so spoiled with gourmet. However, the fast-food amps are very important in the marketplace because there are so many players that could not own a tube amp otherwise.

The new Deluxe Reverb Reissue amps have a circuit board and smaller caps with higher rating (so I am told). Does this make it a problem swapping for Spragues? Is there a detailed layout of the upgrades and modifications to these newer amps in any of your books or any web articles that you are aware? You might include the overall quality difference of the modern design changes on the reissues compared to the original 60s amps.

The only upgrade you need in the Fender Deluxe Reissue is to replace the electrolytic filter caps (I use the Sprague ATOM brand) and you should be good to go. The Spragues will fit on the printed circuit board. The Taiwanese caps that come stock should be replaced immediately. There is another chapter in this book that addresses recapping a vintage amp, however the information will apply to reissue amps as well.

Of course, the stock tubes in the reissues leave much to be desired. If it were my amp, I would change to JJ Brand all the way across the board on the tubes. These are the best new tubes in my opinion. Without looking at one, I can't remember what actual microfarad values Fender used for the main filters (these are the ones that connect directly to pin #8 of the rectifier tube but are mounted under the cap pan), but if they are using two 350 volt caps in a totem pole (series) configuration, use two 220 uf at 350 volt value capacitors. This will help the amp to play in tune when overdriven and the bottom-end will be more robust.

I would like to know how to add reverb to the normal channel of my Blackface Fenders. I have a Super Reverb and a Bandmaster Reverb. What is the "easy mod" to get reverb on both?

There are a couple of different ways to do it. Here is the way I like the best because it will also add vibrato to the normal channel.

You will need:
1. 500 pf capacitor 200 volt or better
2. 10 pf capacitor 200 volt or better

3. 1 Meg resistor ½ watt
4. 3.3 Meg resistor ½ watt
5. Hook-up wire
6. You will also need a soldering pencil and solder.

For starters, remove the chassis from the cabinet and locate the reverb driver tube. This is the third preamp tube from the end in most Fenders. It will be a 12AT7 and the reverb driver transformer will be connected to it. If you look at the socket, you will notice that pins #1 and #6 are tied together, pins #2 and #7 are tied together and pins #3 and #8 are tied together. You want to separate pins #2 and #7. These are the grids. After the two are separated, you will notice one of these pins will have a wire going to the board. On the pin that DOES NOT have anything going to it, connect one end of the 1 Meg resistor, one end of the 500 pf cap, and the piece of hook-up wire. Solder all three to the pin. Now take the other end of the 1 Meg resistor and solder it to ground. At this point the 500 pf cap will have one end soldered and the other end sticking up in the air.

Now look at the first preamp tube and notice that pin #6 goes to a coupling cap (usually a .047 uf) on the board. Notice a wire comes out of the board near the other end of the cap. This wire goes from there to one of the 220K resistors that feed the phase inverter input. (If you don't know where the phase inverter input is, it is the other end of the capacitor that connects to pin #2 of the preamp tube next to the output tubes.) You are going to disconnect that wire from the 220K resistor and move it to the loose end of the 500 pf cap. You may need to shorten this wire to keep everything neat. Don't solder this yet as you will need to add another wire to this connections later.

Next, take the 3.3 Meg resistor and put it across the 10 pf cap. These will be connected to each other in parallel. Locate pin #7 of the fourth preamp tube. There will be a wire already on pin #7. Leave it there but add one end of the resistor/cap subassembly. You want zero lead length so that the other end of the resistor/cap subassembly stands up in the air. Solder this connection.

Lastly, connect a piece of hookup wire to the loose end of the resistor/cap subassembly. The other end of this wire goes to the same 500 pf cap that was installed earlier the reverb driver tube. The airborne end of the 500 pf cap is the correct end to connect and solder the wire to. So basically, you will end up with two wires soldered to the 500 pf cap.

This will give you a little more gain in the normal channel, about the same amount as the reverb channel. If you wish to lose some gain to clean the amp up, you can solder a wire from ground to the loose end of the 220K resistor that you remove a wire from earlier. This will clean it up. If there is excessive noise, you may need to replace all the new hookup wire with shielded wire. When using shielded wire, make sure and ground the shielding but only at one end. The other end of the wire should be clean and possibly shrink-tubed to insulate the shielding.

Is it normal for a Fender Super Reverb to make almost no sound with all the tone knobs at 0 in the vibrato channel? In that condition the three tone knobs become like a volume pot. When I turn up one of them sound comes on. I remember that my Mesa Boogie Mark I behaved something like that.

On the vibrato channel of Blackface or Silverface Fenders that feature a middle control, it is perfectly normal to have the volume go away when you turn all three tone controls knobs off. The tone controls are subtractive. In other words, when the tone knobs are each all the way up, it is the full signal passing through the circuit. When you turn a tone control down, you are cutting out those frequencies within the range of that control! So, if you cut out highs, cut out the mids and cut out the lows, there isn't anything left. In a Fender channel which does not feature a middle control, such as a normal channel, there is a resistor that sets the amount of mids. The exact amount of mids is dictated by the resistor that is substituted for a middle control. You can turn the tone controls down and the volume wouldn't go completely away since the fixed resistor always stays the same thus always having a certain amount of mids.

Is there a way to reduce gain in the reverb channel of a Super Reverb AB763? I must tell you that reverb pan is missing and vibrato pot is disconnected.

The gain would go down considerably if you hooked back up the vibrato intensity pot. Also, you can change the third tube to a 12AY7 or remove the cap that is going to pin #3 of the third tube.

Any of these measures can reduce gain. And if you did them all, you may have reduced it too much. So I would recommend listening after each modification and simply let your ears decide.

I have a one year old Fender '65 Deluxe Reissue amp that sounds great to me but exhibits a quirky behavior. The power lamp goes out when the amp gets hot enough. The sound/performance doesn't seem to be affected. I have already changed the lamp twice, and the new lamps do the same thing. Any insight into what might be causing this behavior would be appreciated.

The problem is minor and can easily be repaired. The lamp socket is corroded and needs cleaning. Unplug the amp first. Remove the lamp. Use a cotton swab and alcohol to clean the inside of the socket. Never attempt to change or remove a pilot lamp with the amp on. One false move and you could short out your power transformer, thus causing irreversible damage.

If cleaning with the swab and alcohol doesn't fix the problem, maybe the socket is not making a good connection. In this case, you could either retention the socket or replace it.

I had a mod kit from Torres Engineering installed by a local amp tech. The amp is a Fender Hot Rod DeVille and it sounds absolutely awesome but it has a very small hi-end buzz when you turn it up past 3. My amp tech keeps telling me that it is probably a characteristic of the amp itself but it didn't do this before; do you have any suggestions or quick fixes for this.

It is a parasitic oscillation. It is caused by improper lead dress, improper grounding, or bad layout. There are only a few options to address such a problem:
1. lower the gain
2. decrease the treble response
3. alter the lead dress
4. change the layout
5. perform a remedial circuit.

These are very hard to deal with and few shops will even touch them. The parasitic circuit is a circuit that occurs, but is not part of the schematic—like a ghost circuit. That is why it is so hard to find. I suggest you complain to Torres engineering. It is their mod that caused it and perhaps they either have a fix for it or will feel compelled to find one.

I am having a little issue with a moderately obvious popping noise when the Vibrolux circuit is in idle, and it is in sync with whatever speed the vibrato is set at. I can't hear it when I am playing though. A friend has the same issue with his Blackface Deluxe Reverb.

I've used shielded cable around the optocoupler and floated the ground point on it, but still have this thing happening.

You have any experience or suggestions about this? I have never observed it on any other Blackface Fenders I have played. Answer from the same person. He found his own answer and emailed me back: Just to close the loop on the popping vibrato noise issue, in case anyone else asks, here's what finally worked on my rig.

On the AA270 Twin Reverb and Vibrolux Reverb circuits, I noticed that Fender added a .022 mfd across the 10 meg resistor associated with that optocoupler. I tried this, and it worked great. The ticking is gone.

I'm restoring a Fender Deluxe Reverb Blackface. The label inside the cabinet says AB 763 Production #4. The rectifier tube is shown as a GZ34. Upon looking at the amp circuit, it is in fact an AB 868. The rectifier tube that came with the amp has no markings on it other than GE. It is about the size of a 6V6. I have to assume it is a GZ34 as it is too small (from the ones I have seen) to be a 5U4, which is normally used in this particular circuit. The voltages are 337-337 A.C. off the power transformer. In your book, *A Desktop Reference of Hip Vintage Guitar Amps,* **you caution against using the GZ34 in the Deluxe calling for a 5U4. Can I safely use a GZ 34 and convert this back to a true AB 763 or is the voltage too high? Or should I use a 5U4 which draws more filament current? I'm confused about what to use. I don't want to stress out the power transformer, if it is in fact designed to be used with the GZ34.**
Either rectifier will probably work. I would not worry about the fila-

ment current because there is enough safety margin built into the design. However, if you are ever concerned about the current rating of a rectifier tube 5 volt winding, there is a simple test to see how the transformer likes the extra filament current. You can take a voltmeter, put it in the A.C. mode and measure the A.C. voltage from pin #8 to pin #2 on the rectifier tube socket. Turn on the amp and check the voltage. If you measure 4.9 volts or better, then the transformer is handling the current well. If the voltage falls below 4.9 volts A.C., then the wire on the winding is too small and cannot carry enough current to the rectifier filament.

But rather than give you a "one size fits all" answer, I think the key here is which rectifier tube does the amp like. Try the 5AR4/GZ34 first. Then, bias the tubes for about 30 mA of plate current. Now change your meter to the D.C. voltage setting and check between pin #3 of either output tube to ground. If voltage is over 430 volts, then the 5AR4/GZ34 is not the best choice. I would change to the 5U4. But remember, since the 5U4 puts out less voltage, the amp will have to be rebiased with the 5U4 rectifier in the circuit to get the best tone from, the amp.

I have a 5C5 Pro amp that really doesn't sound very good. I have had a cap job done, a basic overhaul and new tubes; but she won't give up the goods. What would you do to bring out the best in this amp? Also, if I wanted to change the preamp tube circuit from grid leak bias to cathode bias, would I set up V1 and V2 like V3 with a 2.5K cathode resistor and 25 uf bypass cap?

The 5C5 is not one of my favorite Fenders as a stock setup, but it could be made to really jump through hoops and barrels with only a few minor tweaks. The main thing you need to do is change the cathode resistor that connects to pin #8 of the output tubes. The stock value is a 250 ohm 10 watt. I would recommend changing it to a 560 ohm 10 watt. This will cool off the output stage enough that you will be able to get more gain from the preamp and have the output tubes still respond favorably. Just changing this one component will make the amp have better overall tone with more headroom.

Besides cooling off the output stage to make it more responsive, the preamp gain needs to be improved. There are several ways to do this. A really easy way to get more gain from the preamp is to change from the 6SC7 large base tubes to the miniature #9 pin 12AX7 preamp tubes. You will really put that amp over the top when you change to 12AX7. There are adapters available from Groove Tubes that will allow you to put a 12AX7 into the adapter and then the adapter goes into the octal socket of the 6SC7.

It is a very good idea to change the grid leak bias on the preamp section to cathode bias. You could go the very conservative approach and use the 2.5K resistor with a 25 uf bypass cap, as is the case with V3. Make sure you use 1 Meg grid resistors on the grids of each triode section on both V1 and V2. Also, you must get rid of the series capacitors feeding those same grids. If it were my amp, I would go with either an 820 ohm or a 1K ohm cathode resistor instead of the 2.5K. Also, a smaller cap would give the amp a little more "sting" on the attack of the note. The 25 uf sounds good, but it will be a little boomy. I would try a 1 uf or a 5 uf and do some listening tests to see what it needs. Smaller values have less bottom-end boominess while larger values have less bite and more brawn. There is no one-size-fits all. Try it with your guitar and your playing style and see what sounds best for your situation.

I have a Princeton Reverb that I would like to add a headphone jack. Is there a way to add a headphone jack in place of the extension speaker jack so that when a headphone was inserted the main speaker would be disabled?

That's an easy one, but you want to make sure and use a fairly low impedance headphone. Perhaps an 8 ohm headphone set would be ideal. You will need a jack with a switch on it for the EXT speaker jack. This it the same type of jack that is used on the input jack and the regular speaker output jack of most amps. There are three leads on this type of jack, namely: ground, hot and the switch. Have the hot lead from the output tranny go to the hot lead of the EXT speaker jack. Have the negative feedback wire also go to the hot of the EXT speaker jack. Take the switch of the EXT jack and connect it to the hot of the regular speaker jack. When nothing is plugged into the EXT jack, the regular speaker comes on. When headphones are plugged into the EXT jack, the regular speaker is cut off.

There is already a jumper going from the switch of the regular speaker jack to ground. You want to leave that jumper in place as a safety precaution. If someone turns the amp on without a speaker load plugged in, the jumper prevents transformer damage from occurring.

I am having trouble with a Silverface Fender amp. The treble control does not work on the Tremolo channel. Everything else works fine. I replaced the pot. I checked the solder joints and tested the values of the components in the circuit. I thought the 250 pf cap may be open, so I tried replacing it and no difference.

And worse yet, I also removed the .002 caps on the power tubes as per your Silverface to Blackface modification; and now I have a ringing with the reverb hooked up.

Regarding the treble control, it appears you have a bad ground in the tone stack circuit. The ground for the tone stack will occur through either a resistor soldered to the back of a bass pot or if the amp uses a middle control, it will ground through that pot. Any time there is a ground made from soldering to the back of a pot, as such is the case here, you are depending on the mechanical connection of the pot to the chassis to obtain adequate grounding. If the pot is corroded or if the nut holding the pot is loose or stripped, then the ground will not be adequate and the treble control will malfunction.

I like to remove the pots from the front panel and use a small brass bristled brush to clean the inside face of the pots and clean the brass grounding buss. This assures good connectivity. Make sure the star-washers are shiny because the connection goes through those too. If someone has changed the starwashers to the black anodized ones, such as the ones supplied with some brands of new potentiometers, it will not work. Black anodized coating does not conduct electricity!

With regards to the silver to Blackface conversion, when you get rid of the parasitic suppression cap, sometimes the amp develops a parasitic oscillation. The amp was unstable in the first place and instead of altering lead dress and layout to correct the problem, the genius CBS engineers added the .002 caps as a remedial solution. Never mind that they ground out higher frequency harmonics and make the amp's tone nasally. Removing the cap returned the amp to the way it was. The real problems are in layout and lead dress errors.

I wrote what to do to fix this in my book, *Tube Guitar Amplifier Essentials*. In a nutshell, the solution is to move the tone caps and slope resistor from the component board to the back of the pots. In doing so, you will remove about two or three feet of grid wire!

Secondly, you want to replace the grid wires with shielded wire that connect the volume pot wiper and the signal from the input jacks. As always, the shielding must only terminate to ground on one end which prevents ground loop hum.

Which of the master volumes in your first book would be best for a 5F8A circuit if I want the circuit to be as transparent as possible when the master is cranked to 12?

The circuit most transparent is the circuit using the twin 250K pot circuit to replace the two 220K power tube grid return resistors. When the pot is turned all the way up, the circuit is dead nut stock. I know the resistor is a 220K and the pot is 250K, but there are percentage differences in electronic components and a 250K pot that reads 10% low could easily read 225K while a 220K resistor reading 10% high could read as much as 242K. I don't think there would be any audible difference in this circuit. But if you are going to turn the master all the way up, what's the point in having a master volume control in the first place?

Second, I would never recommend using a master volume control. The master volume circuits printed in my first book are part of a reprint that was written by someone else. I recommend using a power attenuator instead of a master volume.

A master volume control starves the output stage for signal while overdriving the preamp. I don't like the buzzy and homogenized tones of an overdriven preamp. I think the "good" distortion comes from overdriving an output stage. When you overdrive the output stage, you keep the dynamics and the touch sensitivity; but you also get that smooth, organic distortion that only an output stage can get. The power attenuator allows you to turn your amp up, but turn your speakers down.

Is there such a thing as a transformer driven phase inverter? If so, how does it sound? How would I wire one into a Blackface

Princeton type circuit?

Of course there is such a thing as a transformer driven phase inverter. It is simply an interstage transformer with a center-tap on the secondary. You may have heard the transformer driven phase inverter on certain early Gibson amps. I believe the EH150 used one.

There are certain advantages of using a transformer coupled phase inverter. As an advantage, the transformer eliminates the need for a tube. Phase inversion can be achieved without a tube! Also, depending on the turns ratio of the transformer, one can obtain gain by making the phase inversion transformer as a step up transformer, although the turns ratio is rarely greater than 2:1. Sounds good so far, but wait.

There are also disadvantages. An interstage transformer has certain losses and distortions inherent in transformers. For example, the loss in voltage through leakage reactance is greater for higher frequencies than for lower frequencies. The shunting capacitance effect and hysteresis losses also increase with frequency. A really good phase inversion transformer would be very expensive. There are no bargains when it comes to transformers. It would also require more chassis space and there would be a necessity for shielding.

I have serviced amps with phase inversion transformers and every one I've heard sounds muffled and dull—like a cheap guitar cable.

I would not recommend changing to a phase inversion transformer. The Princeton (I am assuming it is not a Princeton Reverb), is one of the lowest gain Fender amps ever made and it does need more gain. If you want to improve the gain of the amp, why not sacrifice the vibrato tube triode section and rewire the phase inverter to the familiar long-tail pair type as used on a Blackface Super Reverb. This would improve gain considerably and the amp would not sound so lackluster—like non-reverb Princetons. The distributive load style phase inverter in a stock Princeton actually loses gain; but the long-tailed pair type as in the Super Reverb, increases gain. Complete instructions on how to perform this modification are in my second book, *Tube Amp Talk for the Guitarist and Tech*, beginning on page 103.

I see in your books that you feel that a Princeton Reverb and a Deluxe Reverb have about the same amount of output (20W). But I notice that the Princeton Reverb uses the same power trans-

former as the Champ. The Champ only has about 360 volts at the output transformer center-tap, yet the Princeton schematic shows 420 volts from the same transformer. Where does this extra voltage come from? My 1968 Princeton Reverb only has about 370-390 volts, depending on what power tubes I use. Is there something wrong with my amp?

I love to see people dissecting schematics and really digging in. It shows you are paying attention. There is probably not anything wrong with your amp. You really can't trust the voltages on Fender schematics. They show the Deluxe Reverb voltages and the Princeton voltages as the same, but in real life every Deluxe Reverb I've ever seen reads high and every Princeton reads low to the schematic!

There are a few points to consider. First, the Fender schematics show the same power transformer part number for both the Princeton Reverb and the Champ, yet they show the Princeton feeding the rectifier tube with 340 volts A.C. while they show the same part number on the Champ feeding the plates of the 5Y3 with 320 volts A.C. This is obviously inaccurate. The actual A.C. coming from the transformer B+ for two identical transformers should be identical. But that is the only legitimate place to compare voltage to voltage because these amps are not using the same rectifier tube. The type of rectifier tube will make a huge difference in the actual D.C. voltage, and D.C. voltage is what you were measuring at the center-tap of the output transformer. Also, the output tube quiescent (idle) plate current will also affect the plate voltage. As the idle plate current goes up, the B+ voltage will go down. This is because a transformer winding has a D.C. resistance inside of it. This resistance drops voltage in accordance with Ohm's law. As more idle current goes through the B+ winding, more voltage is dropped across the winding—leaving less voltage on the plate.

You seemed amazed that a Princeton Reverb is very close to a Deluxe Reverb in terms of power. How could this be when it uses a Champ power transformer? The Champ is configured as a Class A single-ended amp cathode bias. This configuration is about as inefficient as is possible. The Princeton (and the Deluxe) is configured as Class AB push-pull fixed bias, which is the most efficient design possible. The Princeton and the Deluxe both have the same output tubes and the same preamp. The Deluxe uses an optocoupler style vibrato which eats gain, however the Deluxe uses a long-tailed pair phase inverter which adds some gain and compensates for the optocoupler loss. This type of phase inverter has more gain than the distributive load phase inverter of the Princeton. But then again, the Princeton has the bias modulation vibrato which does not eat gain. So the gain is similar, the configuration is similar and the output tubes are similar. The B+ voltage on a Princeton is somewhat lower than the Deluxe (even though the Fender schematics do not show this.) So there is at most a few watts difference. If you listen to the amps, there would seem to be more of a difference because the Princeton uses a 10" speaker and the Deluxe uses a 12". But hook them up to the same speaker and you will hear very little difference in power.

Can a modern amp with 12AX7s (like a Fender Pro Jr. or Blues Jr.) be rewired to use 6SC7 preamp tubes? It's what I have in my 1954 Fender Pro, and I've really gotten used to them. Modern 12AX7 amps sound too bright and too harsh to me. I like the Blues Jr. layout, with reverb and a footswitchable boost, and I am thinking putting the chassis in a 1x15 cab, with the 6SC7s.

To answer your question, yes, a modern amp could be retrofitted to 6SC7 preamp tubes. But I don't think you will like it any better than the 12AX7. It is not the sound of that tube which is the essence of your '54 Pro. The tweed '54 Pro has completely different circuitry and design that is responsible for its tone. I'll bet if you changed the Pro to 12AX7s you would like it just as much. In fact, Fender did just that on some of the late 1954 models. I have seen those amps come into my restoration shop with the chassis tube socket holes punched for the octal 6SC7 type tubes, but with an adapter and a smaller 9 pin socket installed with 12AX7s!

Let's look at the design of the '54 Pro and see how it is unique. For one thing, it uses relatively low plate voltages on all the tubes. This affects the high-end response. When a tube is operated at lower voltage, besides losing gain, it loses treble response.

The Pro used a unique volume control that is different than almost all others. Almost all amplifiers use a voltage divider circuit for the volume control. The voltage divider actually divides off a portion of the signal. On the Pro, the volume control is not a voltage divider. The volume pot simply loads down the signal. That is why the taper is so fast. You can put the slowest taper audio pot you can find in there and it will still go up in volume all at once. That type of volume control has a sound of its own.

The Pro's preamp tubes used grid leak bias instead of the more familiar cathode bias. And it had a simple, one-knob, tone control. Each of these circuits have their own sound.

But perhaps the circuit that affected the tone the most was the paraphase style phase inverter. This style phase inverter circuit achieves phase inversion by taking the signal which feeds one output tube, and runs a portion of that signal back through another preamp tube. When it is run back through another preamp tube, the signal inverts to drive the other preamp tube. You can never get a clean tone from such a circuit because one output tube will have distortion that the other doesn't have. This results in a dirty and grindy tone.

I am working on a Fender Twin and I am not getting a reading when checking the output tube bias using the "transformer shunt method." I have my meter set to D.C. current with the black wire from the meter connected to pin #3 of the 6L6 and the red wire going to what I think is the center-tap of the output transformer. Could you direct me to which wire is the center-tap? Also, when I get the idle plate current milliamp reading of one side, I need to split it in half to get the "current per tube reading" since it is a four-output tube amp, right? This amp is dead quiet until I switch on the reverb. It makes a hum that sounds ground related and is affected by the front panel reverb control. With the reverb control on zero there is no hum but increases as you turn up the level. I should add that in this amp the reverb transformer ground is connected to the ground lug on the reverb output jack; then from that same lug is a 220K resistor going to the positive terminal of the reverb pedal jack. Should this be changed? And does this amp need the reverb and vibrato pedals plugged in to function properly?

To bias by the shunt method, you will place your meter across the output transformer primary. In the case of a push-pull amp, which will have a center-tapped primary, you will connect from one end of the primary to the center-tap. Look at the transformer, there are five wires. Two go to the output jacks. A blue one goes to pin #3 of one set of output tubes, a brown wire goes to pin #3 of the other bank of output tubes and that leaves a red wire which is the centertap. If you can't find the center-tap, on the Fender Twin, you may also use the cold side of the standby switch. This is the same point electrically as the centertap. When using the shunt method on a four output tube amp, remember when you shunt the output transformer, you are actually shunting two tubes and you must divide your milliamp reading in half to get the correct "milliamp per tube" value.

Here's another potential problem. You may have blown the high voltage fuse in your meter—especially if it is a Fluke meter. Some meters have two fuses, one being the fuse you see and the other being a physically larger 600 volt internal fuse that you don't see. This internal fuse sometimes blows when you are checking current and you must take the meter apart to change it. Few people even realize it is there. Yes, the meter could measure voltage OK but not measure current.

Here's a check to see if your meter is working. Put the red lead on the hot side of the standby switch and the black lead on the cold side. Turn the amp on and leave the standby switch in the standby mode. When the tubes warm up, the meter will have all the current that should have gone through the standby switch routed through the meter instead. This will verify that your meter is working. It should read about 110 mA or so.

Regarding the wiring of your reverb circuit, they all are wired like that. I suspect you have a noisy reverb return tube (V4), or the R.C.A. plug patch cables going from the pan to the amp chassis need replacement. You would be amazed at how much difference there is in reverb connecting cables.

Here's how the footswitch works on a Fender Twin: the reverb pedal cuts the reverb off, but the vibrato pedal turns the vibrato on. Without a footswitch you cannot have vibrato. However, you will have reverb with or without a footswitch.

I recently purchased a 1964 Fender Deluxe Reverb. I'm restoring it and ran into a problem with the vibrato. I checked all three disc

caps in the tremolo section and I replaced the filter cap supply. The tremolo is weak at full intensity and if you increase the speed the intensity actually goes away—if you slow the speed down you get some weak tremolo at the slow speed. Also, I found it was originally too fast even when "slowed down" so I swapped out one of the .01 with an .02 cap like your book says, which works nice but didn't fix the weakness problem. I've swapped out tubes with good ones and that is not the problem either. Any suggestions?

More than likely, you will need to replace the opto-coupler. Since the leads going to the light dependent resistor inside the opto-coupler are crimped instead of soldered, there is the possibility that a stray resistance is causing a lack of intensity. Another possible cause would be light leaking into the opto-coupler, which can also make the intensity go away, since stray light coming in will nullify the effect of the neon light source within the optocoupler. In the case of stray light being the problem, there is a possibility that you could repair the optocoupler you have. Try removing the optocoupler and installing either black electrical tape or shrink tube around it. This is to shield stray light. If that doesn't work, it is time for a new optocoupler.

GIBSON

I have a Gibson GA90 amp with six 8" speakers and someone has done a poor job of rewiring them. They have them wired parallel which gave a total of 1.5 ohms on my DVM. Could you please tell me if this is correct or should I wire them differently and what should the total ohms be after rewiring them properly. The output transformer measured .6 ohms, is this correct? Any help you can give would be greatly appreciated.

First, let's make a distinction between D.C. resistance and impedance. A Digital Multimeter (DMM) does not measure impedance. You are measuring only D.C. resistance with your DMM. Impedance includes D.C. resistance, and A.C. reactance (both inductive and capacitive). So you cannot measure impedance with your meter. You can measure D.C. resistance with a DMM.

Secondly, the speakers in a GA90 are supposed to be wired in parallel/series for a nominal impedance of 5.33 ohms.

To wire the speakers correctly, you would take three of the speakers and wire them in parallel. That is to say that all of the "plus" leads connect together and all the "minus" leads connect together. Once the first three are wired in parallel, take the other three speakers and wire them in parallel. At this point, you should have two sets of three—each set being wired in parallel.

Now we are going to take the first set of three and wire it in series with the other set. That is to say, the "pluses" of one set go to the "minuses" of the other set. The "minuses" of the first set would connect to the ground of the amp's output and the "pluses" of the second set would connect to the "hot" of the amp's output.

Once this is done, the D.C. resistance of the six, assuming they are not connected to the amplifier, will be about 3.7 to 4 ohms. To measure this, one would take the two speaker harness leads that would normally connect to the amp and connect them to an ohmmeter or DMM instead. Also, when they are wired correctly, the two speaker harness leads that normally would connect to the amp can be attached to a 9-volt battery instead and the cones should all move in the same direction at the same time. When connected to a 9-volt battery, if any cone is moving backwards while the others are moving forward (or vice versa) then they are wired wrong. Actually, when you connect the "plus" of the battery to the "plus" of the speaker harness, and the "minus" of the battery to the "minus" of the speaker harness, all the speaker cones should move forward.

I am re-capping a Gibson GA19 Falcon, tweed model w/ 7199 at V2. It is wired for a 5Y3 rectifier. I am rebuilding the power supply and am considering a 100 uf / 450 V Sprague as the first cap seen by the 5Y3. The three others were going to be 30uf / 450 V and/or 40 uf / 450 V. I have a friend who cautions against this, based on "soft limits" of the 5Y3.

1. **What do you think about using the 100 uf capacitor in the manner described?**
2. **Would you prefer to see two beefier caps in series as opposed to one large value cap?**

3. **The two cap totem pole is what you describe in your book where you also say to use a 220K / I watt resistor across each cap. When you say across, do you mean (a) tied "end to end" that is, one resistor leg on the plus end and the other resistor leg on the minus end or (b) from the positive end of one cap to the positive end of the other cap?**

4. **Would I have to change the 5Y3 to a GZ34 to use the 100 uf?**

When re-capping any amp, I only encourage bigger caps for the mains and in some instances, the screen supply. I never suggest increasing the cap for the preamp circuit unless one is trying to get rid of bottom-end. You can increase the mains which will help the amp to play in tune when overdriven. In your particular circuit, I would change the 20 uf main filter cap (the main is the one going to pin #8 of the rectifier tube) to a 100uf at 450 volt (or 500 volt). The next filter cap could be replaced with a stock 10uf or, since it is the screen supply, you could increase it to 20uf. But the cap feeding the preamp, I would replace with a stock, 10uf cap value.

Using two caps in series to replace the main filter, what I call a totem pole stack for obvious reasons, is usually only done to increase voltage rating and since the schematic shows 300 volts; I am going to say the voltage rating is fine with one 100 uf at 450 volt cap. So you wouldn't need a totem pole stack in this case.

When we say putting a resistor across a cap, we mean connecting it in parallel such that one end of the resistor goes to the positive lead of the cap and the other end goes to negative. Placing a resistor across each of the capacitors in a totem pole stack is done for three advantages. First and most important, it will cause the voltage across the caps to divide evenly which results in obtaining the maximum voltage rating. You don't want 90% of the voltage to occur across one of the caps and then only 10% across the other. Second, it provides a bleeder circuit that will discharge all the caps slowly when the unit is turned off. And third, although this is not a significant advantage, it will lower the impedance of the power supply some, which results in a more stable voltage supply.

It is fine to use bigger filters for the mains. The tube manuals say don't do it, but the tube manuals also say never to run a 6V6 over 350 volts, so you can't believe everything you read in a tube manual.

Anyway, guitar amps are about tone, not electronic perfection. As long as you hook the positive end of your main filter cap to pin #8 of the rectifier tube, you are fine. This allows the main filter to trickle charge while the rectifier is warming up.

Will it hurt my GA-20 to use a GZ34 rectifier tube instead of the stock 5Y3 tube? The 5AR4 gets the B+ voltage up to 340 volts and sounds quite good. I'm not sure if the power transformer can take it on a long-term basis. What do you think?

Most Gibson amplifiers need help. Remember they were voiced using a Gibson guitar, whose pickups put out two to three times as much signal voltage as a Fender guitar. Since the designers were going for clean, they would deliberately design the amps with low gain and conservative voltages. So generally speaking, with Gibson amps, the voltages are too low and the gain isn't enough to satisfy most players.

With any amp, here is a simple test to see if the power transformer likes using a 5AR4 instead of the stock 5Y3. Simply install the 5AR4. Put an A.C. voltmeter between pins #8 and #2 and read the voltage. If it is 5 volts A.C., the amp likes the tube. If the filament drops below 4.1 volts, the transformer isn't capable of delivering enough filament current to keep the 5AR4 happy.

With the GA-20, the 5AR4 will probably work fine. Here is another little trick to bring the preamp voltages up and ultimately increase the gain of the 6SJ7 pentode preamp tubes. Locate the 47K resistor that is situated between the two 10 uf filter caps. Change this resistor to a 10K resistor and the gain for the preamp will increase somewhat. It isn't dramatic, but every little bit helps.

I own a GA-45 ('58 Gibson Maestro) which is similar to a GA-40 that can be seen on page 180 and 187 of the *Gibson Master Service* book. Originally, the amp used four 8" 4-ohm speakers. I want to change the baffleboard and use one 12" P12R Jensen 8 ohm speaker and one 8" 4 ohm speaker. What would be the best way to wire it?

There are a couple of different ways to wire a 12" 8 ohm speaker to an 8" 4 ohm speaker. You could wire the two speakers in series by connecting the plus of one speaker to the minus of the other and connecting the two remaining leads (one from each speaker) to the amp.

In this scenario, the amp would see 12 ohms which would be a big mismatch if you are connecting this to a 4 ohm amp output. It would probably be very compressed and you would lose volume. However, the power would split where the 12" speaker would get ⅔ of the power and the 8" speaker would get ⅓ of the power.

Another way to hook them is in parallel. To do this, connect the plus of each speaker together and have this go to the plus (hot) of the amp's output and then connect the minus of each speaker together and have this go to the minus (ground) of the amp's output. This would drop the speaker impedance to 2⅔ ohms—which would be a pretty good match for the 4 ohm amp. The down side of this scenario is that the 8" speaker would get ⅔ the power from the amp, leaving the remaining ⅓ of the amp's power for the 12" speaker.

Neither way is optimum, so I think I would just try it both ways and see which way sounds better to your ears.

After reading your second book, *Tube Amp talk for the Guitarist and Tech*, I purchased a Gibson Duo-Medalist (1x12") based on a chapter in that book that states this particular model has all the right components to be a killer amp, but has the wrong tone filtering network and gain structure. Can you help shed some light on what else I can do to improve my anemic sounding Duo-Medalist, and how to install a 3-prong power cord, and 3-prong accessory outlet?

Also, I would like to know what your opinion is on the current production 7591's.

The Duo-Medalist has serious problems with both gain structure and tone shaping. There is too much loss in the preamp circuitry. Looking at the schematic on page 314 of the *Gibson Amplifier Master Service Book*, I can see the schematic of the Gibson Duo-Medalist "A". The tone shaping network that needs to be removed is everything from pin #7 of V1 to pin #5 of V1. This includes the tone controls. Actually the entire preamp from the inputs to pin #2 on V6 needing a complete circuit change. I would change the preamp to a basic Fender Blackface circuit from the input jacks on both channels to pin #2 of V6. Which blackface? You can pick one that doesn't have a middle control, such as the Deluxe Reverb. I would leave the tremolo tube and its circuitry alone and simply hook the .22 tremolo coupling capacitor to one channel or the other. The reverb circuitry would need to be wired like the Fender blackface using V3 as a 12AT7 (and not the 12AU7) reverb driver tube. This would be a mod for an experienced service technician. Care should be taken to have the signal path move from input towards output without crossing back over itself. This will help avoid the possibilities of parasitic oscillations that could occur from performing such a modification.

I would advise against using an accessory outlet for anything that draws much current (another amp for example) as all the current from your amp and your accessory plug would have to come through the A.C. cord on the amp. The schematic I see shows a 2-prong accessory outlet. If you need a 3-prong outlet, why not just use a 3 to 2-prong A.C. adapter—which can be found in any grocery store? This will lift the grounds of the accessories plugged into the outlet. If you are connecting these accessories to the amp, you will already have proper grounding due to the fact that the patchcords will connect all chassis together.

To install a 3-prong A.C. cord, simply hook it up exactly the way it is hooked up now. In other words, connect the white lead where the white lead on the existing A.C. cord is connected and match the black lead of the new cord to the same placed where the old cord's black lead was attached. This will leave you with one green wire left from the new A.C. cord. Simply attach this green wire to the amp's metal chassis and you are good to go.

I think the current manufactured 7591A are a very attractive alternative to paying $200 per pair for NOS 7591A tubes.

I want to do a D.C. heater supply on a Gibson GA5 Reissue, so we're talking about one 12AX7 and one EL84. I was thinking about tapping off the secondary high-voltage winding with a resistor, to a full-wave bridge rectifier, out from there with a couple of 6.2 volt Zener diodes across the line to ground, and the necessary amount of filtering. Will that work? Or do you have a better idea?

That will not work. Filaments need current and the high voltage winding won't deliver much (maybe only a tenth of an amp on that power tranny B+ winding). And the Zener can't hold back the current needed to regulate to 6.2 volts.

I would get a 10 amp chassis-mount bridge rectifier and connect it to the 6.3 volt winding. You need the high amperage rating of 10 amps because the filaments will draw major current at startup when the filaments are cold. With a semiconductor device, even a slight draw of more than the rating will cause a malfunction. 10 amps is sufficient overkill that you will never have a malfunction.

Since you only have a little over 1 amp of filament requirement, with enough filtering, you will be able to make that work. If the voltage is too high, you can bump it down a little with a very small resistance, high-wattage resistor. For example a .1 or .2 ohm, 3 watt or so would probably be perfect, if needed. Maybe you won't need one since there is so little filament draw. The bridge rectifier I am talking about is an epoxy square with markings on each corner and a lead in each corner. In order and clockwise the leads will be marked on the corners: +, ~, -, ~. The two leads marked ~, each connect to a green wire on the 6.3 volt winding. Minus (–) goes to ground, and plus (+) will be your 6.3 VDC. Note: you don't use the center-tap of the 6.3 volt transformer winding (assuming it has one).

I would use a 15,000 uf / 35-volt capacitor, which is a standard value. Since you have a single-ended application here, you need to have the ripple current less than 1 volt peak to peak. The 15,000 uf should get you down to a third of a volt peak to peak.

The 15,000 uf capacitor connects with the plus lead going to the plus (+) lead on the rectifier and the minus lead going to ground. The plus of the rectifier also goes to one filament lead on each tube and the other filament lead of each tube goes to ground.

OTHER AMERICAN

I have a question about early 1960s monoblock tube type amps such as the Dynaco Mark III (60 watts) and tube type P.A. power amplifiers such as the Bogen Challenger (100 watts). Your readers probably see these for sale on ebay and wonder as I do: "Wouldn't one of these make a great guitar amp?" "Is hi-fi really good for me?" "What should I use for a pre-amp?" "Would it be more cost

**effective to sell them and use the cash to buy a Kendrick amp?"
Also, I have no electronics experience. Where would you suggest I
go to get basic soldering and schematic reading skills?**

The 100 watt Bogen was the rocket of choice for one of my favorite
players of all time: Terry Kath! He used a couple of them driving two
Dual Showman cabinets loaded with D130Fs. He later switched to
200 watt Bogens—again two of them! What a sound he had! That
tone was unmistakable.

Terry Oubre, another of my favorite players of all time, for many
years, used a Dynaco amp that his Dad had built a single channel
preamp section using a Twin Reverb design on the same chassis. It
was cleverly laid out. They took some bolts and used them as stand-
offs to add a phenolic component board. It sounded wonderful, and
still does. Yes, he still has that amp. It doesn't sound anything like a
Twin. For one thing, it had only one channel, so the channel merger
220K resistors were missing, which accounted for 3 dB more gain. But
it used the 7199 triode/pentode phase inverter that provided more
gain too, so it was a monster. It used a pair of EL34 output tubes.

To learn more about tubes, find an older University in your area
and visit the engineering library. Most colleges have dozens of books
written in the 40s, 50s and 60s on tube audio.

If you want to learn how to solder, here's the way we used to train
apprentices. Next to our building was a recycling place that routinely
had surplus military tube equipment they sold by the pound. We
would buy old electronic equipment and have the apprentices desolder and disassemble them. After disassembling and completely desoldering a few pieces of tube equipment, one could easily get a feel
for soldering. By using old equipment, if someone screwed something up, it didn't matter. Maybe you can find some old electronic
equipment in a garage sale and disassemble it. Who knows, you might
get some cool parts in the process!

**I recently purchased an old SVT head that had been fully ser-
viced. It worked fine for a month or so; but now, about every
twenty minutes, the power goes off (power light is off). To make
it work again I have to turn it on again and it works for other
twenty minutes and goes off again. No blown fuses. I brought it**

to a tech but he could not find any problems Do you have any hint about the cause of this annoying problem?

The Ampeg SVT is arguably the best tube bass guitar amp ever built. There is an entire chapter on servicing them in my first book, *A Desktop Reference of Hip Vintage Guitar Amps*. Without a crystal ball, I don't know you problem for sure. I assume the problem could not be duplicated for the technician. Did you try a different A.C. wall outlet? Maybe yours doesn't work as well as the technicians. If the pilot light goes off, the problem is happening between the wall plug and the power transformer. If using a different outlet doesn't fix the problem, I would suspect a loose or faulty solder joint or a bad power switch. I'll bet you could find it if you opened it up out of the cabinet and did a very close visual inspection. You could easily kill yourself if you touched any of the high voltage components in that amp while it is operating. The B+ on that amp is around 700 volts or so.

If it were in my shop, I would take it apart, turn it on and take a wooden chopstick and poke every solder joint. When an offending solder joint is disturbed with my chopstick, the electricity will be interrupted and the pilot light will flicker or go off. If that doesn't find it, I would turn it on and play it until it stopped. This would give me the chance to check the switch for malfunction. Without disturbing the switch, I would need only to unplug the amp and measure the power switch leads with an ohm meter. The switch itself should measure zero ohms.

My large-venue amp is a Boogie MK4 head with a 15" JBL speaker cabinet. The head has both 4 ohm and 8 ohm speaker jacks. What general effect on the amp, tubes, and power output does using more speakers in combination, either 4 or 8 ohm, with the amp we have. I would like to understand how to maximize use of these taps and combination (short of frying things).

The idea behind impedance matching it to have a proper load on the amplifier. If you use too high a load, for example a 16 ohm speaker into the 8 ohm output jack, then the amp will lose a little power and the tubes will compress more. If you use too little of a load, for example a 2 ohm cabinet plugged into the 4 ohm output jack, then the tubes will be operating closer towards a short circuit and tube life will be shortened considerably.

On the Boogie amp, the output jacks are wired differently than other amps. There are only two taps on the output transformer; a four ohm and an eight ohm tap. The eight ohm tap is connected to a single 8 ohm jack while the 4 ohm tap connects in parallel to two jacks, each labeled 4 ohm. These are intended to be used with 8 ohm speaker cabinets. If you use two 8 ohm cabinets, the nominal load will be 4 ohms, and you will need two jacks (one for each 8 ohm cabinet). That is why there are two 4 ohm jacks. They are connected to each other internally, so what you really have is a four ohm output transformer tap but with two jacks. When an 8 ohm speaker is plugged into each of the 4 ohm jacks, the nominal load of the two jacks combined will be 4 ohm. If you had a 4 ohm cabinet, you could simply use one of the 4 ohm jacks.

The 8 ohm output jack is meant for using a single 8 ohm speaker load, however, if you wanted to use two 16 ohm cabinets, you could do so, but you would have to make a junction box which would put the two 16 ohm loads together in parallel so that the total load would become 8 ohms. You would hook the junction box to the 8 ohm output jack on the amp.

If you had a 16 ohm and an 8 ohm cabinet, the correct way to hook them both up is to put the 16 ohm cabinet into the 8 ohm output jack and the 8 ohm cabinet into one of the 4 ohm jacks.

I recently found a Supro Thunderbolt Model S6920 "Bass and Organ" amp. The original owner stored it unused for over 30 years. The problem is that, even with two 6L6s, it doesn't get very loud and actually sounds kind of cheesy. I've always thought that Supros were highly desirable and toneful amps; yet, I've read in one of your books that "some two 6L6 Supros aren't very loud." Is mine a "cheesy-bolt" or a tone-monster awaiting resurrection via a tune-up?

Did you say unused for 30 years? While it is true that some Supros are not very loud, it is also true that any amp that has sat unused for 30 years will not be up to snuff. It is unreasonable to evaluate an amp in this condition. Let's compare this to a vintage car. Suppose you tried to start a car that had sat in a garage for 30 years. If it started at all, do you think it would perform as it should? Your amp is long overdue for a complete overhaul. Sure it isn't loud and sure it sounds cheesy. The filter caps and the bypass caps are either ruptured or dried up to nothing and there are probably one or more coupling caps leaking D.C. voltage to the next stage—thus choking the gain. Who knows what other problems occurred from the amp just sitting up?

My advice to you is to get you amp overhauled. Every electrolytic cap in the amp should be replaced with AMERICAN made caps. Not American brand caps made in Taiwan, but real "made in America," American caps. Every coupling cap should be checked for D.C. leakage and any cap leaking more than .25 volts D.C. should be replaced. Either an Orange Drop capacitor or a Mallory 150 or 152 series cap would be best.

I also recommend a complete cleaning and possible retensioning of sockets, jacks and pots. Any resistor that has drifted out of tolerance should be replaced. And last but not least, fresh tubes should be installed. When changing output tubes, it is always best to set bias on a fixed bias amp and at least check the idle plate current on a self-biased amp.

I'm hoping can provide some information. I recently purchased a 1997 Ampeg V4-BH with a quartet of 6L6s. I would like to replace these with 6550s or KT88s for additional headroom. My question is this; does the transformer provide enough voltage to compensate for the additional filament current drawn if I replace the 6L6s with 6550s or KT88s? Of course I understand that I will need to change a bias resistor to get the bias voltage in the correct range. I know that original Ampegs had VERY over engineered power transformers and were quite forgiving but I'm not sure about the SLM versions. Any assistance, or recommendations you may be able to provide will be greatly appreciated.

I haven't seen the 1997 version yet and really don't know the answer. I do know that if the transformer can't take the current, the voltage will drop. So here is the test:

Open the amp up so that you can take a voltage measurement from pin #7 to pin #2 of any output tube socket. Install the 6550s in the amp and put the amp in the standby position. Now check the A.C. voltage on the meter. It should read 6.3 volts. If it reads lower than 6.1 volts, the transformer cannot take the additional current draw of the 6550s. If it reads 6.1 or more, the tranny can handle it just fine.

My Silvertone 1304 was out on the town last night, unsupervised, and the output transformer, among other components, took a spilled beer right into the lower end of the chassis! The fuse blew and when I got the amp apart, the wettest component was the output transformer. Is it possible to rehab it by baking?
You may be able to salvage it, but you don't want to get it too hot. And whatever you do, do not put it in an oven. I would recommend using a hair dryer first to evaporate as much liquid as possible, and then put it in a box that has a low wattage light bulb in it. You may want to leave it in the box for a few days. You don't want to get it too hot—just enough for it to dry out. (**Note:** this person emailed me 4 days later and his amp works perfectly now. The hair dryer and light bulb trick worked.)

I bought a used early 90's Kendrick Model 1000 stand-alone reverb to add some reverb to my beloved 1955 Tweed Deluxe, but I am noticing the dry sound remains too much in front, and the reverb becomes relevant only when Dwell and Mix are around 7-8. The tubes are new and of good quality. What would you would suggest to improve the above two points?
You have a malfunction. That reverb unit should have plenty of reverb and even too much reverb when set the way you are setting it.

Did you unlock the springs? If the springs on the pan are not unlocked, then you won't have much reverb. The early units had a metal push-lock that is accessible from the back at the middle of the bottom. The metal pushes the pan into a sponge and then locks by going to one side. This prevents the springs from moving around during shipment. To unlock, the metal must be moved to one side then pulled back—thus releasing the springs.

The other problem could be a broken spring. If one of the two springs is broken, you will get only half as much reverb. You may have to take off the front to check the reverb pan to see if it still has both springs. This information would also apply to the Fender 6G15 stand-alone reverb.

I own an early 60's Ampeg Jet J-12 amp that has an electrical issue. When played until hot, it starts to smell like electrical

components burning. It actually lost power once a long time ago, but I decided to put it into storage as a later project. I figured I probably killed the power supply or I blew the caps.

Anyway, the amp was recently checked out by an amp repair person, but no problem was found (except a bad input jack). I then played it for about a half hour and smelled it burning again. My goal is to make this amp as reliable and robust as possible without changing the tone too much. Please let me know if this is something that can be done.

I like to see people troubleshooting with the best troubleshooting tools available, the ones that are God-given. If it smells like it is over-heating, it probably is. If you have an amp that is 40 years old, it needs a cap job for sure. You cannot trust a meter check to see if the capacitors are any good, because:

1. A typical capacitance meter runs off of a 9 volt battery and that isn't the same as putting it under a 400 volt application. It may check "good" on the 9 volt meter and then leak like a sieve when installed in the amp.

2. Capacitance meters do not check leakage current which is probably the root of your problem. All electrolytic capacitors leak some current. It is the nature of that type of capacitor. New ones don't leak much. When they get old and leak a lot and there are several in the amp, the sum of all leakage can really tax the B+ winding on the power transformer and cause it to overheat. If you have four caps that are each leaking 10 mA, then four of them together will leak 40 mA. You particular amp uses about a 125 mA rated B+ winding. If the caps are leaking 40mA then the caps are using up ⅓ of the transformer's capability and when the rest of the circuit draws more the 85mA total, the current will cause excessive heat.

3. Nothing lasts forever. An electrolytic capacitor uses an electro-chemical process to perform its duty. This is the same type of process as in a car battery. If you had a car battery that was 40 years old would you trust it?

Another possible cause could be rust on the transformer. Rust causes continuity between adjacent laminates on the power transformer. A transformer uses thin laminates of steel that are insulated from each other. If they become conductive to each other, then the transformer "thinks" it has another "1-turn" secondary on it. Current begins to flow around and around the laminates if the adjacent laminates have continuity with each other. Technically, this current is called "eddy current". When you have eddy current, the transformer simply overheats. You can check to see if your laminates are conductive to each other. Here is how: Take an ohmmeter and put it on one of the lower resistance settings, let's say 200 ohms, for example. Put one test lead on the edge of a laminate on one side of the transformer's core stack, perhaps the left side. Now put the other test lead on a laminate on the other side of the core stack — the right side. If you read infinity ohms (open), then you don't have a problem. If you get a resistance reading, then there is conductivity and the transformer will have eddy currents and overheat after a few minutes of usage. The only cure is a new power transformer or a rewind.

I am preparing to assemble a Hoffman Bassman 5F6A based harp amp. I would like to try for the spongier, browner tone by deliberately mismatching the speaker impedance as described in your third book. My output transformer has 2, 4 and 8 ohm taps. Do I wire the four 8 ohm 10" speakers in series/parallel to get 8 ohms and run them off the 4 ohm secondary? Or should I wire them as a 4 ohm load with a 2 ohm output. Is it possible to wire the four 8 ohm speakers to obtain the 4 ohm impedance? Would it make any sense to mount an impedance selector on the chassis?

Deliberately mismatching the speaker impedance to get twice the load the amp wants to see, will definitely get you the spongier, browner, and more compressed tone you seek. I think you may have the wrong transformer for such an application. I can only assume your motivation in assembling your own amp is that you really want it to sound great. If that is the case, you need a 4 ohm transformer or a 16 ohm transformer and not the tapped transformer you have now.

Using the 4 ohm tap of an 8 ohm tapped transformer is analogous to unplugging the spark plug wires going to two plugs of your V8 engine and using only 6 cylinders. It will work, but it isn't optimum. When a tapped transformer is wound, it is wound for the largest impedance output (8 ohms in this case) and then the taps are added at 70.7% of the secondary winding and 50% of the secondary winding for the 4 ohm and 2 ohm tap respectively. You will not get the optimum tone by using a portion of the transformer. What runs better a 6 cylinder car or an 8 cylinder with two plugs removed?

For the same reason, you would not want to use the 2 ohm tap on your transformer as you would only be using ? the transformer and it wouldn't sound very good even though it would pass signal.

You can only wire four 10" speakers to get three impedances; namely 32 ohm (all in series), 8-ohm (series/parallel or parallel/series) or 2 ohm (parallel). You cannot wire for 8-ohm speakers to get 4 or 16 ohm.

I have Epiphone EA28RVT project amp that has been modified quite a bit. I want to rewire it using a Gibson Falcon as a guide. The schematics are almost identical, but the Gibson tends to have lower cathode to ground resistor values from the pre-amp tubes. I assume this at least in part contributes to the more overdriven sound of the Gibson compared to the Epiphone by increasing the voltage on the 6EU7 plates—correct? I like the sound of the Gibson very much, so I'm planning to use the resistor and cap values from the Gibson schematic instead of those specified on the Epiphone schematic.

1. **The pots are corroded and need to be replaced. The schematic calls for 2¬-Meg audio taper pots for volume and reverb. I'm having a heck of a time finding 2 Meg pots. If I can't find them, what would be the effect of substituting a 1 Meg? Tremelo depth and frequency call for 250K and 500K reverse audio taper pots—what is meant by reverse here?**

2. **When I compare the Epiphone parts to the Gibson, in general they just seem wimpier. This holds for the transformers that are smaller and lighter weight. Would there be any benefit in replacing the output transformer? I'm not sure how to test it out to see how it is functioning, but I do understand that this is a critical part of the circuit. If I decide to replace it, how do I decide what primary and secondary impedance that I need? As you know, this amp uses 2 6V6 power tubes and will be driving an 8 ohm, 15 watt speaker.**

I love to see people just dive into a project amp. There is no better way to learn than hands on. Don't even think about plugging that amp in without using a current limiter. If there is a short in the amp somewhere and you have a current limiter, the light bulb will glow bright and you will know it. If you don't use the limiter and there is a miswire, the amp could go up in a mushroom cloud before you can trip the switch to turn it off.

The lower resistance value cathode resistors do contribute to the gain, but not for the reason you stated. The cathode resistors actually bias the preamp tubes. A large value will cool off the tube and decrease the overall gain of the preamp.

Forget the search for 2 Meg pots. You don't need them. Almost all amps use 1 Meg. It will work just as well, and they are available.

The taper of a pot has to do with how much resistance change occurs with regards to how much the knob is turned. If you turn the knob half way up and the resistance changes by half, then the pot is said to be linear. If you turn the knob half way and the resistance changes by considerably less than half then the pot is said to be audio. And finally, if the knob is turned up half way and the resistance changes by considerably more than half, then the pot is said to be reverse audio. People do not hear linearly. A linear pot in an audio circuit will seem to "come up" all at once. Most tremolo circuits use reverse audio taper pots because that is what is required for the knob to control the function evenly.

I would recommend trying the amp with the stock transformers first and see what you have. You may like it just fine. For the output transformer you will need a 6,600 ohm primary and an 8 ohm secondary. The math works out to a 28.72 to one ratio, so you could choose something near there and it would work. If you are changing the output tranny, I would recommend going with the Fender Deluxe because the type of grain oriented silicon steel used in the core makes a huge difference in sound and we know the Fender Deluxe will sound very good in that circuit.

I have a Bogen CHA20 amp with the 6V6 output tubes that makes a lot of noise. It only has the old two prong style plug. As I put my hands on my guitar strings it quietens down. Would

converting it to a three prong help? Or, should I look for an un-grounded part inside the amp?

I recommend converting your A.C. two-prong plug to a three-prong, grounded A.C. plug. This will definitely help the noise you are having, provided you are plugged into a properly wired three-prong A.C. wall outlet and the outlet is wired correctly. On a three-prong style A.C. cord, there are three wires: a black, a white and a green. The black wire and white wire replace the existing two conductor stock cord. Usually one of the stock leads goes to a fuse and it is preferable to use the black wire of the new cord for that connection. But the third wire, which is green, connects directly to the chassis of the amp. This assures that the chassis of the amp is earth grounded. Your strings also connect to the chassis of the amp via your guitar cable shielding. So, with a properly wired A.C. cord, you should be able to touch your strings without it affecting the noise floor.

I would like to modify my Kendrick 2410 amp so that I can switch easily from tube type to solid-state rectifier. Can this be done?

Yes, it can easily be done on any amp that uses a tube rectifier. The instructions I am about to give will work with any standard 5-volt tube rectifier amp—such as a Super Reverb, Pro Reverb, Deluxe Reverb. You will need only 6 diodes and a SPST toggle switch.

To begin, make two subassemblies, each with three diodes in series. By that I mean the cathode of one diode goes to the anode of the second diode and the cathode of the second one goes to the anode of the third. Do the same thing with three more and you end up with two sets of three diodes in series. Use 1N4007 or 1N5399 diodes.

Here's how to wire to two subassemblies into the amp. The free anode of one set goes to pin #4 of the rectifier tube. The free anode of the other set goes to pin #6 of the rectifier tube. Now you will have the two cathode ends, one from each subassembly, which you will tie together and connect to one leg of a SPST switch. The other leg of the switch goes to pin #8 of the rectifier tube. You need to mount the switch somewhere convenient. On a vintage amp, you may wish to use the extra speaker jack hole—so that you do not drill any holes in a vintage chassis.

You now have the option of solid-state or tube rectifier. Actually, when you choose tube, the switch is open—so the circuit is dead nut stock. When you turn the switch on, the solid-state rectifier shorts the tube rectifier and since the impedance of the solid-state is much less, the current goes through the diodes instead. You are looking at under $10 worth of parts plus however much labor (around $50 or so) to perform this mod.

I am restoring an old Sears Silvertone amp I picked up at a flea market. Everything works but it sounds rough, so I have been systematically going through it. The output transformer primary has a D.C. resistance of 176 ohms from one end to the center-tap and 258 ohms from the other end to the center-tap. Shouldn't these values be close to equal? Does this mean that one side has some shorted laminates? There is some surface rust, but very thin. Should I replace the output transformer? Thanks!

When we hear the word "center-tap" it is only normal to think the center-tap is in the center and therefore the resistance to each end should be about the same. However, the output transformer is not wound for D.C. resistance; it is wound for a specific number of turns. The center tap is the center of those turns. Let's say the primary winding has 400 turns; the center-tap would be placed at the 200th turn. But wait! Just as the inside lane of a track event is shorter than the outside lane, the inside turns of a transformer have less length, and therefore less resistance, than the outside turns. That is to say the D.C. resistance will never match. So how do you test it if D.C. resistance isn't the test?

You need to do a test to see if each side has the same turns-ratio with respect to the secondary. Remove the output tubes and disconnect the secondary of the transformer. Using an A.C. source, such as a Variac or an A.C. wall-wart, put voltage on the secondary. It doesn't have to be much voltage, maybe only a volt or two, but it must be A.C. voltage. Using an A.C. volt-meter, check the A.C. voltage from primary center-tap to one end and record this A.C. voltage measurement. Next, check A.C. voltage from the center-tap to the other end and record this A.C. voltage measurement. The two readings should match. If they don't, it doesn't mean the laminates are shorted, but it means the primary windings are shorted. The short would occur on the side that had the smaller A.C. voltage measurement. If the primary is shorted, you need to replace the transformer.

All About Vacuum Tube Guitar Amplifiers

To check to see if the rusty laminates have become conductive, use an ohmmeter set to a 1K range or more. Touch one meter lead to a laminate near the end-bell on one end of the core stack. Touch the other meter lead to a parallel laminate on the other end of the transformer core near the other end-bell. If there is continuity between the laminates at all, the laminates are conductive and it is time to replace the transformer. A transformer is an expanding and collapsing magnetic field with coils of wire wrapped around an iron core such that electricity is induced in the coils of wire as the magnetic field expands and collapses. The problem with conductive laminates is that, to the transformer, the shorted laminates look like just another secondary—that is shorted! Excessive current is drawn by the primary, and excessive heat builds up as current is induced in the laminates—instead of the real secondaries.

I have a 1969 Sunn Sonaro bass amp that I picked up from a pawn shop. It worked perfectly at the time, but it needed a cap job and new tubes. I replaced all of the electrolytic capacitors and the can cap. I triple checked my connections. I burned in the caps using a current limiter and everything seemed fine. I retubed it and biased the new EL34 tubes and it worked fine, then I replaced the 2-prong A.C. power cord with a grounded 3-prong one and all was still fine. But after about an hour of playing, I blew a fuse. I replaced the fuse and tried the omission trouble shooting technique to see if it was a power tube or the rectifier and nothing seemed to cause the fuse to blow again. So I played a little more, the fuse blew again; this time accompanied by a smell from the power transformer. Now it won't even fire up, the fuse instantly blows. I'm guessing I've fried the power transformer, but I don't know why. Any advice?

If the amp blows fuses without any tubes in it, you most likely have a burned up power transformer. A burnt smell from the power transformer usually means the power transformer is history. The ultimate test is to disconnect all the secondaries of the power transformer and if it still blows a fuse, the power transformer is definitely bad. I don't know what caused this problem without trouble shooting the amp.

When a fuse blows and you cannot determine the cause, the

problem could be intermittent. Until you are sure what caused the problem, always keep the amp in a current limiter. This will prevent too much current from flowing and if too much current tries to flow, the light in the limiter will burn bright instead of the amp blowing up. When a circuit draws too much current and the fuse blows, the power transformer will get very hot. If you install another fuse and the problem persists, it is very possible to overheat and blow the "already very hot" transformer before the fuse blows.

If it turns out the power transformer is burned up, once the transformer is replaced—do not even think about turning on the amp without using a current limiter until you find the problem that caused the power transformer failure in the first place. You don't want to blow the new transformer! You could have an intermittent problem with the bias voltage. If you had bias voltage failure, the tubes would run wide open and you could easily overheat and blow a transformer. Inspect the filament circuit carefully for a short. Inspect and test the rectifier circuit for a short. You could have a small scrap of wire or blob of solder left over from your overhaul that is shorting something intermittently!

MARSHALL

A friend of mine owns a Marshall JCM 800 100-watt head. I performed the usual 6550 to EL34 conversion. The amp has two inputs, a high and a low gain. He wants to use the high gain as the main input and using the low gain input as a dedicated foot switch jack for switching between the two gain levels. Despite the amp only having one channel, is this "channel switching" mod possible? Another project I'm attempting is the conversion of a Silverface Bassman head from 50 watts to 100 watts (or thereabouts). I plan on using transformers from a twin reverb, while replacing the 6L6s with 6550s. Do you recommend this mod? And if so, are there any other stock parts I should replace?
Regarding the Marshall, your friend cannot channel switch the Marshall JCM 800. It only has one channel. The high gain input is

simply a stage in front of the low gain input. The problem is that the low gain input jack has a switch on it that disconnects the high gain circuit when a ¼" plug is inserted to the low gain input jack.

Regarding your Bassman, you can make a really great sounding bass amp doing the modification you describe. I would begin by first rewiring the Bassman to Blackface AA864 specs. In particular, the phase inverter and output stage and biasing circuit could be made to match the AA864.

Use the 6550s with a Twin output transformer, just make sure and run it with an 8 ohm speaker load. This will be correct for use with a pair of 6550 output tubes—given the turns ratio of the Twin output transformer. You should have no problem with the mod, just be sure that if the amp starts squealing that you reverse the brown wire with the blue wire on the primary of the output transformer (blue wire and brown wire each terminate on pin 3 of opposing output tubes).

You may also need to experiment with the negative feedback resistor. Since the Bassman circuit is already 4 ohm and you will be running the new modified version at 8 ohm, you will probably want to use a feedback resistor that is larger than stock. Here's why: An 8 ohm output will have 1.412 times the voltage and .707 times the current of a 4 ohm output. Since negative feedback is figured from voltage, to keep the stock feedback amount, you will need a feedback resistor that is 1.412 times stock value. The AA864 Bassman uses an 820 ohm resistor, so ideally an 1,159 ohm resistor would keep it sounding like an original circuit. A standard value would be a 1200 ohm. Get several and check them with a meter. Pick one that measures low and you are there. Or you could check a few 1000 ohm resistors and find one that measures high to get there.

Generally speaking, using a larger value feedback resistor will make the tone more raw with more distortion. Conversely a smaller value feedback resistor will make the tone cleaner and more even with less distortion.

One last suggestion, beef up the main filter caps and the screen supply filter. I would use two 220 uf at 350 volt caps in a totem pole arrangement for the main filters (series). This would give you a rating of 110 uf at 700 volts when the two caps are arranged as a totem pole. Make sure and use the 1 watt 220K balancing resistors across each 220 uf cap. Also, I would use a 40 uf at 500 volt electrolytic capacitor for the screen supply filter.

I have a late 60s Marshall Plexiglass full stack. Sometimes on the bright channel, I notice a mosquito-like sound on top of the note. It seems like it is worst if I turn up the volume, the treble or the presence. I have tried different speakers and different guitars but the problem persists. What gives?

Almost all late 60s Marshalls are prone to parasitic oscillation, which seems to be exactly what you are describing. Parasitic oscillations are caused when some type of parasitic coupling occurs from a later stage back to an earlier stage. The actual parasitic circuit that is occurring in the amp is one that does not exist on the schematic but is a circuit none-the-less. It is a phantom circuit. It exists, but no one can see it! In your case, it is a higher frequency oscillation that modulates the note you are playing—thus the mosquito-like sound. Sometimes parasitic oscillations occur as a lower frequency that manifests itself as a "motor boating" or amp "shut down." Other types of parasitic oscillations can make an amp sound like a blown speaker on the lower notes.

There are only four ways to cure a parasitic oscillation: modify lead dress, modify grounding, decrease gain or decrease treble response. I like to address the grounding and lead dress and leave the gain and treble response alone. Here are some suggestions to try: On the first preamp tube, I would use a shielded wire going from the input jacks to the grids. These are the wires that go from the input jacks to pin #2 and pin #7 of the preamp tube. I would also recommend a shielded wire feeding pin #2 of the second preamp tube. When using shielded wires in an amp, you want to ground the shielding on one end only—otherwise you can induce unwanted hum.

There is a 5.6K resistor on pin #5 of the output tubes. Make sure this resistor is as close as possible to the pin on the socket. You want the body of the resistor right up against pin #5. The grid wire feeding the resistor should be as short as possible.

The transformers in my 1980 Marshall are rusty, probably because they are badly coated. I heard that these transformers can be cleaned and then painted with a heat-resistant lacquer/varnish.

But how and with what materials do I do that without damaging the transformers. And how do I avoid fluid/lacquer/varnish leaking in the transformers? Can all the transformers be cleaned?

I don't recommend doing any refinishing. If you clean them, you will worsen the problem. The problem with a rusty transformer is the inter-conductivity of the adjacent laminates of steel that make up the core. The transformer is not made of solid steel for a very good reason. It is made from laminates that are lacquered and then placed next to other laminates. The laminates are used instead of solid steel to keep electrical current from flowing across the core. Each laminate is insulated from the adjacent laminate, so current cannot flow around the core. If a solid core was used, the transformer would "think" it was just a thick piece of wire wrapped around the coil! Current would flow round and round and round (this phenomenon is called "eddy currents")—thus heating up the core and robbing power transfer, so it is important that electricity cannot flow across the laminates. When a transformer rusts, the laminates can become inter-conductive and produce eddy currents just as if the core were made from solid steel instead of laminates.

Why not replace the transformers? Those are probably still available and brand new ones won't any have rust to cause inter-conductivity.

I have a 1974 Marshall Superlead. The impedance position markers are illegible. I need to find out which position corresponds to which load. Am I correct in believing that the higher impedance output would correspond to more windings on the OT? If true, I could measure true resistance at the speaker output with each of the three settings. The lowest one will correspond to 4 ohm, the next up would be 8 ohm, and the greatest would be 16 ohm. What do you think?

I love to see people using their gray matter. You are on the right track. Now there are some precautions to take. First of all, you will be measuring D.C. resistance and not actual impedance. The D.C. resistance will be very small. So small that you will need to use a quality digital meter, and it must be set for the 1 ohm range. You will want to short the test leads together and zero out your meter first because the resistance will be very small.

When you measure, plug a short patch cord into the speaker jack and connect your test leads to the patch cord. As you try different settings on your impedance selector, the highest ohm reading will be the 16 ohm tap, the next highest will be the 8 ohm tap and the smallest resistance will be the 4 ohm tap.

I have seen you mention many times in your writings that LCR electrolytic capacitors do not do a good job of filtering out D.C. ripple current. So why then are they in every Marshall amplifier? And it seems you are the only one that says this! I use F&T caps and Sprague and have stayed away form LCR based on your words; but I have taken heat from taking this position. Can you elaborate on the subject some for me please?

Filtering is most difficult when the amp is drawing lots of current. When there is just little current, it doesn't take much capacitance to filter out ripple current. This explains why LCR's sound fine at low volumes. When doing a listening test, do it at loud volumes because more current requires a better filter. At loud volumes (read lots of current), the LCR cap will not remove the ripple current that occurs when A.C. is rectified to pulsating D.C. The 120 Hz pulse will be in the power supply and modulate the signal path. This results in an out-of-tune sound because you will hear the note you are playing along with the note that is modulated by 120 Hz ripple. It will sound something like a ring modulator.

Do a little experiment so you will be able to speak with authority on this subject. On one track of a stereo or multi-track recorder, record the sound of a Marshall amplifier with LCR's. Notice the ugly, out-of-tune sound and the lower pitch, yet non-harmonic note that moves with the note you are playing, but in a non-harmonic way. Pay particular attention to the notes A and B flat. You must turn the amp up to overdrive condition to hear this. When the amp is drawing a lot of current, the LCR's will not filter out the power supply ripple and it is clearly audible as an out-of-tune sound underneath the note.

Now change the LCR filters to F&T and do the same test. Turn the amp up and listen. You can do an overdub recording of this on a second track. Notice how the notes sound in tune when overdriven. Play an A and a B flat. Notice how clear the harmonics are and how there is no out-of-tune sound underneath the note being played. At this point, you should know why I don't like LCR caps. If you aren't convinced, play both tracks into a mixer and mute one track. While it is playing, unmute one track and simultaneously mute the other so you can switch back and forth—while remembering what the other one sounded like. Each time, one track is muted while the other is unmuted. Or, you can turn the balance control on the stereo amp to the left and listen, then to the right and listen. Which one do you prefer? Does the one that is in tune sound better to you? Me too.

Why does Marshall use LCR? I can only guess it is because Marshall is a Brit company and LCR is made in Britain. They have always used British parts unless the parts are not available in Britain. I don't think Brits would want to go out of their way importing expensive German parts over locally produced cheap British parts. With a big manufacturing company, "cheaper domestic" is better than "more expensive imported."

I have heard that a 6550 or 6L6 to EL34 conversion in a Marshall amp requires a change in the attachment of the feedback circuit to the output transformer. Would changing the feedback wire to a different output transformer impedance tap change the amps tone as you outlined in the relationship between feedback resistor value and tone?

Changing the feedback resistor value always changes the tone and character of the amp. Let's look at what negative feedback does and how it works. With negative feedback, a part of the output voltage is fed back into an earlier section of the amp. It is fed back to a part of the circuit that is out of phase with the output. This causes a phase cancellation which reduces gain somewhat. It is used primarily for two reasons. First, any noise produced inside the tube itself will be sent back through the tube in reverse phase, thus canceling the noise. Also, if certain frequencies are louder than others, then more voltage of the louder frequencies are fed back which results in those frequencies being phase cancelled more—thus evening out the frequency response. Here's the trade off. In order to use a negative feedback circuit, one must always sacrifice some gain.

Regarding the Marshall, when changing from 6550s to EL34 one must consider the tube requirements. To drive the tube to full power, a 6550 needs more gain than an EL34. An easy way to increase gain is to reduce the amount of negative feedback. There are a few ways to do this. Perhaps the easiest is by changing which tap the negative feedback resistor is connected to. In an output transformer with multiple taps such as the Marshall, the 16 ohm tap has the highest voltage, the 8 ohm has less voltage and the 4 ohm has the least voltage. If you move the feedback resistor from one tap to another, you are changing the amount of actual voltage you are feeding back.

So to review, you can decrease gain by adding more negative feedback. More feedback can be added by using a smaller feedback resistor OR using a higher impedance output tap. Conversely, gain can be increased by using a smaller feedback resistor or connecting the resistor to a smaller impedance output tap.

Generally speaking, less feedback will make the tone more raw with more distortion and more gain. Conversely a smaller value feedback resistor will make the tone cleaner and more even with less distortion.

I have heard of some people using a different impedance setting on a Marshall JTM 45 than what their speaker setup would call for, presumable to counter a difference in output transformer primary impedance that the JTM 45 has—as compared to the later Plexi Marshall. What is the effect of lower output transformer primary impedance?

The primary impedance of any output transformer is a function of both the turns ratio of the transformer and the speaker load. In other words, the primary impedance (this is what the tubes see) will go up if the speaker load goes up. For example, let's say you had a transformer that had the proper turns ratio to match a 4,200 ohm primary with a 16 ohm secondary. And let's say there were taps on the secondary for 16, 8 and 4 ohms. If you plugged an 8 ohm speaker into the 8 ohm tap, then the primary impedance would be 4,200 ohms. So far so good.

Let's supposed we plugged the 8 ohm speaker into the 4 ohm output. Since we doubled the load on the secondary by using a speaker that was twice what the tap called for, the reflected impedance on the primary would also double. So the tubes would see 8,400 ohms instead of the 4,200 ohms. How would this sound? For one thing, it would be much more compressed. The electrical current from the tubes would be limited because of the added impedance and the response would be more spongy and not as loud. Believe it or not, the output tubes would last much longer because the added impedance would limit the tubes dynamic response thus not allowing the tubes to be driven as hard.

Now let's go the other way. Suppose we hooked that same 8 ohm speaker to the 16 ohm tap. Now the load is half of what it normally should be. If it is half on the secondary, then the primary impedance would also be half. In other words, the reflected impedance on the primary would go to 2,100 ohms. When you consider that a direct short is zero ohms, you can see that 2,100 ohms is half way between the 4,200 ohms and a direct short. How does this affect the sound? It will be punchy, and louder. It will also be hard on the output tubes. The output tubes will be prone to arcing. But even if they don't arc, they will wear out really quickly.

One of the first amps I ever made used four output tubes with a Fender Twin output transformer. Thinking it would be too loud, I used a switch to disconnect two output tubes, so I could play it at either 100 or 50 watts. However, I didn't think about two tubes needing twice the primary impedance of four. I could have easily corrected this by using an 8 ohm speaker load instead of the stock 4 ohm load. I installed perfectly matched tubes in the amp and after about a month checked them and they were still very close. I got a gig and the 100 watts was too loud, so I lifted the two tubes to make it 50 watts and played for about a week like this. I then checked the tubes and could not believe what I found. By running those two tubes at half the impedance of what they should have run, they were almost completely lifeless. They idled at about half the current of the two tubes that were not used!

I use to play through a 50 watt Marshall (early 70s) using the Bass Channel for slide guitar because it wasn't as bright and seemed much smoother.

Does the same phenomenon hold true for the 62 Bassman? Specifically, if a Strat is run through the Bass Channel will there be a noticeable difference in compression/voicing? Is the Bass Channel voiced differently than the normal channel? It seems to compress—like a compressor that has the threshold set low. I have several dealers arguing over this point. Is this normal?

The bass channel is a different circuit than the normal channel. It does have four triode sections whereas the normal channel has two triode sections. The tone circuit has much larger capacitor values, which lower the range of frequencies. The channel should compress more, but I don't know if your particular amp is compressing the right amount unless I hear it. The last stage of gain uses an unbypassed cathode resistor. This degenerative feedback circuit is used to add compression while maintaining clarity. Of course, the Strat played through the bass channel would sound noticeably different given the circuit differences.

The dealers can argue all they want, but if they simply look at the schematic (page 272 of my first book, *A Desktop Reference of Hip Vintage Guitar Amps*) they will be able to "see" the differences in black and white.

OTHER BRITISH

Could a Vox AC30 clone be switchable to use only two of the four power tubes? (Half power switch) Where would you disconnect the two tubes not in use, at the cathodes? I was planning to have 4, 8 and 16 ohm output jacks instead of an impedance selector.

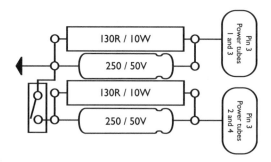

Your configuration looks good except your resistors are backwards. The 130 ohm would be for only two tubes while the 50 ohm would be for all four.

Another way to do the half power switching, and this would be the way I would do it, is to hook the inside pair permanently to both a 130 ohm resistor and a 250 uf at 50 volt bypass cap. Then hook the other pair to a 130 ohm resistor and 250 uf at 50 volt bypass cap. On one circuit, have the ends of the resistor/cap permanently soldered to ground. On the other pair, have the ends going to a SPST switch that goes to ground.

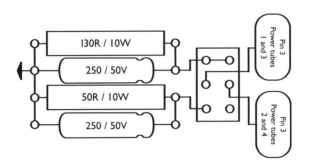

With this circuit, you won't have the "spike" when switching from one to the other. Since one pair is always connected, the current from that pair is constant.

The way you originally drew it, when it was switched, for a moment, there is zero current and then all of a sudden maximum current. I suspect that would make a loud spike during switching. Also the SPST is much simpler than using the DPDT.

When running two output tubes, the impedance selector that was 4, 8, 16 ohms with all four output tubes running becomes 8, 16, and 32 ohm taps. When all four tubes are on, the taps would go back to 4, 8, and 16. When you switch to two tubes, you will need to deliberately mismatch the secondary of the output transformer by 100%. For example, if you have an 8 ohm speaker connected to the 8 ohm output jack and you are using all four output tubes; when you switch to two tubes, you will have to plug the 8 ohm speaker into the 4 ohm output jack. This will correct the reflected impedance on the primary side of the transformer.

Two output tubes want to "see" twice the impedance as four output tubes. If your secondary speaker load is doubled, then the "reflected impedance" on the primary side will also be double. The turns ratio of the transformer is constant, so the only thing you can possibly do to double the primary impedance is to double the secondary impedance (speaker load.)

I have a Sound City amp that I bought at a garage sale, but the impedance selector is missing and the output transformer secondary wires (that went to it) are just hanging there. The wires look weird because they are in pairs (two red, two yellow, two black and two brown). The two blacks are soldered to the groundside of the output jack. Also, there is a gray wire hanging in there that goes to a resistor. I suspect it goes to the impedance selector also, but I don't know for sure. I want to forget the impedance selector and just wire the ¼" output jack. Then I could use it with my Fender 2x15" cabinet (4 ohms). What are your recommendations?

The transformer is most probably wound so that the two yellow wires connect together and count as one lead. The two red wires connect together and count as one lead. And ditto for the two black wires and the two brown wires. The transformer is simply a "bifilar" design. In the "bifilar" design, two wires are constantly held together and treated as one wire as the transformer is being wound. This has many advantages, for example, two wires can carry more current than one wire. And there is more inductance (more inductance helps the bottom-end response.) The ends of each "bifilar" secondary winding connect to each other and count as one wire—even though they are really two wires that were wound as one.

If you just want to wire it without a switch and just use a single output jack, I would recommend re-wiring your 2x15" cabinet from 4 ohm (parallel) to 16 ohm (series). This will give you a 16 ohm load from the 2x15". It is best to use the 16 ohm tap of the transformer. The largest tap of any output transformer always sounds best because when using the highest impedance tap of the secondary, you are taking advantage of using the entire secondary winding.

How do you know which tap is the 16 ohm tap. Easy, since the 16 ohm tap is the entire winding (and not just a percentage of the winding; as in the case of the 4 ohm which is 50% of the winding or 8 ohm which is 70.72% of the winding), you can easily determine this by which set of wires has the most D.C. resistance to ground. Remember the best ground to connect your meter is where the two black wires connect to the ground of the speaker jack.

The resistance you measure will be very small, perhaps a tenth of an ohm or so. Because of this, you must use a digital ohmmeter and

All About Vacuum Tube Guitar Amplifiers

zero out the lead resistance first. The lead that has the most D.C. resistance to ground is the 16 ohm tap, the lead with the least resistance is the 4 ohm tap and the remaining one will be the 8 ohm tap.

The gray wire is your feedback wire. I don't know which tap it was attached to, but attaching it to the 16 ohm will give you the cleanest and most uniform setting. Conversely, connecting it to the 4 ohm tap will give you the rawest and most obnoxious feel. And connecting it to the 8 ohm will be somewhere between those two extremes. So try it on each tap and see what sounds best to you. When you are done, tape off the unused leads so they do not short against anything.

I just got your new book and although it looks pretty good, you say the Vox AC30 is Class A. I know Vox Korg likes to say it is, and that has been the popular conception of it for many years, but it isn't really Class A from what I can understand about it. Do you still think the AC30 is Class A?

We are conditioned from elementary school to think straight A's are better than the AB honor roll. I have discussed Class A quite a bit in my seminars with technicians, design engineers and patent attorneys. The textbook definition of Class A is that the tubes operate on the most linear portion of the curve. This is simply another way of saying the tubes are biased to idle half way between saturation and cutoff. Also, to remain in Class A, the signal voltage should never increase enough to drive the output tubes into cutoff or saturation. Remember, if they are biased exactly halfway between cutoff and saturation and operated push-pull, like the Vox AC30, and enough signal is given to drive them into cutoff, they will reach cutoff and saturation at the same time. After much discussion, although the textbook definition is for the tubes never to go into cutoff, I believe the Class of operation with the Vox is simply a design intention. I think the Brits were attempting to design a Class A amp with the Vox AC30. The tubes do idle nearly half way between cutoff and saturation. And the non-top boost models didn't really have very much gain as it only used one triode section. Certainly with a top boost circuit the output stage could be driven into cutoff which would constitute class AB, but on the other hand, any Class AB amp could have its volume turned down and not driven into cutoff thus keeping it in Class A. Even a single-ended tweed Champ, which is never supposed to be run in Class AB because it is single-ended, can drive into class AB if it is played with a very hot pickup. So I believe with the question of Vox, I am going with the definition of Class of operation being a design intention whether or not the design is 100% valid by strict textbook definition. Incidentally, I prefer Class AB amplifiers. Everything we've ever heard by Hendrix, Stevie Ray Vaughan, Billy Gibbons, Eric Johnson, Robbin Ford, etc is Class AB. The Class AB is louder and more dynamic because with a Class A push-pull amp, the tubes are drawing maximum current at idle.

I am wondering if you could make any suggestions on how to trouble shoot some noise issues I'm having with two HIWATT amplifiers. They are both from early to mid seventies. One is an SA212 and the other is a DR103. They both are in pretty good shape and both are actually very clean. Both units have their original Partridge transformers which look almost brand new. No evidence of worn lacquer or laminate rusting. I am getting noise from these units when they are "on", but it seems to be a mid-frequency buzz that increases as the master volume level control is raised. It is also affected by each channel volume control. It doesn't matter whether something is plugged into the inputs or not. There is a very slight low frequency 60 Hz type hum, but not very offensive at all. It is the midrange buzz that is annoying.

I run these amps through a Furman power conditioner. I'm sure all of the internals are original as well as the tubes. The large capacitors are of the aluminum can type and are wrapped with a plastic shrink skin. Other than the buzzing, these things are FANTASTIC.

So you have two different types of hum. The low frequency hum is from the filament circuit, which has 60-cycle hum from the wall electricity. The higher pitched hum is coming from the power supply ripple current that occurs anytime a wall voltage is rectified to D.C.

The HIWATTs all have the same problem. If you will notice, the grounds from the cathode resistors, bypass caps, etc are all daisy chained together. This is why they all hum. Hiwatt wanted the inside of the amp to have a very clean appearance and they apparently

    All About Vacuum Tube Guitar Amplifiers

thought that since the amp was so loud already that no one would ever turn them up. So they did not consider the hum a problem.

Here's the fix: Install a brass grounding buss on the inside of the amp chassis between the pots and the chassis. The grounding buss is basically a piece of sheet brass slightly smaller than the faceplate, with holes cut in it for the pots and jacks to mount through. Next, you will need to get schematics and locate each component that is connected to ground. Give each and every ground connection (this would include the cathode resistors and corresponding bypass caps, as well as the grid resistors feeding) its own ground wire connection going to the brass buss, the hum will disappear. Here's a tip for finding those ground points easily: The bypass caps will be smaller electrolytic style caps. The minus side of those caps also connect to the cathode resistor. It is the minus side that goes to ground. **Note:** In order for this modification to work, you must remove the daisy-chain grounds that are in the amp now. That's what's causing the hum. This modification is not recommended for a beginner. Unless you know what you are doing, I would bring the amp to a qualified tech with a copy of this page and he should know what to do from there.

But besides the hum, it sounds like it is time for an overhaul. At the very least, you need new filter caps and new tubes. Some of the hum could be a result of dried up filter caps. Components don't last forever and if your filter caps and tubes are original, it is past time to change them.

I have a repair question on a 1986 HIWATT L50R all tube 50w head. It is very strange. The amp sounds GREAT except for one problem. When it is being played it sounds great until you stop playing (amp still on), then only sometimes, it only puts out a weak signal. (You hear the guitar, but very low in volume) Then it will crackle and come alive, sounding GREAT again. It is almost as if playing it seems to break through a dam and the music (and volume) pours forth through some blockage. This seem to occur after it has been warmed up with playing for 20 minutes or so. Then the problem is intermittent. I have replaced the filter caps, and put in new matched E34L tubes. I also did a complete service on the pots, jacks, switches etc. I have

substituted pre-amp tubes, and this does not solve the problem. Any advise is welcome.

There are a number of possible malfunctions, but what you are describing is most likely a bad solder joint. You can have a solder joint that is intermittent, just like a battery cable on a car that is physically connected, but the connection isn't good enough. The best way to troubleshoot this problem is with a chopstick. Put the amp chassis on a bench so that the circuitry is facing you. Connect a speaker load to the output jack and turn on the amp. When it starts acting up, take an ordinary wooden chopstick and lightly tap every solder joint on the amp. When you tap the offending solder joint, you will know it, as there will be a loud popping noise coming through the speaker.

It might not be a bad solder joint; it could also be a bad lead coming from a component. It is possible for a capacitor or a resistor to have a lead internally defective. You would troubleshoot this in the same manner. Simply tap on the leads of every resistor and capacitor in the amp. If you find a cracked or broken lead, you will know it because it will make the same loud popping sound.

If it turns out to be a bad component lead, replacing the component is the cure. If it turns out to be a bad solder joint, desolder the joint with desoldering wick and re-solder using fresh solder.

SPEAKER

I've rewired my cabinet to series/parallel (to get it sounding warmer) with 16 ohms 25 watt reissue Greenbacks, and on my ohmmeter the speaker impedance is indicating 14.7 ohms. Is this OK when my amp is set to 16 ohm output as it is not exactly matched between the cab and the amp? I don't want to over stress my output transformer or is this irrelevant?

Your speaker cabinet and amp are matched, but you are confused about the difference between impedance and resistance. Your meter doesn't measure impedance. It measures resistance! Impedance is not the same thing as resistance. When you add together D.C. resistance, inductive reactance and capacitive reactance, you get impedance.

You are ignoring the capacitive and inductive reactance and only looking at the resistance. Your meter doesn't measure capacitive reactance or inductive reactance. There is no convenient way to measure capacitive reactance and inductive reactance. Inductive and capacitive reactance will actually change with frequency and will be very little. When you measure a speaker for resistance, that number will always be slightly less than the actual impedance. For example, a 16 ohm load might have a resistance measurement of 12 to 15 ohms depending on the actual speaker. Likewise, an 8 ohm speaker could have a resistance measurement of 5 to 7 ohms.

I have a tube amp that has an 8 and 4 ohm output, it is running currently on the 8 ohm going straight into the 12" Celestion (8 ohm speaker) in the combo set up. I have acquired a two twelve cabinet that has Jensen concert speakers and it is set up as a 16 ohm load. Will I damage my tube amp by using the two twelve inch speaker 16 ohm cab? Should I run it from the 4 or 8 ohm output?
You have an ideal situation here. With an amp that has dual impedances such as yours, you can use both taps if you double the speaker load on each. This works out perfectly for you because you can plug the 16 ohm cab into the 8 ohm output and the 8 ohm speaker into the 4 ohm output to correctly match everything. It will work perfectly.

I am trying to decide what speaker or speakers to use which will give me a large sound at a reasonable volume with great brown tone for home use. I am going to have a 40-watt class A amp built and would welcome your input. Any information pertaining to factors which effect speaker tone, efficiency, cabinet style and number of speakers would be greatly appreciated.
The key word here is "reasonable" as in "reasonable volume." For starters, 40-watts may be a bit much for home use unless you have a 2,000 square foot bedroom. Really, 40 watts of Class A is more power than George Harrison used to play the Ed Sullivan Show 40 years ago. Perhaps you should think about using a power attenuator. This way, you could turn your amp up and turn your speakers down. This actually works great because when you attenuate, the first thing to go is the noise; so you end up with a church mouse quiet amp.

For speakers, you need to use something that is highly efficient and can take the power of the amps. An open back speaker cabinet is best, especially for home use. The open back has a much greater dispersion angle, so the sound will seem to engulf the room. Closed back cabinets are highly directional with no sound coming out of the back. Stand in front of the closed back and it is too loud. Stand to the side and you can't hear it. I would say open back is the only way to go.

TRANSFORMERS

You wrote an article once about output transformer taps and their effects on sound but I can't seem to find it. I was trying to explain to my friend that a 16-8-4 ohm multi-tap transformer will sound its best going through the 16 ohm tap, but my friend argued that it should be louder at 4 ohms. I was trying to point out that he is not using the full potential of the secondary winding in the output transformer. He was baffled, and so was I by time he was done with me, so I thought I'd have you confirm what I said to him.

The reason he is confused is because he is thinking in terms of transistors. On a transistor amp, the speaker IS the load. If you run a smaller resistance speaker, the transistor develops more power because when the resistance goes down, the current goes up. Yes, the lower impedance speaker actually changes the operating parameter of the transistors and more current equals more power.

Such is not the case with a tube circuit. The tube is transformer coupled. The coupling is done by winding the transformer to a specific turn's ratio. The taps on the transformer's secondary are placed such that the turn's ratio is larger with lower impedance speakers. When you do the math, the impedance that the tubes see is constant. In other words, each tap has its own turns ratio such that regardless of which tap is used, as long as the correct impedance speaker is hooked to that tap, the output tubes "see" the same load impedance. So in the case of the tube circuit, the operating parameters of the tubes remain constant.

When you use a tapped transformer, the turn's ratio of the transformer goes up (creating more coupling loss) and each tap represents only a fraction of the transformer's full potential. For example, a 16, 8, 4 ohm transformer is first wound such that the full secondary would be used for a 16 ohm load. The 8 ohm tap is placed at 70.28% of the turns of the secondary while the 4 ohm tap is placed at 50% of the turns on the secondary. If one attaches a speaker to the 4 ohm tap, he is actually using only half the transformer which doesn't sound as good as when using the entire transformer. And to add injury to insult, the apparent turn's ratio has gone up considerably when using the smaller tap and so do the coupling losses. A simple listening test will prove this beyond any doubt.

MISCELLANEOUS

I would like to make a line-level output that I can attach to the speaker of my small amp. What parts do I need and how can it be done.

To make a line-out, you need:

1. A ¼" female jack
2. A 100 ohm resistor
3. A 2.2K ohm resistor.

First—Solder the 100 ohm resistor across the jack—one end of the resistor goes to the tip and the other end of the resistor goes to the sleeve.

Second—Solder one end of the 2.2K resistor to the tip lead of the jack. This will already have one end of the 100 ohm resistor connected on the same lead.

Third—This device will connect to a speaker. Decide which speaker you wish to connect it to. The free end of the 2.2K resistor gets connected to the plus of the speaker. The minus of the speaker gets connected to the sleeve of the jack.

Note: You could mount this in the extra speaker jack of a "two output jack type" amplifier such as a Blackface Fender. You would simply use the extra speaker jack as the line out device and instead of hooking it directly to the speaker, the 2.2K would go to the hot of the speaker jack and the ground of the speaker jack would connect to the ground of the line out. **Caution:** Some amps may have a negative feedback loop attached to the hot of the extra speaker jack. If this is the case, remember to move it to the hot of the first speaker jack.

Regarding the power distribution resistors between filter caps, are they OK at ½ watt? I see Hoffman amps using 3 watts and others use 5 watts. How do know what is best.

Here's the way to know what you need.

1. Measure the D.C. voltage across a resistor. That means with the amp on, you would put each end of the resistor to a meter lead.
2. Square the voltage and divide by the resistance. (Example: if the voltage was 20 volts and we were using a 100 ohm resistor, you would square the 20 volts and get 400. Then you would divide by 100 ohms and get 4 as the answer.)
3. Take your answer and multiple by a factor of 2. This is your safety margin, 100%. In the example mentioned, you would get 8 as the answer.
4. Round UP to the nearest standard value. In the aforementioned example you would use a 10 watt resistor, as that is the nearest standard value.

Sometimes, designers overdo it in a bad way. For example, if you overbuilt your screen resistors and the screens were drawing too much current, instead of blowing the screen resistors, you might blow the power transformer! So I am not in favor of going crazy with overdesign. 100% and rounding up is enough for safe operation.

I have seen guitarists with little or no experience that have rolled their own homemade amps and ended up with great sounding amps! Now I am getting the idea of building my own amp but I don't know where to start. What do you recommend?

Roll your own? He who cuts his own firewood is twice warmed and homemade tastes better than store bought.

Starting at the end and working backwards seems to work best for making requirement decisions. You need to look at how you will be using the amp. With that in mind, decide what kind of speakers you will be using and what kind of power you need to drive those speakers. Once you get that far, keep working backwards—from end back towards beginning—and decide what type of output tubes you need.

Then you can decide on the preamp features such as: one channel or two, simple tone control or complete tone stack, how much gain, etc.

Let's use an example. Let's say I rarely play in a gig situation and I want a simple practice amp for my bedroom. It is not necessary to have a 4x12" cabinet for practice. I decide on an 8" speaker to keep it simple.

An 8" speaker can't take much wattage, nor do I need a lot of wattage in my bedroom practice situation. Still working backwards, I decide three to five watts will be plenty. I decide on a 5 watt output tube such as a single 6V6.

Still working backwards, I decide on using a simple preamp circuit with only one volume control and without tone controls. This is starting to look like a tweed Champ.

Next we have to decide if we want to use one of the tube amp kits available on the market or build from scratch. There are advantages and disadvantages to either. For your first amp, the kit is the way to go.

In favor of building from scratch, you have more choices. You can choose different schematics, chassis materials, component board materials, etc. This is a great way to go if you have already built an amp or two, because you have the experience to make some intelligent design decisions.

Transformers will be a big problem when building from scratch. The transformers will have more to do with the sound of your amp than other part. Great sounding transformers cost money and cheap transformers rarely have tone.

Oddly enough, the advantage in favor of building from a pre-designed kit is you have less choices. You are pretty much guaranteed a working amp because the kit has already been designed and the kinks worked out. You don't risk making a dumb design decision. An amp kit has a more predictable outcome. The transformers in a kit are most probably custom made especially for that kit, so you've got the critical part handled. Kits cost much less than buying parts individually. And besides the actual component's pricing, there are hidden costs with building from scratch. If you were to buy parts individually, besides the cost of the parts; you have long distance phone calls, separate shipping charges, extra shopping time, etc. All of this hidden cost is time and money saved when purchasing a kit.

Would adding a 3-prong A.C. plug to an older amp help prevent shocks and, if so, how is it wired? Is there a line conditioner that could be used for safety's sake to prevent shocks?

Adding a 3-prong A.C. cord could stop annoying shocks if the receptacle you are plugging into is wired correctly. You would simply wire it exactly like the 2-prong cord is wired now except the extra wire would be connected to the amp chassis. This extra wire is almost always green. As far as a 3-prong connector eliminating shock hazards, it will only work if the receptacle is wired correctly. A receptacle checker can be bought at almost any electronic supply for $15 to $20. A line conditioner is designed for voltage regulation, spike protection, RF interference filter, and EM interference filter, but will not help for shock protection.

I've been studying up on Tremolo Circuits and have a few questions. On output tube bias tremolo, the oscillator A.C. is passed through a cap to the bias circuit. How many volts A.C. might be put into the bias circuit? I'm guessing that this A.C. is added to the bias, sometimes increasing the bias voltage and sometimes decreasing it. Can the output tubes be damaged during the half cycle where the bias voltage is reduced?

Also, the Tremolux 6G9 and the Princeton Reverb AA1164 circuits are an interesting comparison. The Princeton gets the oscillating voltage from the plate of the oscillator triode, where the Tremolux uses an intermediate cathode follower. Is this superior circuitry? Does one circuit sound better than the other?

The amount of A.C. oscillator voltage that is added or subtracted to the fixed negative voltage of the bias circuit is dependent on how the intensity control is set. I would not worry about damaging the tubes during the half cycle when the bias voltage is reduced. Just as it is reduced, the bias voltage also increases, thus cooling the output tubes. The average will be where the bias would normally be set without the tremolo oscillator on. In other words, if the output tubes were safe before using the tremolo, they would still be safe when using the tremolo.

Regarding the type of circuit that modulates the bias voltage let me say this: The cathode follower is used as an impedance transformer device. Source impedances should be low and destination impedances should be high to achieve maximum stability.

Here is an analogy. Let's say you are going to take a voltage measurement using a voltmeter. Your voltmeter is a very high impedance device and that is how it should be. If the voltmeter had lower impedance than that which it was measuring, then the meter would "load" the source and give an inaccurate reading. It would be like a partial short circuit. In other words, the source impedance should be much lower than the destination impedance. Said another way, outputs should be low and inputs high impedance.

If the oscillator is low impedance (cathode follower circuit), then the way it modulates the high impedance inputs of the output tubes will be very stable.

Now lets say you modulate the bias voltage (without a cathode follower) directly from the plate of the oscillator. You have a high impedance source and a high impedance destination. It will work, but it is not nearly as stable as using the much lower impedance cathode follower. The advantage of not using the cathode follower is that a tremolo circuit could be made by only using one triode section instead of two.

My amp has a triode/pentode switch on it and I am not sure how that works. I notice the amp has less volume in the triode mode but it seems sluggish. Can you shed some light on this design?
Consider the triode/pentode switch a marketing myth/gimmick. I am amazed at how designers will add a switch to an amp to make it lose responsiveness, give it a cool name and then sell it as a feature! The design is touted as taking a pentode and changing it to a triode—which is not possible. The switch basically connects the screen grid to the plate and that is supposed to somehow make a triode. A pentode will never be a triode. There is more going on here than what meets the eye.

Besides the heater, a triode has only three components: the cathode, grid and plate. A pentode has five components: the cathode, grid, screen grid, suppressor grid and plate. For example, in a 6L6 tube, the suppressor grid is internally connected to the cathode. So when you connect the screen to the plate to supposedly make a triode, the suppressor grid is still sandwiched in between the screen and plate and it is still connected to the cathode. Triodes don't have suppressor grids. Even when you connect the plate and screen and count that as a single component, there are at best four components.

If there are four components, could this be a tetrode? No, because in a tetrode you would have the screen grid and not a suppressor grid! When you switch to the "triode" mode, it may distort at a lower volume, but that is only because it lacks efficiency.

You are the guru of tone, hence, I assume you would know the answer to the following question that I have a bet riding on: Do distortion and/or overdrive pedals (in this case a Proco Vintage Rat and an Ibanez TS9 modified to 808 specs) create "synthetic" harmonics via their transistors or op amp chips, or do they overdrive the power tubes in the amp by amplifying the input signal, resulting in natural overdriven tube harmonics?

On certain pedals, it may be possible to set the pedal where you are actually overdriving the amp. For example, if you had a pedal that had clean boost mode such as a Kendrick Buffalo Pfuz or a Klon pedal. In this case, you could use the pedal to boost the signal hitting the amp and if you turned the amp up, you could definitely overdrive the output tubes. That is not the case with the Proco Rat and TS9. Both of those pedals have semiconductors overdriving semiconductors, which flattens out the waveform.

With the TS9 and Proco Rat, depending on how the pedal is set and how the amp is set, you could get both output tube distortion and pedal distortion. Consider this: If the amp was turned down and the pedal was set for distortion, then you would be getting no output tube distortion because the power tubes would not be overdriven. On the other hand, if the pedal was set very clean and the amps volume turned up, you could get output tube distortion. Most people do not set the pedal that way. Usually, pedals are set to add their own distortion to the amp's input signal.

When I connect the negative feedback on an amp I'm building, it causes a loud squeal or howl? I had this occur twice; once on a Deluxe Reverb AB763 clone with a 820 Feedback resister and today on a JTM-45 clone using a 27K Resistor off the 16ohm tap. What could cause this?

What makes a negative feedback loop? The output of an amp is A.C. which is constantly changing from negative to positive. The negative feedback loop is negative only because the signal is out of phase with respect to the part of the circuit it is fed back into. If the negative feedback loop is howling or squealing, it is probably not negative at all. It is a positive feedback loop! If you change the polarity of the primary wires on the output transformer, the positive feedback loop will become a negative feedback loop and the howling will cease. When changing the polarity of a transformer, you must take care to switch the primary wires and not the secondary wires when you have a multi-tap output transformer. If the output transformer has only one impedance output, you could change the polarity of either the primary or the secondary, but with a multi-tap output transformer you can only change the polarity by swapping the primary leads with each other.

I learned this lesson back in the eighties when I built an amp that had a switch to change the circuit from the 5F6 Bassman to the 5F6A Bassman. I wanted to be able to switch back and forth and see which circuit I prefer. The main difference between those circuits is where the feedback is injected back into the circuit. On the 5F6, the feedback goes to the bottom of the tone stack. One the 5F6A, it goes to the bottom of the phase inverter. When the switch was turned on my project amp, one position always howled and that was because those two points were out of phase with each other. I learn an important lesson that a feedback loop is only negative when it is injected into a part of the circuit that is out of phase with the feedback loop.

Next time you build a clone amp that uses a negative feedback circuit, I suggest you connect but not solder the primary wires of the output transformer until you can test the amp and see which polarity is correct. Once you have determined the correct polarity, then you can solder the wires in place.

I recently purchased a Kendrick amp kit, which is a clone of a 5F1 Champ. There are only a few resistors in the kit, but how do I know the value of each?

Its time to learn the resistor color code. Each color represents a number. There is a sentence that has been used for almost a century now to help us remember the resistor code color sequence. The first letter of each word in the sentence is the first letter of a color and each color represents a number. Before you read the sentence, remember that electronics has been mainly a male dominated field and what was politically correct in the 20s may seem distasteful today. Here's the sentence: Black Boys Rape Our Young Girls But Violet Gives Willingly. Now we have the modern women's lib, politically correct version: Black Boys Race Our Young Girls But Violet Generally Wins. Use whatever works for you to remember the sequence of colors.

Black   = 0
Brown   = 1
Red   = 2
Orange = 3
Yellow = 4
Green   = 5
Blue   = 6
Violet   = 7
Gray   = 8
White   = 9

Here is how to use the code. The first two colors are simply the first two digits of the resistor value. The third color represents how any zeros (called the multiplier) you add to the end. For example, 1 Meg would be Brown Black Green. That would be (1, 0, plus five zeros), hence 1,000,000 = 1 Meg. Let's do another. Lets say we were looking for a 68K (same as 68,000 ohms.) Six is Blue, Eight is Gray and three zeros translates to Orange. So the resistor would be blue, green, orange.

Sometimes there is a fourth color band, either gold or silver that would indicate the precision of the ohm rating. For example: If there is no fourth color, the resistor is considered 20% tolerance. If the fourth band is Silver the tolerance rating is 10%. And if the fourth band is Gold the resistor is considered within 5% tolerance. So a 20% tolerance 100,000 ohm resistor (100K = brown, black. yellow) could be measured and any reading from 80K to 120K would be within tolerance. If the same 100K resistor had a silver band for the fourth color, then it would be rated 10% tolerance and any actual measurement from 90K to 110K would be within spec. Likewise, if the fourth color band were gold, the resistor would be a 5% tolerance rating. An actual measurement of 95K to 105K would be correct for this resistor.

Occasionally you will see gold or silver as the third banded color. Remember the third band is the multiplier. In this rare case, the silver multiples by .01 and the gold multiplies by .1. So if we needed a 5.6 ohm resistor, it would be green blue gold. Similarly, a .98 ohm resistor would be White Gray Silver. In the case of silver or gold for the multiplier, there still may also be a fourth color of silver or gold to indicate the percentage tolerance.

Now my question is, if I put a 47K resistor instead of the 100K resistor for the negative feedback resistor, will I loose distortion or grind in the sound. Someone told me that it would cut a bit of gain at the power amp output.

I'm a little confused on the definition of gain versus distortion, could you explain it to me quickly!!

The smaller value feedback resistor will allow more negative feedback. Negative feedback is simply a small signal coming from the output of the amp that is injected back into the circuit but at a point that is opposite phase with respect to the output. This has the effect of "Phase canceling" some of the signal (gain). The negative feedback resistor connects the output of the amp back to an earlier place in the circuit.

The intention behind negative feedback can be made clear. Certain notes are inherently louder than others. When a louder note is played, it would have a tendency to have more output. When it has more output, more (negative) feedback is sent back to that earlier section of the amp and this extra feedback phase cancels some of the signal. This causes less volume (less volume = less gain) in that particular frequency. Now let's say there is a frequency that is not as loud as the other notes. Relatively speaking, that particular note will not have as much feedback and therefore, it will not phase cancel as much as the other notes. You get the idea. Negative feedback cuts overall gain in order to make the amp play more evenly. Using a smaller feedback resistor (47K) will allow more feedback to get through. The smaller value does not "resist" the flow of electrons as much as the 100K value, so more feedback gets through. With more feedback through, the overall volume of the amp is decreased and the frequency response will be at a more consistent level.

When a small signal is amplified to make it larger than it was, it is said to have increased in gain. The guitar may put out ⅛ of a volt, but after the signal is amplified through a 12AX7, it may come out somewhere near 12 volts. This is gain. There is a larger output voltage than input voltage. Gain by itself is not sufficient to drive a speaker. It takes power to drive a speaker, so the gain is fed into a power amp section (your power tubes) and is transformer coupled to a speaker.

Distortion is anything that is in the output that was not in the input signal. In other words, if your output tubes are saturating, that was not in the input so it is by definition—distortion!

Let's say the output stage is not saturating and the sound is totally clean—without any distortion. It still has some gain (the output is greater than the input), but not distortion. What gets confusing is the fact that to drive the output tubes to saturation requires a very hot signal voltage, which means you need a lot of gain.

I have heard the term "sag" when referring to a tube amp's tone. I am not sure I know what that means. Could you please explain?
When you play a note, the amp will respond to the attack of the note. At first, immediately after the initial attack of the note, it will compress slightly (some amps more than others). Then, as the circuit begins to recover, the volume will rise slightly. This is happening as the string is beginning to die out, so you get a singing quality here. The envelope of how any particular amp does this is loosely referred to as sag. If the sag is very pronounced, it is sometimes referred to as "meow." This natural sonic envelope is similar to the Attack Decay Sustain Release envelope of a synthesizer, except it cannot be easily adjusted as in the case of the synth. An amp will usually have more sag when it is played hard than when it is played at lower volumes. Deliberately changing the impedance of the amp's power supply can modify the sag. Such impedance changes could include changing rectifier tubes, going from tube to solid-state rectification, increasing or decreasing main filter cap values. Sag or, "note envelope" can also be affected by output tube bias. In general, as the tubes are biased hotter, the sag will become more pronounced.

Cathode biased amps, such as the tweed Deluxe or Fender Champ, will have more sag than if the same amps were fixed bias. Tube rectifiers will produce more sag than solid-state rectifiers. Sometimes, amps sag too much and end up compressing the front off the note. Fast players generally don't like much sag because if they are fast, they can play faster than the amp can recover. Conversely, slow pickers or slide players usually prefer a little more sag because the sag can contribute to sustain—even at lower volumes.

I have heard the terms "non linear" and "clipping." What is that all about?

A tube can be made to draw more current or less current. If it draws so much current, that no matter how positive the grid goes, the tube cannot pass any more current, then the tube is saturating. If a greater amplitude of input signal is applied, instead of the tube drawing more current, the top of the wave is "clipped" off. Instead of the waveform being rounded on top, it is more flat. This is called clipping and it is called such because the top of the wave gets clipped off. It can also happen if the tube draws too little current. If you keep drawing less and less current, you will eventually get to the point of no current. This condition is called "cutoff." If you reach cutoff and a greater amplitude of signal is applied, then the bottom of the wave is "clipped" off and clipping is said to have occurred. Depending on how your output tube bias is set, clipping could occur at the top of the wave, at the bottom of the wave, or at both top and bottom simultaneously.

The tube's response is the most linear when it is operating exactly half way between saturation and cutoff. When the tube is operating in the most linear part of its characteristic curve, it is said to be operating in Class A and it will have an output that is a faithful reproduction of its input. When it operates into the limits of cutoff or saturation, it is operating in the most non-linear portion of the tube's operating curve. In this condition, its output will not be a faithful reproduction of its input because clipping will occur.

I recently purchased your second book and had a question about removal of two output tubes on a four-output tube amp. Between the pages of 375-377 there are two questions on this subject. The first answer states you have to match the impedance by unplugging one of the speakers (in a Fender Twin) resulting in an 8 ohm load. Is the normal output of a Twin 16 ohms? The answer on the second question states you should double the speaker impedance after removal of the tubes. Also another book I have states that I

should reduce the speaker load by half, i.e. If I had an 8 ohm speaker I should now use a 4 ohm speaker, but in the case of a Marshall amp with the switch at 8 ohms and using two 16 ohm cabs I should simply switch the switch to the 4 ohm position, but wouldn't that result in be doubling the impedance between the amp and speaker? I am very confused and maybe more confused after reading my own question I hope you can shed some light on this for me.

The book that says you should reduce the speaker load in half is wrong. You should have the actual speaker load be double what the impedance selector is set for. A transformer is an impedance matching device. It consists of two coils of wire wrapped around an iron core. The ratio of one coil to another determines the primary and secondary impedance. Yes, these coils are wound to a very specific ratio. The primary winding is what the tubes see as they are connected to the primary; and the secondary impedance is what the speaker sees as the speaker load is connected to the secondary winding.

A four-output tube amp wants a primary impedance of approximately 2K. A two-output tube amp wants to see approximately 4K. So two tubes need twice the impedance of four tubes.

If you take a transformer wound for 2K primary and hook it up to the speaker load it was made for, then the primary is in fact 2K. If you were to double the speaker load by hooking the secondary to a speaker load that is twice what the transformer was originally wound for, then the primary becomes 4K.

A Twin is a 4 ohm output and the primary is approximately 2K. When you hook it to an 8 ohm load (one speaker instead of two), then the primary becomes approximately 4K. Two tubes want to see 4K, while four tubes want 2K.

In the case of the 100-watt Marshall, when you remove two tubes, you simply would connect a speaker that is twice what the impedance selector is set for. Example: Using a 16 ohm cabinet with the impedance selector set to 8 ohm when switching from four to two output tubes. Or as your one book stated: using two 16 ohm cabinets (8 ohms total nominal load) and setting the impedance selector to 4 ohms.

Zero ohms is a dead short. If you remove two tubes and don't deliberately mismatch the speaker to the output, then you are running two tubes at approximately 2K which is half way between a direct short and what the two tubes would like to see (4K). It will work, but the tubes won't last very long and if you play it loud, your output tubes may arc—thus ruining them.

I have a Fender Blues DeVille I would like to apply a nice varnish to the tweed, as you do on your amps. Can you recommend a varnish and how it might be applied?

There are many different types of sealers that could be used, but I use Polyshades by Minwax. I like the Honey-pine matte finish. Apply it with a rag or other lint-free cloth. Use three or four coats and sand between coats with 000 steel wool. Use an appropriate tack cloth to remove sanding debris before each new coat. You want to let it dry and sand between coats so that the "feel" will be right. tweed has a roughness to it because it is a textile product. When you apply the lacquer, you are sealing the pores of the weave. When you let a coat of lacquer dry and then sand it, you are leveling out the tweed. After several coats with sanding in between coats, the feel will become smooth and actually feel like an original.

The Polyshades is sold at Walmart, Sherwin Williams, or Home Depot; and is available in pints. One pint will be plenty enough to do one cabinet. Expect to pay a little less than $9 for a pint and remember to add lint free cloth, tack cloth, and 000 steel wool to your shopping list.

Caution: There are some guitar cases covered in polyester tweed instead of typical cotton tweed as used on amps. Do not use Polyshades polyurethane lacquer on a polyester tweed guitar case. The polyester from the tweed interacts with the lacquer and makes it turn green.

How do you calculate the proper resistor wattage when putting them on a board (when to install a .5 or 1 or 2 watt resistor?

On most circuits, ½ watt will be just fine for everything except the screen resistors (usually 5 watt is fine), power resistors (1 watt values are almost always enough for these) and if the amp is cathode biased, the cathode resistor will usually be quite large (5 to 25 watt range). So I recommend using 1 watt resistors for the power resistors, 5 watters for the screen resistors and ½ watt resistors for everything else and then I would recommend calculating the actual wattage after the unit is working.

Here is how to determine the best wattage rating for a resistor.

With the amp on and in the play mode, put a voltmeter across the resistor in question and measure the D.C. voltage across the actual resistor. Once you get the D.C. voltage measurement, square it and divide the product by the resistance. This will give you the actual wattage in the circuit. Then you must DOUBLE IT FOR SAFETY.

Example: You have a 10K dropping resistor between two filter caps. You check the voltage across it and measure 60 volts D.C. Square the 60 volts to get 3,600 and then divide by the 10K. You get .36. You double that for safety and get .72. Rounding up to the nearest standard value, you see the 1 watter is the way to go.

I'm moving from the USA to New Zealand soon. I own several Kendrick amps and would like to wire them for New Zealand voltage. Nominal voltage in New Zealand, as it turns out, is 230 A.C. at 50 Hertz.

All Kendrick amplifiers have export design primaries that can be wired to take any voltage in the world. To hook up the amp for 230 volts, it is a matter of changing a couple of transformer wires going to the fuse and the switch.

All Kendrick power transformers have two primaries. For 120 volt operation, the primaries are hooked in parallel. For 240 volt operation, the primaries are run in series. There is a tap at 110 volts on one of the primaries, so if we use that tap and run both primaries in series, we will have a 230 volt setup. The primaries of the transformer are connected to the fuse and the switch. Right now, for the stock 120 volt setup, you are using black and brown together as one 120 volt line lead and you are using gray and white together as the other line lead. One set goes to the fuse and the other set goes to the switch.

To change from 120 volt operation to 230 volt operations:

1. Disconnect the brown wire from the black and attach it to the violet wire. (The violet wire is not currently being used). The brown and the violet wire should be soldered and taped up or held together with a wire nut.

2. You will not be using the white wire that is currently connected to the gray wire. Disconnect the white wire from the gray and tape it off by itself.

All About Vacuum Tube Guitar Amplifiers

3. Now it is time to check your work. The black wire and the gray wire are now the connection to the 230 volt line with one going to the fuse and the other to the switch. The brown and the violet wires are attached to each other but not to anything else. The white wire is taped off and not being used. The blue wire is also taped off, as it was before, and not being used.

4. You are now ready to connect the amp to 230 volt 50 Hertz power.

What features do you recommend for a multimeter for someone that builds amplifiers as an occasional builder? Which features are "must-have" vs. "nice to have"? What ranges and resolution/accuracy would you recommend for things like capacitance measurements (i.e., is it necessary to be able to measure pF caps or something more like 0.001 to 100 uF?)

Most basic multimeters have common denominator features. For example, almost all will have ohms, D.C. volts, A.C. volts and D.C. current. All of these features are must haves. If you want to use the meter to bias amps using the shunt method, you will need to get an American made meter—preferably one that uses an internal 600 volt fuse. These type units have two fuses: one that is easily accessible and then the 600 volt fuse that requires disassembling the meter to change.

Besides the regular, common denominator features, it is really nice to have gain measurement in decibels with a relative function. The relative function allows you to zero out the meter. This is really useful to measure dB of a stage. For example, let's say you are troubleshooting an amp with extremely low gain. You can put a signal generator on the front of the amp and attach your meter to the input. At that point, you adjust your signal generator to 100 millivolts and then set your meter to dB. You then hit the relative button to zero out the dB signal. Then you place the meter on the output of the tube and you can measure the relative dB gain. If you do this on a 12AX7 style tube and the gain is 35 dB or so you know that stage is working. If it is less than 30 dB, you know there is something amiss. I would stay away from a meter that checks capacitance. Instead, get a separate capacitance meter. There are some really nice ones available for less than $60. I recommend getting the kind that has both test jacks and test leads. The jacks are the most accurate way to connect a capacitor, but sometimes you need the leads.

What is the maximum plate dissipation for a 6V6? When I check different sources, I get different answers—varying between 14 watts and 12 watts. I know it will vary by manufacture, but what do you use as a rule of thumb?

It will vary from manufacturer to manufacturer of different tubes because the GE manual gives the wattage rating for a GE tube, while the R.C.A. gives the correct rating for the R.C.A. version. I like having a tube manual for general information, but you can't really hold them as the last word on guitar amp design. For example, a tube manual will tell you not to run the 6V6 over 350 volts and yet almost all 6V6 guitar amps run them over 350 volts. A case in point is the Fender Deluxe Reverb. These typically use 425 to 440 volts and yet the tube manuals all agree that the 6V6 shouldn't be run over 350 volts! You must remember: a tube guitar amp is not about electronic perfection; it is about tone! Anything that sounds good to you that does not cause a fire, blow a circuit breaker, or produce an annoying mushroom cloud is acceptable.

If you want to know how I would determine the plate watt dissipation of a tube, here it goes. This may sound unscientific, but I forget the math and go the empirical method. Using the maximum preamp signal that will be used with the amp, decrease the output tube bias voltage until the plate just begins to glow cherry red. This is really easy to get just right if you are in a darkened room, so turn down the lights. As soon as that plate just starts to glow cherry red, you have just barely exceeded the plate watt dissipation.

Now, turn the bias voltage the other way, backing it off till the plates no longer glow red. You have reached, by definition, the maximum plate dissipation.

I am just checking the voltages on my 5E3 clone I just built. I have a dual-wire heater system as per your book and it works great, but I was wondering if I should get a voltage reading from pin one of the 6V6 tubes to ground? If so, how much? Right now I am getting zero volts. Is that how it should be? I understand the 6V6s do not have to be grounded when using a two-wire heater set-up? Why did Fender tie pins 1 and 2 together originally? What plate voltage should I be looking for on tweed Deluxe, if I want more headroom and reasonable tube life? Also, what B+ plate voltages would be ap-

All About Vacuum Tube Guitar Amplifiers

propriate for a vintage warm sound with early breakup?

There is no pin #1 on a 6V6 or a 6L6 tube. I know the socket has all eight pins, but, pick up the tube and look at it and you will see pin #1 is missing on the bottom of the tube! Actually, pin #6 is missing also. If that is the case, why did Fender wire pin #1 to pin #2 on the original tweed Deluxe amp? Fender was using a daisy-chain type heater wiring which used a wire on pin #7 while pin #2 went to ground. At that time, it was standard practice to ground pins that were not being used and since pin #2 was already grounded vis-à-vis the daisy chain filament circuit, they simply tied pin #1 to pin #2 because pin #1 was the closest ground.

In the two-wire fixed biased amps such as the 5F6A Bassman, they used a two-wire heater circuit and attached pin #1 to pin #8 because pin #8 was the cathode ground in that amp and the closest available ground. In the tweed Deluxe circuit, the output stage is not fixed biased, so pin #8 goes to a cathode resistor instead of directly to ground.

And later they found a use for pin #1. They used it as a mounting post to mount the grid resistor of the Blackface amps!

The statement you make about grounding is too general to be conclusive. Regardless if you have a two-wire or single-wire filament there are certain ground circuits that are essential. For example, if the cathode isn't grounded, the tube won't work. It must be grounded directly for a fixed bias operation or grounded through a cathode resistor. The heater circuit must have a ground reference too, lest the amp hum. That reference could be in the form of a center tap, or one side grounded, or an artificial ground (two 100 ohm resistors, each going from one side of the heater winding to ground).

I am getting strange readings on my 5E3 clone amp. From pin #8 of the rectifier tube socket to ground reads 6.4 VAC and pin #2 to ground reads 7.1 VAC. What might be the cause of this imbalance? The plate voltage reading from pin #3 of the 6V6 socket to ground minus the voltage across the cathode resistor reads only 300 volts. I have tried several known good rectifiers, but the voltage only changes minimally. What can be the cause of low plate voltage?

On the 5 volt rectifier heater circuit, there is no ground reference;

therefore, any reading you take from the 5 volt heater to ground is invalid. Consider the imbalance as operator error. The way to check the 5 volt heater is to measure from pin #8 to pin #2. Make sure your meter is set for A.C. volts. That reading should be 5 volts. If it is high, either your wall is higher that the voltage rating of the primary, or the primary is too low for the wall (i.e. a 110 volt transformer plugged into a 120 volt wall).

Next, take the 6V6s out of the socket and measure the B+ voltage (pin #3) to ground. How much higher did it go? If it went up quite a bit, it is possible the particular set of 6V6s was drawing too much current and as a result, they were dragging down the transformer. If you are using EH 6V6 tubes, they are for sure drawing too much current, because that type 6V6 always draws more than a regular 6V6 in a cathode-biased amp. If that is the case, you can either use a different set of tubes or use a larger value cathode resistor.

Concerning the low plate voltage, there are four things that can cause low plate voltage:

1. Excessive current being drawn through the power supply. This results in a larger voltage drop across the internal resistance of the transformer winding. The more current, the more drop. A bad output tube or incorrect biasing could cause too much current to be drawn. Or, certain brands of tubes draw excessive current in a cathode-biased amp. Any voltage lost in the transformer winding takes away from the B+ voltage.

2. Bad filter caps or filter caps that whose value is too low. The filtering keeps the voltage up there. If the filter capacitors are no good or too small, the voltage will drop.

3. Weak rectifier tube. If the rectifier has a larger than normal drop across it, then that drop is subtracted from the applied voltage — thus leaving less plate voltage.

4. Too small a current rating on the transformer B+ winding. The rating of a B+ winding is related to the gauge size of the magnet wire used to wind it. Smaller ratings are done with thinner wire. Larger ratings use larger diameter wire. If the wire is too small, the resistance will cause a large voltage drop across the internal winding of the power transformer and that voltage will be subtracted from the applied voltage — leaving less B+ voltage.

All About Vacuum Tube Guitar Amplifiers

I read your article on testing transformers with a variac. Most of the variacs I've seen have a 3-prong standard electrical plug for a connection to it. How do you rig something to connect to the output jack of an amp to supply the 1 volt described in your article?

I use an ordinary electrical cord with a regular A.C. plug on one end and a couple of alligator clips on the other end. Just take a cheap, two-conductor extension cord and whack the female end off. Attach alligator clips, one clip for each conductor and you are there.

Is it really bad on a class A tube amp to run it with a bad preamp tube? What kind of damage could it cause?

Damage to an amplifier is almost always caused from too much current flowing. Preamp tubes do not draw much current anyway. Even if one was shorted, there is enough resistance on the plate resistor to limit current into the low milliamps range. The other way a preamp tube can go bad is too much noise, low gain or loud microphonics. None of these problems cause damage to an amplifier. Even if a tube shorted internally, there are resistors connected to every component in the tube (cathode, grid, plate). These resistors would either burn up, acting as a fuse; or they would limit current. Whether an amp is Class A or not, running it with a bad preamp tube would not cause any major or irreversible damage.

I would like to make an amp using a 12AX7 preamp tube as the power tube. Output transformers with a 50K primary impedance really aren't a stock type of transformer, right? I did some preliminary research on this configuration and stumbled upon an article about how Vox Korg had to design a "proprietary" circuit to get the 12AX7 to be able to drive a speaker. Maybe this is because they wanted to use a stock transformer? Where can I even find a tranny with impedance that high?

Remember this: tubes all have some voltage output and some current output—it is just that when the tube is designed, certain aspects are favored over others. On a power tube, the design emphasis is on current amplification. On a preamp tube, the design emphasis is on voltage gain.

A power tube needs a lot of gain (input voltage) to drive it and therefore requires a preamp tube. The preamp tube doesn't require much input voltage.

Now a 12AX7 is low current (read high impedance!) and to drive a speaker, it must be much higher current (read low impedance). So an impedance matching transformer must be used that matches a 12AX7 to the speaker. You normally run one side of the twin-triode 12AX7 at 100K, so here is the math for what kind of transformer you would need.

This is not hard if you read it slow and use your noggin. Turns ratio equals the square root of primary over secondary. Let's break that down. Primary is 100K, secondary is, let's say, 8 ohms. 100K/8= 12,500. The square root of 12,500 = 111.8 So you a 112 : 1 turns ratio would work great. Kendrick makes a transformer called a 100A that is a 112:1. It is used in their reverb driver that matches a 12AX7 to an 8 ohm reverb pan.

If you ran both sections of the 12AX7 together, you would double the current and therefore half the impedance. That's tech jargon for saying; "You could run both sections together if you run it into a 50K primary."

Here's the math to find turns ratio: 50K/8 = 6250. The square root of 6,250 = 79.05 So a 79:1 would be correct. I do not know of any 79:1 transformers that are available off the shelf.

Notice that dividing the 112 turns ratio of the 100K primary transformer by the square root of two could also get the 79:1 answer. Notice, when you double or half the primary impedance, you change the turns ratio by a factor of the square root of two.

A standard Reverb driver transformer for a Fender Twin Reverb, for example, has a turns ratio of 56 to 1. It is configured for a 25K primary (both triode sections of a 12AT7) and an 8 ohm secondary (reverb pan load).

The Fender Reverb driver transformer could work fine with a 12AX7 provided you configured it so that both sections of the 12AX7 were in parallel (you are running it at 50K) and you used a 16 ohm speaker load. Yes, if you ran it at 16 ohms and used both sides of the 12AX7 together that would work perfectly.

Here's the math: 50K/16 = 3125. The square root of 3125 = 55.9, so the 56:1 will work fine in that configuration.

I am assuming that pre-amp stage gain is the result of the relationship between the plate load resistor, cathode resistor, cathode cap and B+ voltage. Please, correct me if I am wrong. Is there a mathematical equation that will allow me to calculate the stage gain based on the components involved?

Yes, of course, there is a mathematical equation for gain, but you don't need it. It's too complex and it is not relevant to guitar tone. It deals with square roots, pi, and fractions three tiers tall. It is very complicated because it is frequency dependent. Even the capacitive reactive of the coupling circuit and bypass cap would have to be included in such an equation.

Guitar amps are not about electronic perfection, they are about tone—so forget the math. I think the essence of your question is really about how the various parameters and components affect tone. Here are the relationships:

- When the plate voltage is increased, so is the gain and headroom. As the plate voltage is lowered, you decrease gain and decrease headroom (breaks up quicker).

- Lower plate voltages don't produce high-end as well as the higher plate voltages. So the lower plate voltages won't have the shininess of the higher plate voltages.

- When the size of the cathode resistor is made larger, it will cause the tube to be biased colder (lower gain but with more headroom). As the cathode resistor is made smaller, the tube is biased hotter, such that it idles with more plate current. If the cathode resistor is made too small, the amp will hiccup when you stop playing.

- The size of the cathode cap in relationship to the size of the cathode resistor dictates how much bottom end or lack there of. The standard cathode circuit component values for a 12AX7preamp is a 25 uf cap across a 1.5K resistor. As the resistor goes up, the cap has more effect. As the cap value is made larger, it is easier for bottom-end to be amplified. As the cap is made smaller, it is harder for bottom-end to get through it, so the smaller cap will filter out some of the bottom-end.

- If you remove the bypass cap, you increase headroom, increase compression but lose gain.

How does the value of the coupling caps affect tone? (i.e. In my Marshall PA, the coupling caps in the output stage are .1uF while in the Marshall Lead model they are .022uF.)

Generally speaking, the larger cap allows more bottom-end to get through the circuit whereas the smaller cap filters out the boominess and allows for more highs and mids in the mix. From all the listening tests I have done, I generally prefer the big sound of the .1uF caps to couple the phase inverter to the output tube grids. In earlier preamp gain stages, the bigger cap may sound a little too boomy, and perhaps the smaller value would work better. For example the .1 uF may work great in the early preamp stages for harmonica or perhaps bass guitar, yet may sound muddy and boomy for guitar. Just remember this rule: there are no rules. Amp voicing should be done by trial and error with listening tests. That is the only way you will know for sure what you like in that particular application.

If I change transformers in my four output tube amp from 100W to 80W would it be possible to use 6L6G instead of the 6L6GC? Would the plate voltage be lower, or can I lower it to 350-360?

6L6GC tubes have a max plate watt dissipation of 30 watts. 6L6GB tubes have a maximum plate watt dissipation of 25 watts and the 6L6G has a maximum plate watt dissipation of 19 watts. The most you can get out of a 6L6G quartet is 76 watts; however it is the circuit itself and not the wattage of the transformer that dictates whether you can use a less powerful tube. To use a quartet of 6L6G tubes, you need a circuit that is designed for 76 watts or less. Reducing the preamp gain would be one thing you could do to decrease the actual wattage output. Reducing the B+ plate voltage or even biasing the output tubes colder could also reduce wattage.

The wattage rating of a transformer just means how much it can take before it overheats. It has to do with the size of wire the transformer is wound with. Thinner wire will overheat if too much current tries to pass through it. Thicker wire can handle more current just as a larger water hose can deliver more gallons.

You could still use 100 watt transformers as long as the circuit didn't run the tubes over 19 watts each. In fact, that is a trick designers use to squeeze every ounce of performance from an amp.

For example, if you put a transformer that is rated for more wattage in an amplifier circuit, since the wire is thicker, it is easier for the electrons to get through the transformer. The result is a punchier amp. You've seen Jim Marshall do this on his Park Amplifiers. He took a 100 watt power transformer and used it in a 50 watt amp. I did the same thing on the Kendrick Texas Crude Gusher. I used a 100 watt power transformer on a 50 watt amp. The original Bedrock amplifiers, as designed by Brad Jeters in the late 80s, also used massively overbuilt transformers to achieve uncommon dynamic response. Laney too used this same concept to achieve an ultra efficient design.

I have heard that you can do severe damage to a tube amp by running it without a load or speaker hooked up. Also, they say you can damage the amp by using a guitar cord, and not speaker cable to hook to the cabinet. When buying a used amp, what would be some signs that this could have happened by the previous owner?
Yes, operating a vacuum tube amp without an appropriate speaker load is a recipe for disaster. That is why it is so important not to keep playing on your guitar if you turn on the amp and there is no sound. If you ever turn on an amp and there is no sound, stop playing immediately. You should check to make sure the speaker is plugged in before you pick another note.

If an amp is operated without a speaker load, the output transformer lacks a circuit in the secondary. With only a primary connected, the transformer "thinks" it has become an ignition coil and the self-inductance of the coil causes ultra-high voltage to occur. This voltage is so high that it will arc—as in miniature lightning bolts. The arcing will cause irreversible damage as the heat from the arcing carbonizes the insulation, creating a path that encourages arcing in the future; even when the speaker load is later connected properly.

In the case of using a guitar cord for a speaker cable, similar damage can occur. The guitar cable is designed for high impedance while the speaker cable is designed for low impedance. The output of the amp puts out several amps of current. If you use the high impedance guitar cord in place of the speaker cable, the electrons cannot get through the cable fast enough and arcing could occur as the electrons are bottlenecked in the output transformer. It is like trying to push a river through a drinking straw. Something has to give.

Telltale symptoms of these two problems could include any evidence of past arcing, such as carbonized insulation, intermittent popping sounds, or a carbonized path in between two leads of one or more of the output tube sockets. Usually the carbon is between pin #2 and pin #3 of one or both of the output tube sockets. In extreme cases, the carbonization isn't seen on the outside, but it tunnels through the socket, internally! This is nearly impossible to see, but the intermittent popping will easily be heard when the amp is played loud.

Will the tone or performance of a reverb pan be negatively affected if it's mounted vertically on the side of an amp instead of flat on the bottom?

Yes, it does make a difference. Remember gravity is constantly pulling downward on the springs that are part of the reverb pan. All of this is taken into consideration and the reverb pans are designed with a particular mounting style in mind. There are six possible styles in which a pan could be mounted. If you are reading the seven digit part number on the pan, the last digit, which is the 7th digit, will tell you how the pan is intended to be mounted.

DIGIT #7 signifies the Mounting Plane

A = Horizontal Open Side Up

B = Horizontal Open Side Down

C = Vertical Connectors Up

D = Vertical Connectors Down

E = On End Input Up

F = On End Output Up

I have recently completed a single-ended, Class A, amp kit (similar to the 5F1 Fender Champ). It fired right up the first time. It sounds good at low volumes; however, when the volume is past 5, it starts to distort a lot and doesn't sound very good. I tried it on two different speakers, a 12" Bogen speaker and a 10" Gibson/Kalamazoo speaker. It seems really loud and may be overpowering the speakers. Is there something wrong and if so what should I do?

It sounds like your may have so much gain that it is driving the Class A amp into Class AB—which will give the tone and characteristics you describe. Perhaps you forgot to hook up the feedback loop? If the feedback loop is disconnected, the gain will go up to the point of driving the amp into Class AB. It will be really loud, but since the bottom half of the wave is clipped off, the resulting tone will be buzzy and homogenized—like what you are describing. I would suggest you carefully go back over the layout and check everything to make absolutely certain the wiring is 100% correct. Double check the feedback resistor value. If the value is too high, the amp will perform as you describe.

Will a 6L6 output transformer work in place of a 6V6 output if you still intend to use 6V6s? For example, could I remove the old 6V6 output transformer, replace it with a 6L6 style output transformer and still use the 6V6 tubes? Or, should I use 6L6 tubes? Will there be any gain or loss?

It won't work well. You might get some sound out, but the impedance matching is wrong. The output transformer's primary load impedance will be too small and your tube life would be severely shorted. The output transformer is like a transmission in a car. It takes a high impedance source and matches it to a low impedance load. When you do a mismatch like that, it is like hopping in your car and taking off in 5th gear. It could probably be done, but it wouldn't work well.

You can, however, make a 6V6 output transformer work with 6L6 tubes if your power transformer can take the extra filament current. I'll tell you how to check that later, but if you put 6L6s in a 6V6 socket that is non-self-biasing; the tube is going to draw too much current. Therefore, you must pre-adjust bias before putting 6L6s in the amp. In fact, I would start with no output tubes in the amp and set the actual voltage between pin #5 and ground of the output tube socket to -50 volts. This will be enough negative bias voltage to tame down the 6L6s from the start. You don't want to put a pair of 6L6s in an amp and have them go into runaway while you are scrambling to quickly adjust the bias voltage before they blow.

After you set the approximate bias voltage on pin #5 to -50 volts and before you actually fine tune the bias setting, place the output tubes in the sockets and set your multi-meter to A.C. volts. You should measure from pin #2 to pin #7 of either output tube socket. You are measuring the filament voltage. It is supposed to be 6.3 volts. Anything from 6 volts to 6.8 volts means the transformer is handling the extra filament current of the 6L6 well. If you check the filament voltage and it is less than 6 volts, the power transformer cannot take the extra filament current required of the 6L6s and the 6L6s should not be used. Once you have determined that the transformer's filament winding can handle the tubes, you should now reset the bias of the output tubes using whatever method you normally would use.

I'm starting to build a clone of the Fender Pro Amp, according to the layout as stated in your book, "A Desktop Reference of Hip Vintage Guitar Amps, (pages 318–321). What's your opinion about this amp? I like its simplicity. I'm still in doubt of which model to build, the 5D5 or the 5E5. The differences I see are the phase inverter tube and the feedback resistor. Is it best to build the 5D5 or the 5E5, and for what reason? Or is there a reason NOT to build this amp at all?

Is it necessary to add the 1.5K ohm grid resistors at the power tubes and the 470 ohm screen resistors on pin 4 of the power tubes, which both are not in the schematic yet present on all later Fender amplifiers?

Also, some Fender schematics show a switched type output jack so that the speaker output jack gets shorted to ground when not connected, and some don't. Is there a special reason for that? Is it better to do this on all tube amps?

There is a great reason for always using a shorting type jack with any amplifier. It is a safety circuit which prevents you from blowing up the output transformer should someone accidentally operate the amp without a speaker properly connected! A tube amp must have a proper speaker load or else arcing will occur in the output transformer and across the tube sockets. I would recommend always using that circuit as a safety precaution because when a speaker plug is plugged into the jack, the switch opens and from a circuit standpoint, it is not there. But on the other hand, should someone forget to plug a speaker into the jack and the amp is accidentally operated, then the shorting jack prevents you from doing serious damage to the output transformer.

In my opinion, the 5E5 sounds better than the 5D5. That paraphase inverter on the 5D5 simply will not get a clean sound at all. The

5E5 is more versatile in that it can get a beautiful clean tone and a great lead tone.

I would recommend building it just like the schematic and layout except for a couple of things:

For the first filter section coming from the rectifier, instead of using the the two 16 uf 450 volt parallel wired main filters, use two 100 uf 350 volt capacitors in series for the main filters. The circuit would be similar to the way the AB763 Super Reverb was done in the 60s except the Super Reverb used two 70 uf 350 volt in series for the first filter section. Also, in order to keep the voltage evenly divided across these two caps, a 220K ohm 1 watt resistor should be used across each of the filter caps. This will assure you that the voltage is evenly divided across each filter cap. It is interesting to note that the AB763 Super Reverb used those same 220K ohm resistors across each of the main filter capacitor.

I wouldn't use the 1.5K grid resistors on the output tubes unless the amp has a parasitic oscillation problem. The 470 ohm 1 watt screen resistors are a good idea. They keep the tubes from drawing excessive screen current and this makes the amp have more power and the tubes last longer. However, if you change the circuit to add the screen resistors, you may wish to make the 2500 ohm (on the E) or 10K (on the D) about half the stock value. This will bring up the voltage feeding the screens and will compensate for what you lose by adding the 470 ohm screen resistors.

I just put power to the 1964 Vibroverb clone amp for the first time to tweak this amp in. It is plugged into a current limiter with a 100 watt bulb, then the current limiter is plugged into my Variac in which is set to exactly 120VAC. I got my 8 ohm speaker load plugged in. All the tubes were pulled from a working 1964 Vibroverb. Here is what happens: I flip the "on/off" switch to the on position and the 100 watt bulb in the limiter glows for a few seconds until the tubes and filter caps are drawing their current; then the bulb drops to a very dim glow. This is normal. Then I switch the "Standby" toggle switch to "play" mode and the bulb glows brighter and drops a small bit. All of a sudden I get this loud pitched squeal starting low in volume then gets very loud. The

bulb is also getting brighter to about 50% of the bulbs brightness. Then it starts chirping, and then comes a motor boat sound. It is horrible! Of course I hit the OFF switch. The bulb is sucking power but not enough to glow 100%. I have gone over my wiring several times and it is correct. What could be the problem?

You either have a parasitic oscillation or the output transformer is in backwards. And if will make you feel any better, the first amp I ever built had the same problem. Here's the remedy. To check if the output tranny is wired backwards, set up the amp and make it squeal. There is a wire going from the speaker jack to the board. This is the negative feedback loop wire. While it is squealing, disconnect the feedback loop wire. You can just snip the wire. The problem will either get better or worse. If the problem gets worse, the transformer is installed correctly and the problem is a parasitic oscillation. If snipping the feedback wire stops the problem, your transformer primary wires (blue and brown), that go to the output tubes must be reversed. After you reverse them, hook the feedback loop up and try another test.

On the other hand, if the transformer is installed with correct polarity, then you have a parasitic oscillation. The basics for getting rid of a parasitic oscillation include shortening every grid wire (pins #2 and #7 on a 12AX7 or 12AT7) in the amp as short as possible, then use shielded wire for the input jack to grid wire and the volume control to grid wire. Make sure you only ground the shielding at one end because if you ground both ends, you will be causing a ground loop and mucho hum. At this point, the amp should be fine.

I bought a Kendrick ABC Box for hooking up three amps at one time. The tremolo from the Fender brown Deluxe is somehow being sent to the GA-8 when both are switched on together. You actually hear tremolo from the GA-8 too, as if they are daisy chained. How could this be? I thought the brown Deluxe's tremolo modulated the output stage and not the preamp. Can it be remedied?

Oh Yes. The tremolo design of the brown Deluxe is such that the tremolo oscillator modulates the output stage. The ABC Box connects to each amp, but every output is buffered and no signal can get from the A output to the B or C and vice versa. So how does tremolo get to the GA-8? The tremolo doesn't really. What you are hearing is phase cancelation. The brown Deluxe amp is apparently out of phase with the GA-8. When the tremolo on the brown Deluxe goes louder and softer, more or less phase cancellation occurs with the GA-8, so the tremolo appears to be coming out of the GA-8. You could probably minimize this effect by changing the phase of one of the amps. (Changing phase is fancy talk for swapping the speaker leads on one of the amps). For years, I ran three amps and a Leslie and when I clicked on the Leslie, it sounded like all the amps were spinning.

I purchased what were supposed to be 2 NOS matched pairs of R.C.A. 6L6GC blackplate tubes from various sellers on ebay, in hopes that I could hear what all the fuss is about in my 1967 Fender Blackface Bandmaster. It's the AB763 circuit with bias adjustment, not hum balance adjustment. All 4 tubes do have matching emission ratings on my Sencore Mighty Mite IV tester, and test perfect for shorts and grid leakage. However, when I bias them in my amp using the transformer shunt method (reading the bias current on pin #3 of the output tubes) they read as much as 14 ma apart. I've tried them in all combinations and can never get any closer than about 5 ma apart. The other thing that I noticed about these tubes in the circuit I thought was odd, was the fact the current reading always seemed to drift higher slowly the longer I continued to measure them in the circuit. Both sellers said they he had matched them in their Fender Super Reverb amp. I contacted the second seller who said the tubes shouldn't hurt my amp. But, I've always thought that matched output tubes means they match on plate current readings. So how many milliamps apart is considered safe when you have 440 volts on the plates? And, if they aren't matched, does it put any undo strain on the OT? Is it safe for the amp to use a pair of these tubes, or should I not use them in a push-pull output circuit?

It's not unsafe to run a pair unmatched. If they aren't matched, they just won't sound as good. I draw the line of whether the tubes are matched at 5 mA of idle current difference. If the difference is over 5 mA of plate current, then I would not consider them matched.

The current will go higher if the tubes get hotter. That is why it is a good idea to let them stabilize before taking a reading. In my opinion, neither guy knew what he was doing. A regular old style tube tester doesn't work because it tests the tubes at low voltage (150 volts or so). A tube will perform differently at low voltage than high voltage. And the guy that matched them in his Super Reverb may not have known enough about matching to get accurate readings. For example, when you are measuring one tube in a circuit, the other tube in the circuit will affect the current draw and plate voltage measurement of the tube being measured. The only way to really test them in a Super Reverb is to do them one at a time without any other tubes in the amp. I'll bet he didn't.

I have a Kendrick Power Glide Power Attenuator that I would like to run with my 100-watt head and two 8 ohm cabinets. Do I run the head at 8 or 16 ohms for two 8 ohm cabinets?

There are two versions of the Power Glide; the 16/8 ohm and the 4/2 ohm. Which one do you have?

If it is the 16/8 ohm Power Glide, the optimum setup is to run the speakers in series and run the amp at 16 ohms. This is possible only by using a series box. You can make a series box with three plastic (Marshall-style) ¼" phone jacks. You must use plastic type jacks or if you use Switchcraft style jacks, you would have to use a plastic box for this to work, as these jacks cannot share any common grounds.

Here is how to wire it. Take the hot from one jack and connect it to the ground of the next jack. Then take the hot of the second jack and it will go to the ground of the next jack and the hot of the third one to the ground of the first one. This way, all three jacks are in series with each other.

When you get the series box made, you would plug your cabinets into it and run the output of the Power Glide into it. It doesn't matter which jack you plug the cabinets and Power Glide into as they are all in series anyway. Then you would run a speaker cable from the amp to the Power Glide's input. This setup would allow you to take advantage of the fact that your output transformer is wound for 16 ohms. When you use the 16 ohm output of the amp, you are using the entire transformer and will get a much better sound.

If you don't have the Marshall-style plastic jacks handy, you could go to Radio Shack and buy three regular jacks but use a plastic enclosure box. You should be able to get everything there for under $15.

If you have the 4/8 ohm Power Glide, you would simply plug both cabinets into the Power Glide and set the head to run at 4 ohms.

I am building an amp whose design normally calls for a 5AR4 rectifier. I want to experiment with using two 5U4GBs. Is that safe or will they burn up the tranny?

You can try two rectifiers. Before hooking up the red B+ winding wires to the sockets, place two rectifier tubes in the sockets and read the A.C. voltage across pin #2 and #8. If you get 5 volts or higher, you can do it.

I have converted my tweed Twin clone to cathode bias. My amp is too loud. Can I pull a pair of output tubes.

If you pull a pair of tubes, you must double the cathode resistor value and double the speaker impedance. Use an 8 ohm load instead of the stock 4 ohm load. You must double the cathode resistance because two tubes draw half the current of four.

I'm in the process of building a replica of a Silvertone 1392 amp and I'm at the stage to start installing the various caps. I purchased a number of Mallory 150s and noticed the caps in my original amp appear to be electrolytic caps as they have a solid band of color around one end of the cap. Does this indicate a "positive" or "negative" end on the original caps? The Mallory's don't appear to have a positive or negative end, as there are no marking that would represent this. The caps I purchased are: Mallory 150 Series: .02uf 630V, .047uf 630v and .01 uf 630V. Am I missing something?

The Mallory 150 capacitors are not electrolytics. When a cap is made, two pieces of foil separated by a dielectric are rolled up like a cigarette. One of the foils will be the outermost and for the least amount of hum, should be attached to the most grounded part of the circuit. In the old days, a black line was a marker to show which lead of the cap was the outermost. With Mallory 150s and with Orange drops, when you read from left to right, the right side is the outermost. When they built the Silvertone, they probably did not even pay attention to this. If you have the outermost foil to the grounded-most part of the circuit, then the hum level will be at a minimum. All of those coupling caps have one end going to the plate circuit (high voltage) and the other end going somewhere else (this will always be the grounded most part because there is no high voltage on this end.) The coupling capacitor will go from the plate of a tube to either a load resistor, tone stack or volume control.

Just mount them with the left side (as you are reading the lettering on the cap) to the plate of the tube and you will be doing it the best way. I would not pay attention to the way Silvertone did it because they were not paying attention to such detail and ala Murphy's Law, they will probably have it mounted incorrectly anyway.

I have a "custom built" harp amp that is driving me insane with a 'tic' at the end of a note, during a long, low note it will tic repeatedly. It's worse when the Bass control is turned from 3 o'clock to fully up.

For the past few days I have been suspecting some sort of parasitic oscillation and looking at the lead dress to the power tubes and tried to dress wires out as close as possible to a Fender. As only one power tube (V5) ticks (I found this out by running I power tube only, none of them tick in V4).

Points that may be of note are that the output trans is situated right at the other end of the 17" chassis from the power supply and leads are relatively long (in this respect it can't be much different to many Premier and smaller Ampegs).

The 'manufacturer' is suggesting that I fit a 50 pf grid to ground cap in each power tube, but I'm concerned that this is just masking the symptoms not curing the problem.

I'm about to take the amp into the back garden with a tin of lighter fuel and a box of matches and give it it's own Viking burial unless you have a better suggestion.

Your amp is suffering from motorboating which is the nice word for parasitic oscillation. This is a real problem when someone designs an amp. You can't really copy Fender or Ampegs, because the Fender amps were plagued with parasitic oscillation problems. And the Ampegs didn't have enough gain for the bad layout designs to show negative results. Grounds sometimes overlap on the chassis such that one could modulate the cathode of some previous or later stage; grid wires could capacitively couple to some previous or latter part of the signal part, thus causing oscillations.

One very obvious mistake is the output transformer near the input of the amp. That is the worse possible design and you are asking for oscillations by having the output transformer so close to the input. The output transformer could easily couple with an earlier stage either magnetically or capacitively. That coupling could amplify back into itself and you end up with an oscillation. But who knows? It could be a combination of things, such as some grounding problem mixed with some capacitive coupled phantom circuit!! Yikes! This is where you really got to get that lead dress right.

Ampeg got away with murder in their horrible lead dress because their amps had little gain and were fairly dark sounding. Gain and treble response affects oscillations. Reduce the gain and the likelihood of a parasitic oscillation is lessened. Likewise, losing highs—which is what the shop was trying to get you to do with the 50 pf grid to ground cap, will help the symptom, but is just masking the real problem. The real problem is the layout is not happening. Try removing the transformer from the chassis (you could use long primary wires to get it out of the amp's chassis) and hooking it up and see if the motorboating goes away. If it does, move the transformer around until you can find a place on the chassis it will work and still be mounted on the chassis.

Fender had really bad layout on their phase inverter. The leads going from the plate coupling caps to the grid of the output tubes were too long. That wire should be very short. We are talking about the wire going from the phase inverter's output capacitors (usually .1uf on a Fender) to the grid circuit of the output tube (usually a 1.5K resistor). In fact, in your case I would shorten it AND use shielded wire. Remember the rule of using shielded wire: when using shielded wire inside of an amp, always ground the shielding only on one end.

The 1.5K grid resistor should be mounted such that there is zero lead length between the body of the resistor and pin 5 of the output tube socket. If the lead is too long, the resistor will not suppress parasitic coupling! Shorten it so there is no length between the body of the resistor and pin #5.

I have given you enough information to solve your problem. If you do everything I recommend and the ticking is still there, I would be very surprised. But if what I said doesn't work, then we need to start looking at the grounds to make sure everything has its own ground and two grounds are not overlapping or sharing the same wire at any time.

I have a few general questions that I have wondered about and I suspect others have also. Explain what the number's on tubes mean (12AX7 style numbering). Is it OK to move the negative feedback send from the 16 ohm send to either the 8 or 4 to lessen the effect? Do you know of any vintage HiFi style or organ speakers that work with guitar? Could you explain how the physical elements of a speaker effect the sound?

There are four different naming systems commonly used to identify tube types. Among these four coding systems you have the American common names, the American industrial/military names, the European common names and the European industrial/military names.

Let's start with the American common names such as a 6L6 or a 12AX7. The first number of the name will indicate the filament voltage of that tube. For example, the 6L6 uses a 6.3 volt filament. A 12AX7 uses a 12.6 volt filament. A 5AR4 for example, would use a 5 volt filament. With the American common names, the last number will indicate how many elements are inside the tube. For example, the 6L6 has six components. It has cathode, grid, suppressor grid, screen grid, plate and heater. On a 12AX7, there are seven elements inside: two cathodes, two grids, two plates and a heater. A 5AR4 would have four components. In this case, since it is a full-wave rectifier tube, it has a heater, a cathode and two plates. There are also suffixes that are used after the last number. The following suffix letters are in common use in tube designations and have the indicated significance:

G signifies a glass bulb and an octal base, for example a 6L6G.

GT signifies a straight-sided glass bulb and an octal base.

A, B, C, D, E, and F; assigned in that order, signify a later and modified version which can be substituted for any previous version but not vice-versa. The assignment of a suffix in this series does not convey any information as to the nature of the modification incorporated.

X signifies a base composed of special low-loss material

Y signifies a base composed of special intermediate-loss material

Is there a book or something that a novice can learn a little

more about amps and how to understand some of the technical language in your articles. I find them very interesting, but I don't know what all the fancy words mean.

There are many books you can get that will help you learn about tube amps. One of the best books on electronics, in general, is *Basic Electronics* by Navpers. These are almost always available on eBay as it was the *Electronics Training Manual* for the Navy. It is designed to explain everything easily enough so that an 18 year old that just enlisted in the Navy, could understand electronics. Do not get the *Basic Electricity* unless you want to learn about wiring breaker boxes. You want *Basic Electronics* as this will have important tube information. It won't have anything specifically on guitar amplifiers, but you can learn a lot about amplifiers, Class A, Class AB, phase inversion, preamps, gain, push-pull, single-ended, rectification, power supplies, etc. This is one of the first books I ever read on electronics and I can say it really puts everything in perspective. There are different versions of *Basic Electronics* by Navpers. Anything written or published in the 60s would be best. The later versions briefly touch on tubes and talk more about transistors. The 60s versions talk almost exclusively about tubes.

When you are ready to learn more specifically about guitar amps, I would recommend my first book, *A Desktop Reference of Hip Vintage Guitar Amps* as the most important book you can read. It talks about certain classic amps, how they changed circuitwise from year to year and how that affected tone. It is interesting reading and you will learn how tube amps work by reading stories about amps and not studying math. When that book was written, I gave the manuscript and a high-lighter to my wife with instructions for her to highlight any word or phrase she did not understand. Then I rewrote those passages and added a glossary to the back of each chapter. When a word was used that one would not use in everyday conversation (for example, one does not say the word "cathode" all the time), it appears in a glossary at the end of the chapter. "Cathode" is just a fancy word to describe the part of the tube where electrical current enters.

If there is a University in your area, try the engineering library. Assuming the University has been there a while; there will be all sorts of interesting books in the engineering library. I used to go to the engineering library at UT in Austin and dig up all sorts of cool stuff. I found the 1953 Audio Engineering magazine where Tung Sol announced their new tube, the 5881! There were racks and racks of tube related books from the 40s, 50s and 60s. Get yourself an "off campus" library card. They may charge a few bucks, but you will have opened up a wealth of knowledge.

If you blow a speaker while playing your tube amp, will this result in damage to the amp? Should a person immediately turn off the amp? How many times have you encountered an amp that needed repair from someone running it without a speaker hooked up, or without a load on the amp? How long would it take before the transformer would blow and what other parts would it take with it? Any irreversible damage when that happens, tone wise?

It could be a big problem if the speaker blowing causes the amp to lose its load. You never want to operate and amp without a speaker load. If there is only one speaker in the amp, or if the speakers are wired in series, blowing a speaker could spell disaster if you keep playing. I recommend to immediately stop playing. Once you stop playing, you can take your time turning it off. It is OK to have the amp on without a speaker load as long as there is no signal going into the amp's input. That is because it is the expanding and contracting magnetic field that causes the self-inductance and arcing, so as long as there is no signal, there is no expanding and contracting magnetic field. How many times have I encountered this problem? Well, it doesn't occur often. If an amp has parallel speakers as in the Twin Reverb, or Super Reverb or if it has series/parallel speakers such as a Marshall; blowing a speaker would not leave the amp without a load. Also, almost all Fenders and many other brands use a shorting jack such that the hot is grounded out if no speaker cable is plugged in.

You should stop playing immediately and check the speaker and speaker wire connection should you ever turn an amp into the "play" mode and not hear sound coming out. When an amp is damaged from running it without a speaker, what usually happens is the amp is put into the "play" mode; the player plays, but doesn't hear sound; then the player turns the amp's volume up and plays really hard — thinking if he plays harder it will somehow start making sound, then the transformer and sockets arc. Yes, the arcing will also occur on the output socket between pin #2 and #3. This is the high voltage arcing to ground via the heater connection. It will damage the socket and once the socket or the transformer arcs; they will want to arc again in the future. So the damage is pretty much irreversible.

Sometimes the carbonization that forms on the socket during arcing can be scraped off the sockets, but it is always best to replace them. The arcing could cause shorts and opens in the transformer as well. The higher voltage could possible take out the main filter caps as well.

I know two 8 ohm speakers wired in series make a total impedance of 16 ohms because 100% of the current has to pass through each speaker. Therefore, the 8 ohm impedance is doubled. I also know that the same two speakers wired in parallel would become 4 ohms because the current has two paths instead of only one, thus the impedance of two paths is half the impedance of one path. But my question: does this same relationship hold true for ceramic caps?

Actually the opposite relationship exists for capacitors. Two same value caps in series are half of one's value. Two same value caps in parallel become twice the microfarad value of one. Capacitance value is affected by the surface area of the conductor and the space between the conductors in the cap. If the surface area of the conductors is increased, so does the microfarad value. If the space between the conductors is increased, the microfarad value goes down. When two equal value caps are in parallel, the surface area is in effect doubled, therefore the overall capacitance would be twice the value of a single cap. On the other hand, when two equal value caps are in series, the space between conductors is actually doubled so the capacitance is halved. Series/parallel connection of four equal value capacitors would be like the speaker example—the overall capacitance would be the same value as an individual cap.

I want to build an amp and I'm still not sure which circuit would work best. I notice the phase inverter style as the main difference in the different model tweed amps. Can you explain the differences between the phase inverters used in the Fender Pro 5D5,

the **Fender Pro 5E5 (or Bassman 5E6-A), and the Fender Pro 6G5 (also used in all newer Fender models, also Marshalls etc).** I would recommend using the phase inverter circuit as found in the 6G5 Fender Pro. This type of phase inverter has a few different names. It is sometimes called the long-tailed pair phase inverter or the grounded grid inverter. It will perform well, it sounds good and you can get a range from clean to dirty with it. There are reasons why it was used on almost every Fender amp after the 6G5. It sounds good and it produces gain.

The phase inverter circuit used in the 5D5 Fender Pro is called a paraphase inverter. It takes advantage of the fact that running signal through a tube will invert phase 180 degrees. It takes some of the signal driving one output tube and runs it through another triode section to invert phase. The new inverted phase signal drives the other output tube. You will never get a good clean tone with this type of inverter. It sounds ratty no matter what else you do to the circuit. The inverted signal will have distortion that the uninverted signal will not have, so half of the waveform will have distortion that the other half will lack. This will never get very clean and that is why harp players love this one.

The 5E5 Fender Pro features a split-load type inverter that uses only one triode section. This is also known as the distributive load inverter. It is the only type of inverter that uses a single triode section. This type is also the only type that actually loses gain. It sounds very good, but you will need to boost it with a stage in front.

As a long time guitar player, your article on "Resurrecting A Dormant Tube Amp" struck home with regards to potentiometers; both amp and guitar-types. You indicated in your article to clean the pots by spraying a cleaner inside the pot and working the shaft of the wiper back and forth, to clean the contact surface. How does one get the cleaner inside the potentiometer?

All my guitars and amps have scratchy pots and your recommendation to clean the pots could not be more correct, except the pot is surrounded by a press-on metal backing "can." Once the ears of this can are lifted, the back of the pot then can be removed exposing the wiper area for cleaning, however I believe you will agree I will never get the cover back on again. Is there a secret to cleaning the inside of the pot without removing the back housing?

Look at the potentiometers in question. There will most likely be a small opening where the three leads come out. I use De-Oxit Tech Spray which has a small "soda straw" that comes with every can. The idea is to insert the little "straw" in the spray head so that the spray comes out of the end of the "soda straw." It is a simple matter of putting the "straw" in the open space where the leads come out.

If you have sealed "J" type pots, there is no good way to clean them, but then on the other hand, there is no way for dirt to get inside either. Sometimes pots become scratchy when there is D.C. on them and no amount of cleaning will help. This condition can be caused by a bad preamp tube or a bad coupling cap. If D.C. is leaking out of the grid of a preamp tube, D.C. will appear on the guitar's volume pot. If a coupling cap is leaking D.C., it can make a volume pot or tone pot scratchy. I recommend cleaning it and seeing if the scratchiness goes. If it still scratches you can use a voltmeter to determine if there is any D.C. on the pot. If there is, you will need to either replace a preamp tube or replace a coupling cap, depending on where the D.C. is coming from. And then on the other hand, sometimes a pot is bad and no amount of cleaning will rid the scratchiness.

I'm considering building a couple of tweed Vibrolux 5F11 clones to sell. Do you have any thoughts/cautions about this circuit? You never see the real ones or clones around, and there may be a reason for this. Any thoughts would be appreciated

Both the tweed Vibrolux and the tweed Tremolux are prone to parasitic oscillations and instability problems. Even the originals have to sometimes be modified to even work. I think these problems can be remedied with an improved layout and using shielded wire on the grid circuits. Basically, the tweed Vibrolux is a one channel (Instrument Channel) tweed Deluxe with Tremolo and a 10" speaker. The Tremolux would be a better choice as it is a tweed Deluxe with Tremolo and a 12" speaker and a slightly larger cabinet. Either way, you will benefit greatly by taking the .02 coupling cap between pin #1 and pin #7 of the phase inverter tube and move it off the board to the tube socket. This will eliminate much grid wire. You will still need the 1 Meg resistor from pin #7 to ground. Also, use shielded wires on the input jack to grid circuit and on the volume pot to grid circuit.

I am unfamiliar with a couple of terms used by amp guys. What is the "screen supply filter" and what is the "main totem pole"? I have the filter component board raised in my Fender Super Reverb amp, but I am not sure which part is the screen supply filter. What is the screen?

If you look under the cap pan on your Fender Super Reverb amp, the first two caps, nearest the output transformer, are the mains. The stock value is 70 uf at 350V. In some amps and, in your case, it is a pair of capacitors in series and with a 220K ohm resistor across each capacitor. This configuration of two caps and two resistors in series is what I am calling a totem pole. The filter capacitor next to that is the screen supply filter. The reason it is called the screen supply filter is because it is connected on the positive end to the screen resistors on the output tubes. It actually supplies voltage to the screen resistors.

What is your opinion on star grounding? a) Is it worthwhile? b) Or is it better to simply reduce the number of ground points from the board and not isolate the pots and inputs?

Star grounding works well, but if you use a star ground, there will be so much extra wire in your amp that it will look like spaghetti. All this extra wire is unnecessary. I like to use a star ground only for the screen supply filter capacitor, the main filter capacitor, the cathode circuit of the power tubes and the B+ center-tap ground. This star ground is best located as far away from the input jack as possible or at least on the other end of the chassis from the input jack near the power tubes. I ground the filter that supplies the preamp tube plates near the input jack. And I ground the other filters and the associated cathode circuits of the tubes they supply near each other in the order they occur from input to output. In other words, the filter that supplies the phase inverter tube and the cathode resistor of the phase inverter would both ground together near the middle of the amp. This works well, is totally stable and hum free.

I might also add that I am a supporter of brass grounding busses. Over a decade ago, I was building a batch of Kendrick 118 amps (these are small amps similar to a 5F1 Champ). Even though there are only a few grounds in one of these, we had always used a brass grounding plate in their construction. Well, I ran out of brass plates and built up 10 of them without grounding busses. The preamp components simply grounded British-style to a wire on input jack. While I was testing these 10 amps, I noticed a residual hum but it didn't seem that bad. Then the brass plates came in and I changed one over to compare it to the others. Everyone was taken aback by the extra quietness that the brass buss provided. After that eye-opening experience, I became 100% sold on brass grounding busses and have always used them on everything.

With regards to filter capacitors, can I use different type capacitors besides electrolytics, such as Solen, Musicaps or oil filled caps and would they perform as well or better?
Absolutely. The reason everyone uses electrolytic capacitors for filter cap applications is because they have lots of microfarads in a small package for a relative low price. For you to get the same microfarads in a non-electrolytic cap, the package will be very large and much more expensive. They will probably work better because almost any type of cap will have less current leakage than an electrolytic type and the other types will last much longer.

Solen is a private brand of the Canadian distributor Solen Electronique and sold by many dealers, including Parts Express and Antique Electronic Supply. There are a couple of different types, but the type that would have enough microfarads for a filter cap would be the Solens FAST CAP.

Hovland's MUSICAPS are one of the original high-end capacitor designs. A polypropylene film and foil design with high-end tweaks, they come in values up to a 10uf, so perhaps this would not have enough microfarads for most applications.

Military tube equipment usually uses oil filled capacitors because of the long-term dependability associated with them.

I built an amplifier based on a Blackface Deluxe circuit using a modern plastic bobbin transformer. I was told by the maker that it didn't matter if the transformer had a plastic bobbin. The transformer was cheaper than the paper bobbin ones so I thought I would save a few bucks. I thought the amplifier sounded pretty good and I was very proud of what I had built. But then my friend built a similar amp with the same circuit and splurged on a paper

bobbin transformer. We A/Bed the two amps and his ate mine alive. Do you think it could be the transformer and if so why?

If you've read any of my books, you probably already know how I feel about paper verses plastic. When I was just starting out in the 80s, I ordered a batch of custom-wound transformers from a trans-former company. The designed called for a paper bobbin with the exact interleaves and number of turns of a 1959 Fender 5F6A Bassman output transformer. When the transformers finally came in and I built up an amp with one of them, the amp sounded OK until I compared it side by side with my own 5F6A. At the time I didn't know a plastic bobbin from a paper bobbin, so I called the company to talk to them about it. As it turned out, they changed the design that I had worked so hard to reverse engineer and gave me a plastic bobbin. They told me it was an upgrade over paper and that no one used paper bobbins anymore. Worse yet, they refused to give me my money back or replace them with paper because according to them, and I quote, "the transformers exceeded the spec I had requested and they were in business to make money." I did end up using two of them. I used one for a door stop and another in a tube tester I made — just so I would have a load for the tubes. But I learned my lesson the hard way. The moral of the story is simply that you cannot expect to get a vintage paper-bobbin sound with plastic. It doesn't work and the only people that will tell you otherwise are the ones that have an interest in selling them.

I have an old Montgomery Ward Airline, and would like to know the speaker impedance. It has two 12" speakers, wired in par-allel. A friend of mine put his ohmmeter to the point where one of the lead wires connects to the speaker, and we got a reading of 2 ohms. This means I have two 4 ohm speakers, right? I thought that seemed slightly unusual since most amps these days have either 8 or 16 ohm speakers, but it's a pretty old amp (early 60s). What do you think?

I am not sure how your friend measured the speakers. In the first place, you must disconnect the speakers from the amp to get a valid reading. The output transformer secondary is a coil of wire that connects to the speakers in parallel, so if he measured with the speakers connected, the reading is invalid. Not only that, but it is always a good idea when measuring anything with low impedance, to zero out your meter before you take a reading. To zero out the meter you simply short the two meter leads to each other with the meter in the proper range. There will be a trim adjustment and you adjust it until the meter reads zero ohms. This way, you will not be measuring the resistance of the actual meter leads.

But even if the speakers were disconnected, you cannot measure impedance with an ohmmeter. An ohmmeter measures D.C. resistance. Impedance is defined as the sum of everything that impedes electron flow, so a valid reading would include D.C. resistance, A.C. capacitive reactance and A.C. inductive reactance. Whatever is measured by the meter is only the D.C. resistance and does not include any A.C. reactance. The A.C. reactance is not very much, so you can still use a meter to measure the resistance and just be aware that the D.C. resistance measurement will be slightly less than the actual impedance. For example, an 8 ohm speaker may measure somewhere around 5 or 6 ohms. A 16 ohm speaker will measure anywhere from 10 to 12 ohms. And the 4 ohm speaker will measure around 3 ohms.

I see on eBay where folks have assembled kits and are selling them for a paltry $75 to $175. They cite: 5E3 clones, 5F1 clones, Single-ended 6V6 Champ clones and the like. I'm not looking to review any specific offer, but for a hand-wired and already assembled amp; how hard is it to go wrong, especially for the price?
It could be really easy to go wrong. If those amps were built by inexperienced builders, there is a very good chance the amps are tar babies. Think about it: if the amps really sounded great, would someone want to sell them for almost nothing? The 5E3 Deluxe layout is particularly prone to parasitic oscillation, motor-boating and general instability. When I first started building that circuit nearly two decades ago, I copied the original layout and found about 3 amps out of every 20 built that sounded like there was a blown speaker when the low E string was plucked. You could change the speaker and get the same result. It turned out that the amp's layout is very prone to parasitic oscillations.

Not only that, but most kits on the market do not address the other problems with the original designs. That is to say any design flaws in the original designs are ignored. An original tweed Champ transformer is wound with wire rated at 30 mA and yet the output tube typically idles at 45 to 55 mA. They always run hot and are problematic. Kit sellers that are merely copying vintage designs are usually unaware of recurring problems of specific vintage designs and simply copy the flaws as well. The parts are usually from the lowest bidder (almost never the best sounding). Many of them use off-the-shelf Hammond transformers which are made in Canada and wound for a 115 volt primary. When you plug it into a 120 volt socket, all the voltages are too high and the heaters that are supposed to be 6.3 volts are run so hot that the tubes sound bad and wear out quicker than normal.

There is an old saying that is true today as it was 100 years ago: When you see a deal that looks too good to be true, it probably is.

I have a late 50s Silvertone with a single 12" speaker and a pair of 6V6 output tubes. I need a new speaker, but the old one is blown and I am not sure of the impedance of the speaker. I was told it should be a 4 ohm speaker, but I have never seen or heard of a 4 ohm 12" speaker. Is there a way to know what kind of speaker load it requires?

I do not know for sure and I don't have an original here to look at, but I can give you a test to perform that will tell you if it is a 4 ohm or an 8 ohm. I am not familiar with anyone using 4 ohm 12" speakers and I would guess it to be 8 ohm, but here is the test.

You will need a variac which is basically just a variable A.C. auto-transformer. With the variac, you can dial in a specific A.C. voltage. Remove the output tubes from the amp. Remove the chassis from the amp cabinet. With the amplifier unplugged and the output tubes removed, connect one end of the variac to pin #3 of one of the 6V6 output tube sockets. Take the other lead from the Variac and clip it onto pin #3 of the other output tube socket.

Next, take an A.C. voltmeter set to a low range (perhaps the 2 volt range will do nicely). And connect it to the output of the amp. This is where the speaker was originally connected, however you want to disconnect the speaker and connect the voltmeter instead.

To perform the test, begin by advancing the Variac voltage going to pin #3 of each output tube socket. You want to advance the voltage on the variac while you are monitoring the voltage reading on the voltmeter. Keep advancing the Variac until you see the voltmeter go to exactly one volt. At this point, the voltage across the primary will be the turn's ratio of the transformer. Now, without disturbing the variac, take the voltmeter off the output and connect it across the Variac. Record the voltage. That is your turn's ratio.

If the voltage coming from the variac measures anywhere from 28 to 36 volts, then the intended speaker load will be 8 ohms. If the voltage on the variac measures anywhere from 40 to 50 volts, then the intended speakers load will be 4 ohms. These figures are only valid for a two 6V6 amp, but they should tell you what you need.

I managed to acquire a Champ style kit. It was a blast to put together and I learned a lot in the process. When I turn it up above about 10½ however, I start to get the really squealing, fragmented, high distortion on the low notes. Is this just a quirk that I'll have to get used to? I'm just curious what would cause that problem. Ideas? What you are describing is common with a Champ style amp using higher gain tubes. You are driving it into Class AB, which is basically clipping off the bottom of the wave. There is a small capacitor on the end of the board near the chassis. It is a 25uf at 25v. Disconnect one end of this and the problem is solved. Fender sold the 5F1 Champs sometimes with and sometimes without this capacitor. With it, when you have a good gainy preamp tube, you can drive it into Class AB. Without it, the gain will come down a little and keep it in Class A.

I have a problem with this weird sub note that is always there. It doesn't matter how loud or how soft the volume control is set. From having the volume so low that you can barely hear the amp, on up to overdrive; nothing I have done has changed it at all. That is why I am considering the problem to be output transformer related. Also, often at low volume, if I let a chord sustain, I get this sound like vibrating tin cans or glass. But it seems electronic, and not a vibration. And it only does it at a certain point during the decay, towards the end. When there is an ugly sub note that only occurs at loud volumes, it is usually a filter cap problem. But when it happens even at soft volumes, it is almost always a speaker voice-coil rub. I know for sure that your case is a voice-coil rub because you have a classic telltale sign of voice-coil rub. When you play a single note and let the note sustain until there is nothing left, at the very end of the sustain, the sound like vibrating glass is a dead giveaway.

This could be caused by the speaker frame becoming bent or improper tightening of the speaker mounting screws. Just as a drummer wants the drum head to have equal tension all around the perimeter of the drum, the speaker needs equal tension around its frame. Uneven tension could compress the paper gasket on the perimeter of the speaker and cause the frame geometry to become non-symmetrical, thus causing the voice-coil to be just enough off-center to rub against the pole piece or top-plate. There is one possible long-shot remedy that works about one time out of twenty. If the voice-coil rub was caused by improper tightening of the speaker mounting screws, you may be able to experiment with loosening some of the mounting screws and tightening others to manipulate the frame geometry of the speaker. This is best done with one person playing a single note and waiting for the "vibrating glass" sound at the end of the sustain and the other person using trial and error tightening of the speaker mounting screws while listening to the vibrating glass sound. If you happen to manipulate the frame geometry so the "vibrating glass" sound is gone, then it should work fine at loud volumes.

There is no repair save for a speaker reconing or replacing the speaker. The recone may not work if the frame of the speaker is bent.

I am thinking of playing with a flea market SUNN PA head with two KT88s and turning it into a Bass head for a friend. 6550s re-biased would be cool in this, no? Or should I keep the Genalex Gold Monarch KT88s?

You can try different tubes and let your ears decide. I don't know how old your Genalex KT88s are. If they are worn out, it may make more sense to replace them. Aside from spending $400 for a NOS pair of Genalex KT88s, I like the JJ brand KT88 tubes. These sound great, are reasonably priced and I have been using them in all three of my personal Leslies and never had a problem. Of course, the 6550 tubes are nice to use for

All About Vacuum Tube Guitar Amplifiers

a bass amp also. They just won't develop as much power as the KT88.

I am building a kit amp and I plan on using a 6V6 tube. I am confused regarding putting a jumper between pins #1 and #8. When this is done, am I restricted to only using the EL34 power tube? In other words, once you jump pin #1 and #8 are you are stuck with the EL34? If not I would like to install the jumper and maybe switch to an EL34 at a later time and listen to different tubes and the tones they produce.

Jumper pin #1 and #8 together and leave it, then you can use almost any type of tube. Look at the bottom of a 6V6; you will notice there is no pin #1. The same is true of a 6L6, but the EL34 has a pin #1 and it should be connected to pin #8, so connect pin #1 to pin #8 and use whatever you wish. When a 6V6 or 6L6 is used, it won't connect to the socket on pin #1 because the tube itself doesn't have a pin #1!

I have a AB763 based amp that has a problem with unwanted / unnatural distortion on the low frequencies when turned up past seven. It isn't real present all the time but comes into the signal every now and then. I have visually checked all the connections and changed tubes to no avail. It seems to only be on the reverb channel. I am asking for some sage advice on this one.

You have a parasitic oscillation. You said it is an amp based on an AB763, but is the layout similar? Spacing and placing of components is critical in a high impedance circuit. You do not want the circuit to cross back over itself.

For starters, I would use shielded wire for the input jack to grid connection and the volume control to grid connection. You want to ground the shielding only at one end of the wire; otherwise you will end up with ground loop hum. Perhaps grounding the end nearest the front panel will be best.

Secondly, I would take the tone caps off the board and hang them on the pots. This would eliminate about 2 or 3 feet of grid wire in the tone circuit. You can hang the slope resistor off the back of the tone caps, so you end up with only a short wire going to the plate resistor on the board. If you do those two things, there is a 98% chance the problem will disappear.

I have converted amps from two-prong ungrounded to three-prong grounded by replacing the 2-wire (two-prong plug) power cord with an earth grounded 3-wire (3-prong plug.) Someone wrote on the Fender forum about the power trannies being isolated from the chassis so it didn't matter which side you connected the "hot" lead to. When I converted my own amps, I always tested the chassis to the grounded outlet with my voltmeter on A.C. and made sure that when all was hooked up there was no A.C. showing on the meter. I once saw some A.C. on a chassis if the wires were reversed. If I did see A.C. on the chassis (which went away when the leads were reversed) how can the Fender power tranny be a pure isolated or floating tranny?

Best not to jump to conclusions. All you know is what you measured, what you don't know is how the A.C. got there. The primary of a power transformer is designed to be totally isolated electrically from the chassis. That is one of the purposes of having a transformer: to keep A.C. wall current off the chassis. Not all amps used a power transformer. There were powertransformerless designs that were not isolated ("amp in the case" Silvertone for example), and one could easily end up with 120 volts on the chassis! I can't explain your "chassis hot" situation without doing the testing myself, but I can assure you the voltage on the chassis was not coming from the A.C. primary winding—lest the transformer was shorted (i.e. primary to the case of the transformer.) There are other possible explanations. Perhaps the chassis was picking up A.C. magnetically. The magnetic field of the transformer could have been interacting with the steel chassis such that A.C. was induced. Perhaps the ground capacitor was shorted. Or perhaps the wall A.C. receptacle was wired wrong. If the third prong of the A.C. wall receptacle was not grounded, then the amp would act like a 2-wire design!

I recently finished building another amplifier from scratch. Like the first one I built, which was later solved, I'm having humming problems. With the guitar volume down I still hear the noise. When I turn the amp volume down, it goes away. Also, when I pull the first pre-amp tube the noise also stops. Could I be getting hum from the filaments?

On both projects I've used Hammond transformers and chassis.

It's weird, due to the fact that when I built the first one using the same filtering and transformers in a salvaged black Bassman chassis, I didn't get any hum or noise whatsoever.

If it stops when you pull the first preamp tube, it is either a problem with the particular preamp tube, or the grid wires feeding the preamp tube should be shielded.

You can even see which side of the tube it is coming from by using a little trick I use. Isolate each side of the preamp tube by using a jumper wire. First, I would jumper from grid to ground. This will tell you which side the hum is coming from. Or if shorting either grid to ground does not stop the hum, you will know it is not coming from the grid circuit and it is probably the preamp tube itself that is humming.

Also, it is important to have a ground reference on the heater filament circuit. If the filament winding has a center-tap, you need only ground the center-tap. If it doesn't have a center-tap, you can make an artificial center-tap by using two 100 ohm resistors. You would connect one end of each resistor to one leg of the filament winding and connect the other end of each resistor to ground. This will balance out the hum induced by the filament.

I'm am using a 1965 Blackface Bassman circuit. I'm getting a hum through to the first preamp tube in the normal channel. Which points in the schematic do I test to see if there is A.C. coming through where it shouldn't be? I want to test the circuit to see how much A.C. is coming through the filter cap bank.

Suppose I pull the first pre-amp tube and the amp stops humming, could this indicate A.C. leaking through the 20 uf 500 volt cap going to the top of the two 100K resistors? How could I block this A.C. so the amp stops humming?

There are two types of hum that you can experience in a guitar amp: 120 cycle and 60 cycle. The 60 cycle is coming either from the filaments or induced from magnetic fields produced by the 60 Hz A.C. line. The 120 cycle hum is coming from the power supply ripple current. Ripple current occurs when the 60 Hertz 120 volt A.C. is rectified to D.C. Since virtually all amplifiers use a full-wave or full-wave bridge rectifier, the ripple current is twice the frequency of the 120 volt wall supply, making it 120 Hertz.

If you have a true RMS meter, you could simply set it to A.C. volts and check the voltage across the filter cap. This will give you the RMS A.C. ripple voltage (120 cycle). Usually, hum from the first preamp tube is not coming from the power supply. Why? Since the power supply of a Fender amp has several stages of filter caps, each filter cap reduces the ripple current more and more so that by the time power is applied to the first preamp section, all ripple current has already been filtered to pure D.C. voltage.

It is more likely the hum is coming either from the filament, or the grid is picking up hum from not being shielded. Try grounding out the grid and see if the hum stops. If it does, you must need a shielded grid input wire or the grid wire is too close to a filament wire.

Also, make sure you have a ground reference on the filaments. There are several ways to do this depending on whether your power transformer has a center-tap on the 6.3 volt winding. If it does not, a pair of 100 ohm resistors, one on each side of the winding and the other end to ground, will make a nice, hum-canceling, artificial center-tap.

Is there any way I can use a rheostat to safely attenuate between the speaker and output? How would I accomplish this using a Blackface Bassman? How do I figure out which parallel and series resistance to use?

The rheostat could be used if you used a fader circuit. In a fader circuit, you would use a 100 ohm rheostat in parallel to a 4, 8 or 16 ohm, high-wattage load resistor (The exact value would depend on the output of the amp). The 100 ohm rheostat across the load resistor wouldn't change the overall load much. You would then connect the amp to the load resistor/rheostat assembly. Your speaker would connect between the wiper of the rheostat and the ground of the amp's output. This would approximate the correct impedance and the rheostat would be your volume control.

Another possibility is to use an L pad (available at Radio Shack or Parts Express). You would need a 100 watt L pad for a 50 watt amp (safety margin). The L pad is like a Rheostat except it is constantly correcting the impedance.

Other than that, you could use the circuit I use in the Kendrick Power Glide Power Attenuator. I have a 6-position double pole rotary switch that adds a resistor in parallel with my speaker and a resistor in series with the speaker/resistor parallel combination. This circuit keeps the impedance correct on every setting. The resistor in parallel provides an alternate path for current, so it drops the current actually going through the speaker. However, when you place a resistor in parallel with the speaker, you drop the nominal impedance. So to correct the impedance, another resistor is added in series. The series resistor has the added benefit of dropping voltage. The exact values to use will depend on how much attenuation you need and what impedance the amp wants to see.

For example, let's say you wanted to attenuate 6 dB and you are playing a 4 ohm 50 watt amp. You would need a 4 ohm resistor to put in parallel with the speaker. This would drop the nominal load to 2 ohms. The current will be cut in half because it will now have two paths—through the speaker or through the resistor.

Next, you will place a 2 ohm resistor in series with the speaker/4 ohm resistor circuit. This will do two things. First, it will correct the nominal impedance back to 4 ohms because the 2 ohm resistor plus the 2 ohm (speaker/4 ohm) assembly makes 4 ohms. And secondly, it drops the voltage in half.

The 2 ohm resistor will need to be 25 watts minimum (I would double it for safety), because it will take half the power. The 4 ohm resistor must be at least 12.5 watts (Again, I would double it for safety to 25 watts) because it is taking up a quarter of the power and your speaker is getting the other quarter of power. This gives you a total of 6dB attenuation.

The Kendrick Black Gold 5 is listed as a 5-watt amp, with a KT66 biased for 50–60 mA of plate current. Okay, plate voltage times plate current equals power in watts, right? Doing the math with these numbers gives a plate voltage of 100 or 83.3 volts; these values seem really low. Help me understand. Does adjusting the plate current actually increase/decrease the power output of the amp, OR does this little puppy bring more to the show than 5 watts?

You can't get there from here. Plate voltage x plate current is not output power! You are looking at D.C. plate voltages and D.C. idle current. The power output on an amp is A.C. power. It's A.C. voltage and A.C. current that provide output power.

Adjusting the plate current only changes the idle point (bias) of the tubes. If it is set too low, it can affect power (not as loud). But it does not necessarily increase or decrease power.

It is the changing voltage times the changing current, both of which are A.C. that gives you output wattage. But even still, let's not ignore the fact there are losses in the output transformer. Output power is the power that is driving the speaker.

Here is a test. While having someone play your amp, measure the A.C. voltage across your speaker with a true RMS meter. Square that number and divide by the ohms of the speaker. That will give you the actual power.

I have many obsolete tubes designed for a 150 volt plate supply. I would like to make a power transformerless amplifier with them. Can I rectify the wall voltage without using a power transformer? Should it be a full-wave rectifier? And what about the grounding? And how would I connect the heater circuit?

Yes, you can rectify the wall voltage but you must choose tubes whose filament current is equal to each other and whose filament voltages add up to 120 volts. If you look in the back of the R.C.A. Tube Manual, there will be example circuits of this design. It is usually referred to as an A.C./D.C. amplifier because since there is no power transformer, it can be operated with either A.C. or D.C. This design is almost always operated as a full wave rectifier because that is the only way to get the 120 volt wall to produce 150 volts D.C. On this type of design, one must select tubes that draw the same amount of filament current. The filaments are wired in series and the sum of the filament voltages will add up to 120 volts or slightly less. If the sum of the filament voltage is less than 120, a ballast resistor is used the "eat up" the remaining volts to equal 120. If the filament of any one tube goes out, then all tubes stop lighting up.

If you've ever seen the "Amp in the Guitarcase" Danelectro/ Silvertone amp, you have already seen this design in action. Many 50s and 60s tabletop radios and phonographs used this design. I would advise against using this design due to the severe shock hazards associated. One side of the 120 volt A.C. line must be grounded to the chassis. Therefore, no matter how careful you are, the possibility exists to have a live chassis with 120 volts A.C. on it. Considering the string-ground on your guitar, you could easily have that same 120 volts on your guitar strings. Ouch.

You could use an isolation transformer between your guitar input and the amp to protect you from electrical shock, but if you are going to use a transformer anyway, why not just use a power transformer?

I have a 5F1 Champ clone. What if I load it with 6L6 or any other octal power tube?
It depends on the output transformer. If you have an original, or a copy of a Champ transformer, it will work for a while but eventually overheat and blow. The Champ transformer is wound with thin wire that is only rated to handle 30 mA. Most Champs idle at slightly more than that, so they are always running hot, even when using the stock 6V6. It is all about wire size. If the wire is too thin for the amount of current the wire is carrying, the electrons will overheat it like a heating element and eventually burn it up.

When I was building Champ clones, the Kendrick Model 118, I designed my transformers with thicker wire so they could take almost any octal tube with the same pin out as a 6V6. If your clone is using the Kendrick 118AHD output transformer, you can use almost any output tube, including a 6550, 7581A, 5881, EL37, 6L6, 6V6, EL34, etc.

What idle current bias setting do you recommend for a four output tube (6L6) amp? Most books I've read indicate 35 mA per tube with a two 6L6 amp. Does this mean I should go for 70 mA per side when biasing my tweed Twin clone?
Any setting you use that sounds good to you and does not make the plates of the tubes glow cherry red is correct. I will tell you from personal experience that I like biasing a Twin at between 50 and 55 mA per side. This works out to only 25-27 mA per tube. I spent an entire afternoon once doing listening tests with Jimmie Vaughan and trying many different bias settings on Jimmie's Kendrick 4610 amps which have four output tubes configured very similar to a

tweed Twin. The 50-55 setting will allow you to turn up without over compressing the notes.

Your situation may be different. If you are playing in a bedroom and can't really turn it up, perhaps you would like setting bias such that the idle current is much more. This will make the amp sound better at a low volume and it will break quicker.

Here is the acid test to determine how much current your particular amp can go without exceeding the plate watt dissipation. Start decreasing bias voltage. This will make the idle current go up. Turn the amp up to the volume you will be playing it and play the amp while watching the output tubes. You are going to idle the tubes hotter and then play. Keep advancing the idle current until you get to the point where the plates of the tubes just start to glow cherry red. At that point, you have just exceeded the plate watt dissipation of the tube. Now back it off a little and you will be at the maximum the tubes can possibly take in that particular amp. Make sure and play the amp and keep watching those plates. You never want to have the amp biased where the plates glow cherry red for any length of time. To do so will invite malfunction. Also, if you set it up with the volume low and then play the volume loud, you may start to glow the plates. That is why it is so important to set it up by playing at the maximum volume you will normally play.

My amp is showing high voltage on the tube plates (530 volts). The diagnosis is that the power transformer may have a winding or two cooked. The options considered so far are to get another transformer or to run with a higher rated tube. First, I know the tube option sounds wacky but I've run the amp single-ended with just one 6L6 for a lot of years and a whole lot of gigs. I expect you're familiar with this and might say it's a senseless thing to do. But the "approach" came to me via some of the most experienced harmonica players out there. I was real skeptical myself but when Rod Piazza and William Clarke say they do it, then that's good enough for me. Well, maybe now I'll have to pay the price. The tube suggested so far is a KT90 or something similar to handle the voltage. I'd like to hear what it does but I'm also concerned about the amp. What I'm doing isn't exactly in a

manual (as far as I know) and I'd like to keep the sound I am getting using the single output tube. Any information and suggestions would be much appreciated.

Here is a simple test to check to see if your power transformer primary is indeed shorted. Place an A.C. voltmeter across the 6.3 volt winding. On most amps, the easiest place to connect the meter will be across the pilot lamp. If the primary is shorted, the 6.3 volt reading will be high (7 volts or more). If the 6.3 volt reading looks normal, then the primary is not shorted.

There are other reasons the plate voltage could read high. For example, if the output stage is biased cold, the plate voltage will go up. I would not recommend changing the tube type to solve this problem. That would be treating the symptom instead of the problem.

The problem may be related to you running only one output tube in an amp designed for two output tubes. A single-ended transformer must be designed with an air gap in the iron core so that the core doesn't become magnetized. If the core becomes magnetized, then signal cannot pass through the transformer. Your amp is designed as a push-pull amp and therefore the output transformer does not have an air gap. A push-pull output transformer, when used with two tubes, doesn't have a problem with becoming magnetized because there are opposing polarities in the center-tapped primary that would cancel any magnetizing effect. On a single-ended design, the current is flowing in only one direction and therefore the air gap is mandatory. If you desire to run a push-pull output transformer in a single-ended fashion, it could be done if you regularly change which socket has the tube. You could put the single output tube in one socket for one gig and then change to the other socket on the next gig. This would help prevent the output transformer core from becoming magnetized.

Is there a trick to getting a solder joint to hold and stick on the back of the pots? I have seen many pots in the last week or so and some say "solderable backs" on the description. Should I sand a spot on the pot to bare metal? I noticed the shafts on most pots being plastic. Is this better technology, rather than

having the metal shafts that are in all my Fender amps?

Pots made from aluminum, aluminum alloy or stainless steel cannot be easily soldered. Almost all others can. It is always best to rough up a spot on the back of the pot to "help" to solder to stick.

The shafts in the pots are non-conductive nylon. This is better from a grounding/noise perspective. Actually all CTS pots (like the ones used in vintage Fenders) are now made with the nylon shaft.

Looking at the tube filaments as they progress from where the parallel wiring feeding them starts at the tranny, they get brighter and brighter as they progress away from the source. Is that normal?

Different brands or types of tubes sometimes look brighter than others. Instead of looking to see if they are going brighter and brighter, check the voltage across the filament leads of each socket to see if the voltage across the socket is a consistent 6.3 volts. This will tell you if anything is amiss. You can't really measure brighter and brighter, but you can measure voltage. You want to set your meter for A.C. voltage. To check the filament voltage on the 12AX7, one lead of your voltmeter goes to pin #9 and the other to pin #4 or pin 5. On the 6L6, 6V6, KT66, EL34, 6550, 7581A, or 5881; the filament leads are pin #2 and pin #7.

INDEX

All About Vacuum Tube Guitar Amplifiers

All About Vacuum Tube Guitar Amplifiers

U

V

W

Z

Tube Amp Talk
For the Guitarist and Tech

Foreword by Billy F. Gibbons
Trainwreck Pages Part 2
**537 pages of questions and answers, troubleshooting, mods,
tweaking and much more**

In this book you will learn:
- How to slow down the speed parameter of your vibrato (page 77)
- The 26 basic rectifier tubes and their characteristics (page 287)
- The top ten tricks for steel guitar players (page 95)
- The twenty tricks to stabilize the unstable amp (page 61)
- The four steps to correctly apply tweed (page 50)
- The five simple troubleshooting techniques (page 26)
- What is involved in a cap job (page 31)

So, from the glitz and glimmer of Hollywood, to the Texas rude, crude and super-blue'd, New York City 'cross to loudness-in-London, and easin' through Euro-town to ampin' up in Africa, turn to your favorite section here and let it rip.
Red Beans and Ricely Yours,
      —Billy F. Gibbons

Price is $34.95

www.kendrick-amplifiers.com

Check your local book dealer or contact
Kendrick Books (512) 932 3130

All About Vacuum Tube Guitar Amplifiers

"Tube Guitar Amplifiers Servicing and Overhaul" DVD:

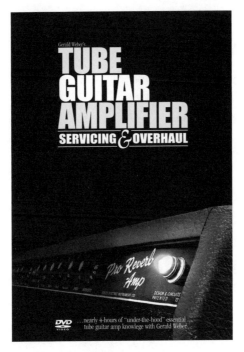

On this DVD you will learn:
- Different types of Silverface and how to convert each to Blackface
- What wiring mistakes can be found in 70% of all SVT amps
- Common problems found in Valco manufacturer amps
  (National, Gretsch, Supro, etc.)
- Checking coupling caps for leakage
- Checking output transformers
- Identifing leads of unknown transformer
- An easy way to determine output wattage
- Basics of servicing and overhauling
- Selection of rectifier tube for tone
- How to perform a cap job
- Common sense troubleshooting techniques that can be done without
  any special equipment

Price is $69.95

www.kendrick-amplifiers.com

Check your local book dealer or contact Kendrick Books (512) 932 3130

"Understanding Vacuum Tube Guitar Amplifiers" DVD:

Gerald Weber's second DVD is presented for all experience levels as a path to experience a "transformation" in their experience of vacuum tube guitar amplifiers. In short, you will never see tube guitar amplifiers the same again.

Four hours of essential vacuum tube circuit concepts such as: resistance, capacitance, reactance, gain, preamps, output stages, tube types, power supplies, tone circuits, biasing, and all the basic concepts needed to really understand what is going on in your amp. An in depth analysis of several classic vintage circuits – component by component – shows what each component does, how it affects tone, and what will happen if you alter the component's value. Organized by topic, with point-and-click navigation, you can use this DVD as a handy reference anytime you wish to review. The DVD was filmed live, during an actual "Tube Guitar Amplifier" training seminar, so you will hear real questions asked by real people and you will learn the real answers. Whether you are a guitar player wanting to learn about tube amps or a seasoned tech, you are certain to find enormous value in this book and DVD.

Price is $69.95

www.kendrick-amplifiers.com

Check your local book dealer or contact Kendrick Books (512) 932 3130

All About Vacuum Tube Guitar Amplifiers

Texas Twist

AND

2 Proud
2 Beg

*Red Hot Blue is
Gerald's band, a
Blues Rock Power Trio.
Their music was
described by Recording
Magazine as "ZZ meets
Stevie Ray at Lou Rawl's
house for some good
homecookin!"*

Texas Twist
Some songs featured: Texas Twist • Member of the Race • Ladies Don't Go that Fass • Killin' Time • Nineteen Year Old Body • Tin Pan Alley • Song for My Father

2 Proud 2 Beg
Some songs featured: 2 Proud 2 Beg • Josie • Sunrise Interlude • Slippin into Darkness • Secondary Lover • Installments • Man's World • Beijing Sunset • Hip Hug Her

Red Hot Blue is:
Gerald "G-Man" Weber: Guitar, Hammond B3, Piano, Kurzweil sampler, Vocals
"Red River" Dave McCLure: Bass Guitar
"Bunky" Yates: drums, percussion

Price is $15 each

www.kendrick-amplifiers.com

Check your local book dealer or contact
Kendrick Books (512) 932 3130